EAST MEADOW PUBLIC LIBRARY
3 1299 00981 2000

East Meadow Public Library
1886 Front Street, East Meadow, NY 11554
(516) 794-2570
www.eastmeadow.info

D1226404

running science

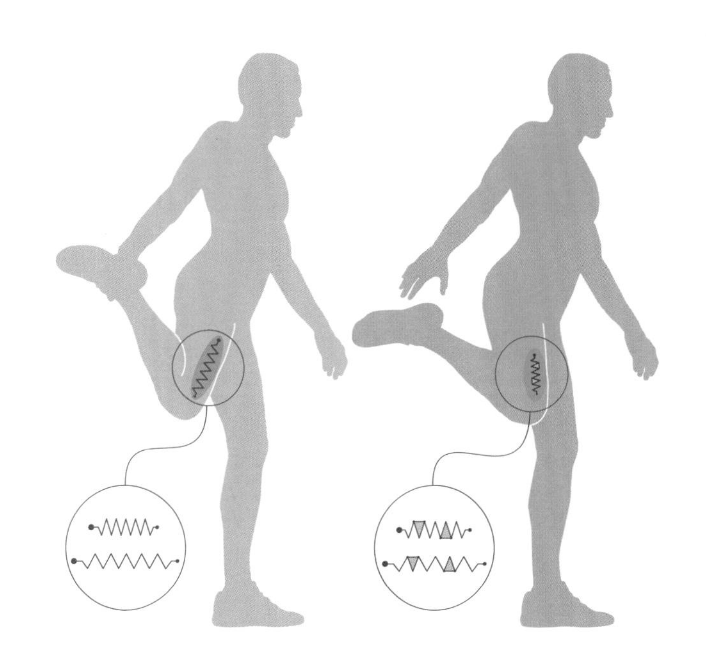

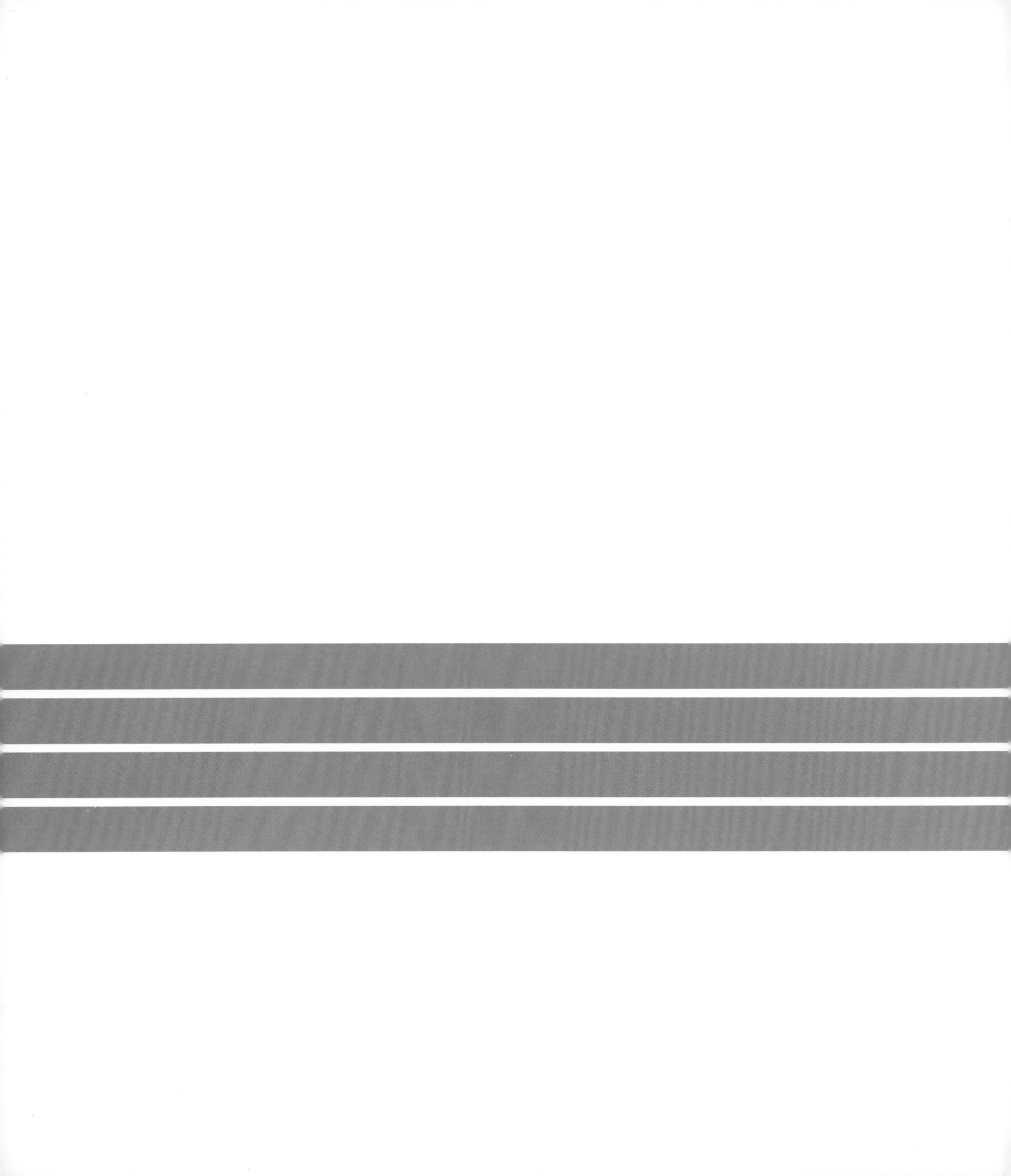

running science

Consultant editor **JOHN BREWER**

optimizing training and performance

Contributors
ANNA BARNSLEY
JOHN BREWER
LAURA CHARALAMBOUS
DANIEL CRAIGHEAD
JAMES EARLE
IAIN FLETCHER
JESS HILL
ANDY LANE
PAUL LARKINS
BOB MURRAY
CHARLES PEDLAR

THE UNIVERSITY OF CHICAGO PRESS
Chicago and London

The University of Chicago Press, Chicago 60637
The University of Chicago Press, Ltd., London
 Published 2017.
Printed in China

26 25 24 23 22 21 20 19 18 17 1 2 3 4 5

ISBN-13: 978-0-226-22399-5 (cloth)
ISBN-13: 978-0-226-22404-6 (e-book)
DOI: 10.7208/chicago9780226224046.001.0001

Library of Congress Control Number: 2016051234

This book was conceived, designed, and produced by
Ivy Press
An imprint of The Quarto Group
The Old Brewery
6 Blundell Street
London N7 9BH
United Kingdom
T (0)20 7700 6700 **F** (0)20 7700 8066
www.QuartoKnows.com

Publisher Susan Kelly
Creative Director Michael Whitehead
Editorial Director Tom Kitch
Art Director James Lawrence
Commissioning Editor Jacqui Sayers
Design JC Lanaway
Illustrators Nick Rowland and Rob Brandt
Copy Editor Gina Walker
Development Editor Steve Luck
Assistant Editor Jenny Campbell

Note from the publisher
Information given in this book is not intended to be taken as a replacement for medical advice. Any person with a condition requiring medical attention should consult a qualified medical practitioner or therapist.

10

Contents

Introduction

The human body evolved to allow us to run. Along with walking, it is our preferred form of locomotion and, in the past, our ancestors had to run either to catch their food or to avoid becoming the food of predators. On the face of it, running is a simple action—it involves placing one foot in front of the other to generate forward momentum, and differs from walking because there is a point during each step phase when both feet are off the ground. But the human body is an advanced mechanism, and the science that underpins running is both extensive and complex.

Since the era of the ancient Greeks, running has evolved from an essential means of movement into a competitive sport and a recreational pastime. In more recent times, it has also provided scientists with a means of better understanding how the human body reacts and adapts to stress, whether as a consequence of exercise or environmental conditions. For many years, there was very little interaction between runners on the roads, tracks, and trails, and scientists in their laboratories studying the effects of running. But as the scientific understanding of running increased, so, too, did the demand to apply this knowledge to the enhancement of human performance. Scientists started to step out of their laboratories and began working alongside athletes and coaches to improve the way in which athletes trained and performed. Initially, this focused on elite runners and, in particular, on those involved in endurance running, where simple predictive correlations were found to exist between laboratory measures such as maximum oxygen uptake capacity and performance times. But over time, the extent of the support from scientists increased, with nutritionists, sports psychologists, and biomechanists all finding themselves able to provide insights and knowledge that contributed to better preparation and performances.

▶ ***Running and science*** *While elite athletes work with scientists to maximize their performance, runners of any standard can help to fulfill their potential and run more healthily through the lessons learned from research.*

USA
TDK
GATLIN
BEIJING 2015
USA
TDK
GAY
BEIJING 2015
JAMAICA
TDK
BOLT
BEIJING 2015
USA
TDK
RODGERS
BEIJING 2015
TDK
TDK

At the same time, the sport of running has shown phenomenal growth, with many millions of people across the globe now taking part in a sport that for many years was confined to highly trained, elite runners. In particular, endurance running has become the domain of the "normal" person, with marathons and even ultramarathons now featuring individuals who in the past would never have competed in such events. This growth in mass participation running is fertile ground for running scientists, because these "recreational" runners still want to do as well as they can, even if they might not be standing on the podium at the end of their race. Many will not have a coach, but most will want to learn and appreciate how best to prepare and compete, and how to get the very best they can from their bodies. This is where science can help. It may not be able to turn a recreational runner into an Olympic champion, but science can help any runner to get more from their performance, stop them making certain mistakes, and, above all, improve their enjoyment of the sport.

The scope for science to impact on running is immense. It can contribute to all running distances, from sprints to ultramarathons, and there are few, if any, areas of running where a better understanding and application of science cannot make a genuine and worthwhile difference. This book takes a closer look at all of these areas, firstly by better understanding the science of running, and then by demonstrating with practical illustrations and examples how this understanding can be used to enhance running performance.

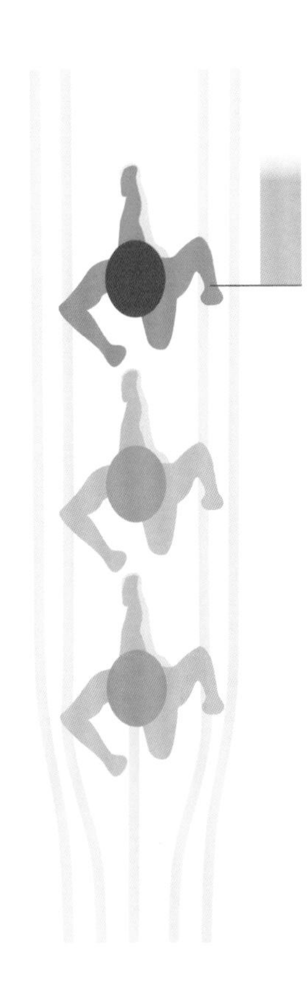

The structure of our bodies, including bones, muscles, tendons, and ligaments, is our anatomy, whereas physiology is the manner in which the body functions to create a living organism. The human body is a complex organism that constantly adapts and regulates itself to support life, with many systems interacting to ensure that it functions effectively. While human beings have evolved in a manner that means all of us have a similar generic structure, each individual differs as a result of their inherited genes. Science, however, has shown that we can make changes to our bodies, particularly through exercise. The body can adapt to the stimulus of running through positive changes to many of the anatomical structures and physiological processes that support life, and these changes can not only improve running performance, but could also create long-lasting health benefits.

chapter one

the runner's body

John Brewer

What is running economy?

Why do I get out of breath so easily?

In order to run, the body has to produce energy. This occurs as a result of the breakdown of either carbohydrate or fat within the muscles, and for low- and moderate-intensity running this process always uses oxygen. Approximately one-fifth of the air that we breathe into our lungs consists of oxygen—the rest is mainly nitrogen, along with small amounts of other gases such as carbon dioxide. Oxygen is transported across the membranes of the lungs to attach to hemoglobin in the blood. With each beat of the heart, oxygen-rich blood from the lungs is pumped to the exercising muscles, where it combines with fat or carbohydrate to produce energy. At the same time, blood that is low in oxygen is pumped away from the muscles and back to the lungs.

The amount of energy needed to run at any given speed will vary from one person to another, and this in turn determines how much oxygen the exercising muscles require. Someone who uses a lot of energy—perhaps as a result of poor technique or a high body fat percentage—will need more oxygen, and be less efficient, or economical, than a runner with good technique and low energy expenditure. This is known as "running economy" and it has a real impact on performance.

Efficient runners have good running economy. They use less oxygen, save energy, and suffer less fatigue. On the other hand, runners with poor economy have to use more oxygen at each speed, resulting in an increase in breathing frequency, heart rate, and fatigue.

A simple analogy is to think of two automobiles—if one requires more fuel than another while traveling at the same speed, the one with the highest fuel consumption will stop first. Runners are no different. A runner with poor economy will stop before a runner with good economy when running at the same speed.

Common areas for improvement

▶ ***Running economy*** *Running style and weight affect running economy. Sprinters have large amounts of muscle and a powerful action, which enables them to run quickly, but at high energy cost. Middle-distance runners rely on fast, sustained speed, which creates an "oxygen debt," often sacrificing efficiency for speed. Long-distance runners have low body weight and expend minimal energy so they tend to be efficient, economical runners.*

Oxygen uptake for runners with different disciplines

1 Weight Carrying extra weight—particularly body fat — requires more energy. Consequently, runners with a high body fat percentage need more oxygen and have poor running economy.

2 Extension of leading knee A straight knee will produce resistance when the foot lands on the ground, which needs to be overcome with extra energy before forward momentum is generated.

3 Foot strike Feet landing too far ahead of the body cause a braking motion that must be overcome before moving forwards. Avoid over-striding by planting the foot slightly ahead of the body.

4 Height of trailing foot Efficient runners with good running economy tend to keep their feet close to the ground, and avoid wasting energy with a high follow-through of their trailing leg.

5 Rotation Over-rotation wastes energy and makes it hard to run efficiently in a straight line. Rotation has to be stabilized with counter movements, requiring more energy and poor running economy.

6 Rear foot action Swinging the rear foot outward while bringing it forward for the next step is a common problem. This creates instability and a loss of forward momentum, which requires additional energy and oxygen.

7 Angle of body Relaxing and leaning slightly forward creates the optimum trunk angle and displacement of the body's center of gravity, making forward momentum easier and minimizing braking forces.

8 Bounding Vertical displacement does not help forward motion, so raising the body vertically should be avoided. Too much vertical displacement uses extra energy and results in poor running economy.

How can a runner's maximum potential be measured? → Will I ever win the Olympic Marathon?

The faster you run, the more oxygen that is needed to sustain the supply of energy to the muscles—a process known as aerobic metabolism. As running speed increases, the body responds by increasing both ventilation rate (the volume of air entering the lungs per minute) and heart rate to pump blood and oxygen around the body more quickly. This results in an increase in the rate of oxygen uptake, known as VO_2—the amount of oxygen extracted from the air in the lungs each minute—which is measured in milliliters of oxygen, per kilogram of body weight per minute (ml/kg/min).

In an ideal world, runners would be able to increase their oxygen uptake to match the rate at which energy is required. But unfortunately this is not the case, because there comes a point when it is impossible to supply any extra oxygen to the muscles. At this stage, the body has to obtain energy without using oxygen—this process, known as anaerobic metabolism, causes fatigue. When a runner has reached the limit of their oxygen uptake capacity, they have achieved their "maximum oxygen uptake" value, or VO_2 max. The higher that a person's VO_2 max is, the faster they should be able to run before experiencing fatigue. Scientists around the world use an athlete's VO_2 max value as the definitive way of assessing their capacity for endurance exercise.

Scientists have shown that VO_2 max can be improved by training, resulting in physiological adaptations that include the utilization of more alveoli in the lungs (small air sacs where oxygen diffuses into the bloodstream), enhanced cardiac output (the volume of blood pumped by the heart each minute), and a greater density of capillaries surrounding the muscles. However, even with training, there is a limit to how much VO_2 max can be improved, because it is largely determined by genetic factors. So, if you want to be an Olympic marathon champion, you need to choose your parents carefully.

Improvements in VO_2 max from different types of training

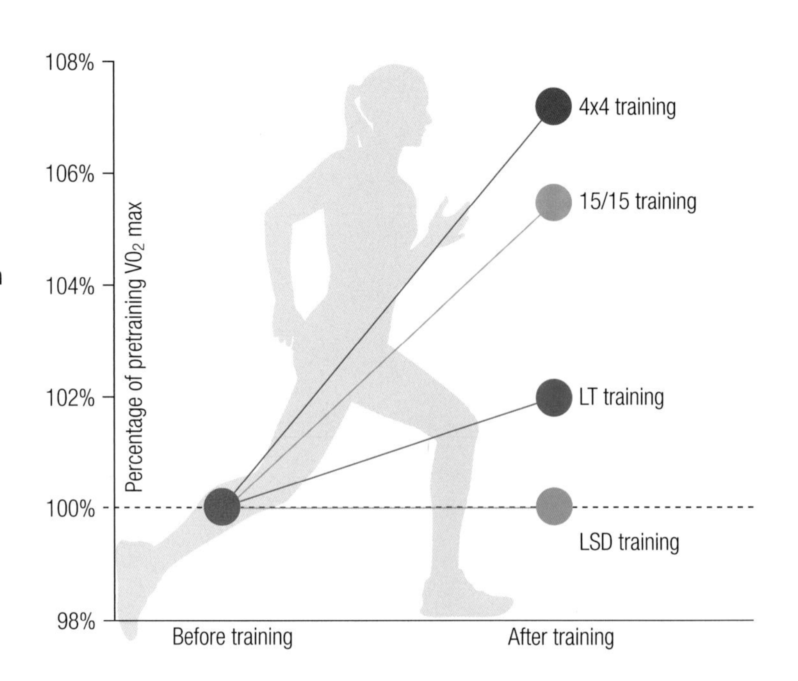

Need to know

VO_2 can be calculated using the Fick Equation:

$$VO_2 = Q \times (C_aO_2 - C_vO_2)$$

($C_aO_2 - C_vO_2$) is also known as the arteriovenous difference, where:

Q = cardiac output

C_aO_2 = arterial oxygen content (traveling from the lungs to the muscles)

C_vO_2 = venous oxygen content (traveling from the muscles to the lungs)

▶ ***The human body and exercise*** *The body's responses to exercise differ significantly from when at rest, and show the capacity that humans have for exercise. Increases in ventilation, heart rate, and cardiac output combine to deliver more oxygen to the muscles, ensuring that energy can be produced at a rapid rate.*

◀ ***The effects of different types of aerobic endurance training*** *Although to a large extent VO_2 max is genetically predetermined, it can be improved with training. However, there is an upper limit beyond which an individual cannot increase VO_2 max, no matter how much or how hard they train, and this limit will vary from one person to another. In individuals who have done only a modest amount of exercise, studies have shown that VO_2 max can be increased by around 10–15% after six–eight weeks of training. More recently, studies have shown that VO_2 max can also be increased with high intensity interval training (HIIT), where bursts of high-intensity exercise are repeated after a short recovery period. There is evidence that gains from HIIT can be as great, if not greater, than those from endurance training.[1] The graph shows the results of one study that compared long slow-distance training (LSD), lactate threshold training (LT), interval training with fifteen seconds of running followed by fifteen seconds of active rest (15/15), and interval training with four sets of four minutes of running followed by three minutes of active rest (4x4).[2]*

▼ ***Reaching the plateau*** *Elite runners can achieve higher speeds than non-elite runners before they reach their VO_2 max—the plateau where the body is no longer capable of taking in enough oxygen to meet the cost of running at a faster speed.*

Oxygen uptake plateau for elite and non-elite runners

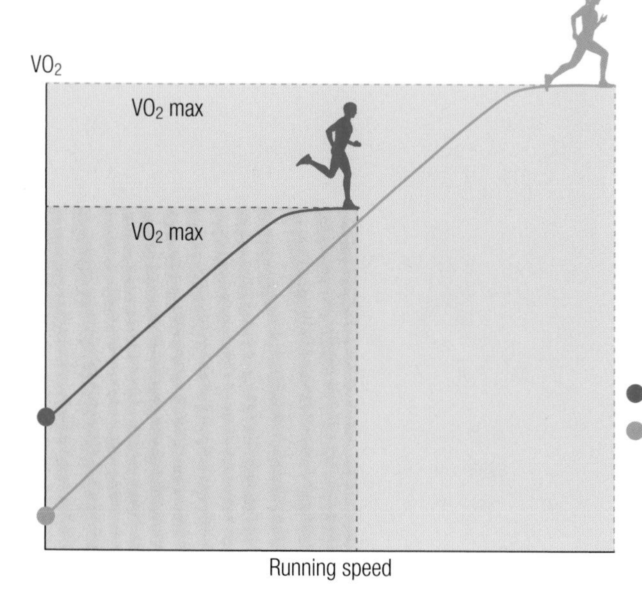

Typical physiological values for an elite athlete

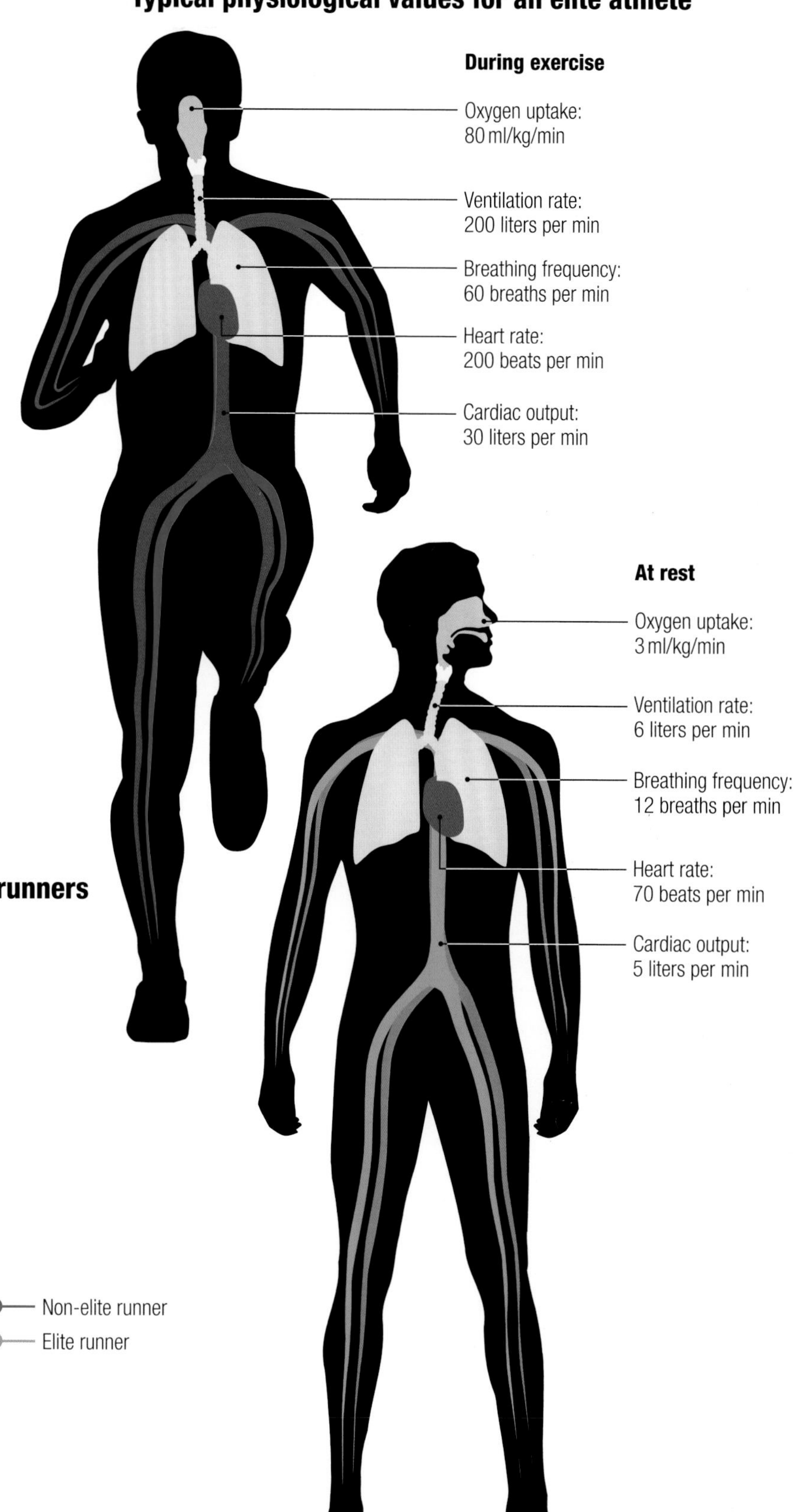

What affects recovery rate after intensive exercise?

Why can't I catch my breath after I stop running?

At slow speeds, the body's demand for energy is low, so the rate at which oxygen is supplied to the muscles to help break down carbohydrate or fat for energy is also low, and can easily be met from oxygen in the air that is breathed into the lungs. The equal matching of energy demand with oxygen supply, resulting in the release of energy, occurs during aerobic respiration. This process produces energy at a sufficient rate for low- and moderate-intensity exercise, with minimal fatigue, meaning that it can be sustained for prolonged periods of time. However, as running speed increases, so, too, does the body's demand for oxygen. When the rate at which oxygen can be supplied to the muscles to produce energy no longer meets the rate at which energy is required, an additional means of providing energy needs to be found.

When faster running and high-intensity exercise cause this situation to arise, the breakdown of carbohydrate occurs without oxygen being present—a process known as anaerobic respiration. Unlike aerobic respiration, anaerobic respiration does not involve the breakdown of fat—this fuel can only provide energy through aerobic respiration.

It is simply not possible for anyone to supply oxygen at the rate required for very high-intensity exercise. The higher a person's maximum oxygen uptake (VO_2 max), the more likely they are to be capable of meeting their energy needs through aerobic respiration, whereas individuals with a low VO_2 max quickly have to resort to fatigue-inducing anaerobic respiration when running speed increases. One of the main factors that differentiates elite from non-elite endurance runners is that those at elite level can run at higher speeds and exercise intensities using aerobic respiration alone to support energy provision.

At the cessation of high-intensity exercise, the body will have produced an amount of energy without the presence of oxygen, thus incurring an "oxygen debt"—the difference between the amount of oxygen that the body required, and the amount it was able to take in. This debt needs to be repaid when exercise stops, and is the reason why people breathe rapidly and deeply after intensive exercise. Carbon dioxide is also exhaled during the rapid breathing. Elite runners can increase their oxygen uptake rapidly in response to an increase in work rate, so they develop less oxygen debt and their recovery time afterwards is shorter.

Aerobic and anaerobic contributions for different distances

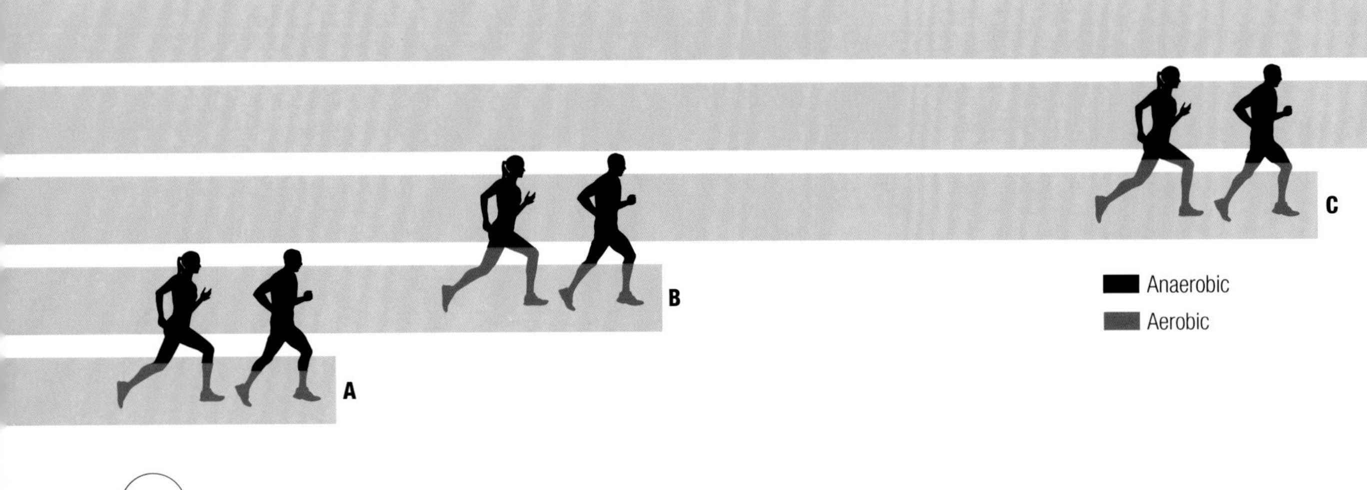

Oxygen debt

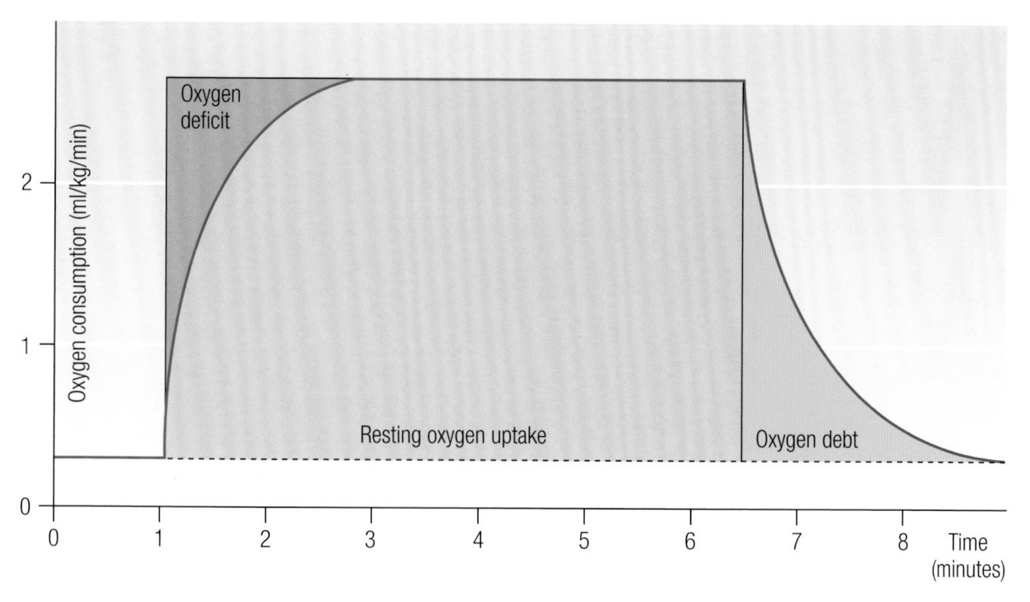

◀ ***Deep breaths*** *When exercise starts, there is a rapid demand for oxygen. The body takes a short period of time to react to this extra demand, so consequently some energy is produced anaerobically—without oxygen—until the rate of oxygen supply matches the rate at which energy is required. When exercise stops, the body "repays" this oxygen debt through an increased supply of oxygen during the early part of the recovery process.*

The oxygen deficit is the difference between how much oxygen is taken in compared to how much is needed

The oxygen debt is additional volume of oxygen taken in at the end of exercise to restore all energetic systems to their normal state

Aerobic and anaerobic respiration compared

Aerobic

- Slow to moderate rate of energy production
- Uses fat for energy
- Requires oxygen
- Slow to fatigue
- Low lactic acid

- Use glycogen for fuel
- Moderate- to high-intensity exercise
- Some lactic acid and oxygen debt

Anaerobic

- Rapid energy production
- Rapid fatigue
- Energy without oxygen
- Energy from creatine phosphate
- High levels of lactic acid
- Oxygen debt

◀ ***Different systems*** *The provision of energy during exercise relies on the aerobic and anaerobic systems. Both have the same outcome: they support biochemical reactions in muscles that result in muscle contractions. Some aspects of each system differ, while others are similar, and the extent to which one or the other dominates depends on the duration and intensity of exercise that is undertaken.*

	A	B	C	D	E
Men	**100 m**	**200 m**	**400 m**	**800 m**	**3,000 m**
Aerobic energy contribution	20%	28%	41%	60%	86%
Anaerobic energy contribution	80%	72%	59%	40%	14%
Women	**100 m**	**200 m**	**400 m**	**800 m**	**3,000 m**
Aerobic energy contribution	25%	33%	45%	70%	94%
Anaerobic energy contribution	75%	67%	55%	30%	6%

◀ ***Energy sources*** *When running speeds are fast, there is greater reliance on the anaerobic system for energy provision, while over longer distances, the aerobic system dominates. One of the by-products of anaerobic running is lactic acid, which quickly causes fatigue, and is the reason why 100 m running speeds can only be sustained for short periods.*[1]

What is lactic acid, and why does it build up during exercise?

At the end of a sprint, why do my legs feel like jelly?

While anaerobic respiration results in the rapid provision of energy for high-intensity exercise, it also produces a fatiguing by-product called lactic acid. If running speed remains high, and anaerobic respiration and the consequent production of lactic acid continues, the muscle cells become acidic, which inhibits the metabolic pathways that break down glucose to produce energy. This is actually a safety mechanism, since it helps to prevent the body from harming itself during extreme exercise, and causing injury or permanent damage. As a result, the rate of energy production is forced to slow due to the build-up of lactic acid.

Elite athletes are known to be able to tolerate higher levels of lactic acid than non-elite athletes during exercise, and are also able to clear lactic acid from their blood and muscles more efficiently when exercise has stopped. Their well-developed respiratory and cardiovascular systems mean that they can quickly supply high volumes of blood to the muscles to deliver oxygen and remove the lactic acid and carbon dioxide. The length of time that breathing rate remains elevated after exercise will depend on the intensity of the exercise and the fitness of the individual, and an active recovery, consisting of light jogging and stretching, has been shown to result in a more rapid recovery than stopping all movement completely.

Contrary to popular belief, there is little scientific evidence to suggest that lactic acid causes the muscle soreness that is often experienced one or two days after rigorous exercise. However, it does cause a painful "burning" in the muscles during high-intensity exercise, making rapid, coordinated movement difficult and forcing runners to slow down, and sometimes even causing an unsteady, "jelly-like" feeling. One of the reasons why sprinters cannot sustain their pace over longer distances is due to the build-up of lactic acid in their muscles and blood.

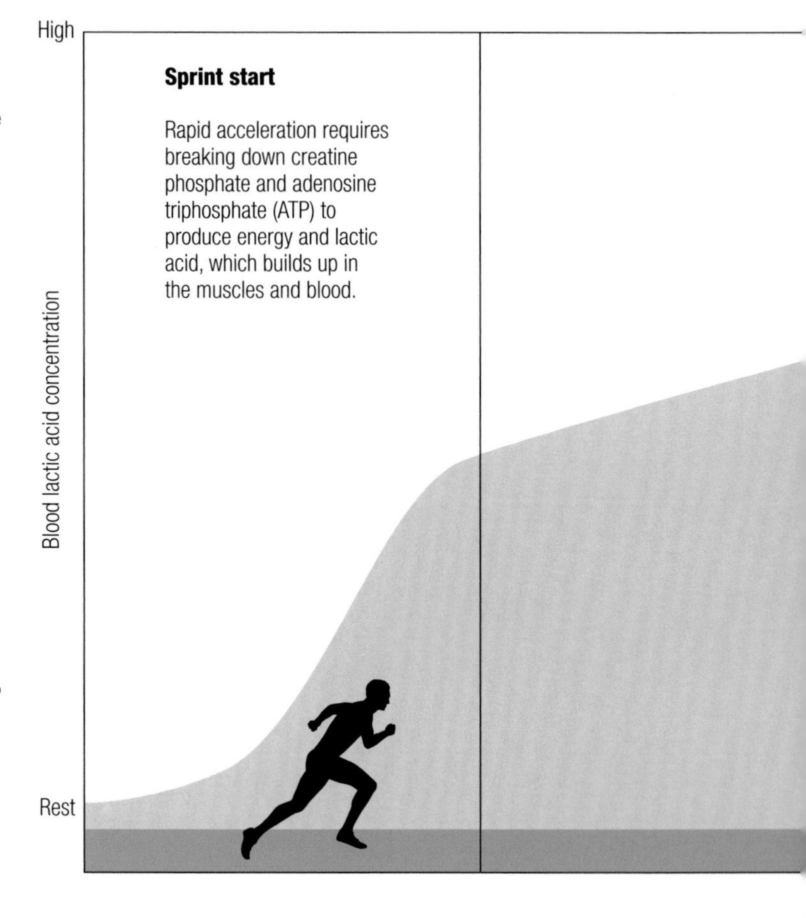

▶ ***Lactic acid*** *Running a 400 m race is all about tolerating high levels of lactic acid while maintaining step frequency and step length. Elite runners sustain near maximal sprinting speeds for the duration of the race, whereas recreational runners are more likely to demonstrate a gradual decrease in speed as the race progresses. Toward the end, lactic acid in the blood and muscles affects coordination and the muscles' ability to contract properly, and even with a warm down, lactic acid levels will take some time to return to normal.*

▶ ***Recycling lactic acid*** *Some of the lactic acid in the bloodstream returns to the liver, where it is converted to glucose through a process called gluconeogenesis, which plays a vital role in the maintenance of blood glucose levels. Lactic acid that is transported to the liver is converted back to pyruvate, and then to glucose, through a complex process called the Cori Cycle. Some of this glucose can be stored as glycogen to be used for future energy production, or to regulate blood glucose levels.*

The Cori Cycle

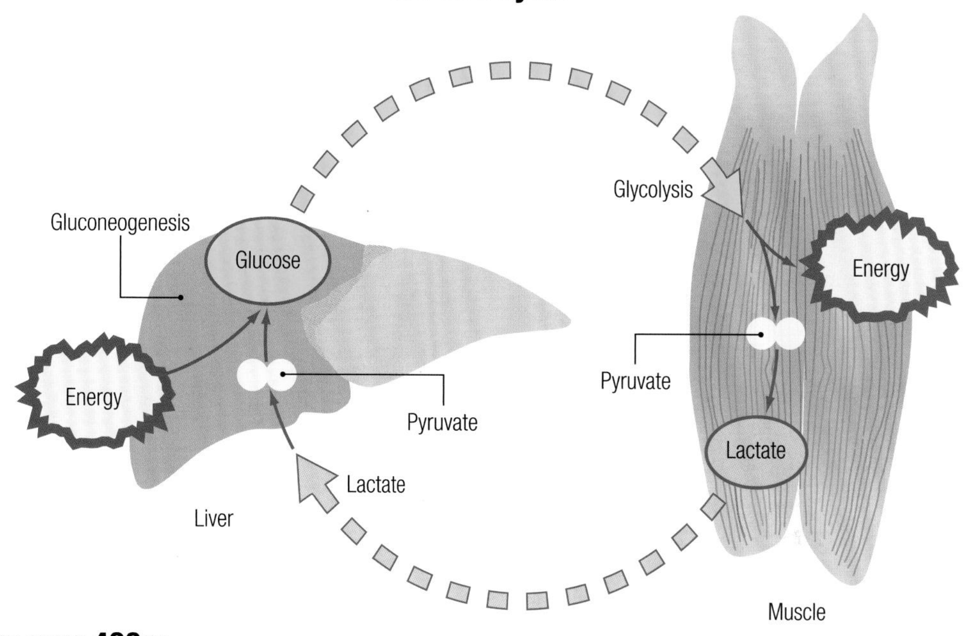

Blood lactic acid concentration changes over 400 m

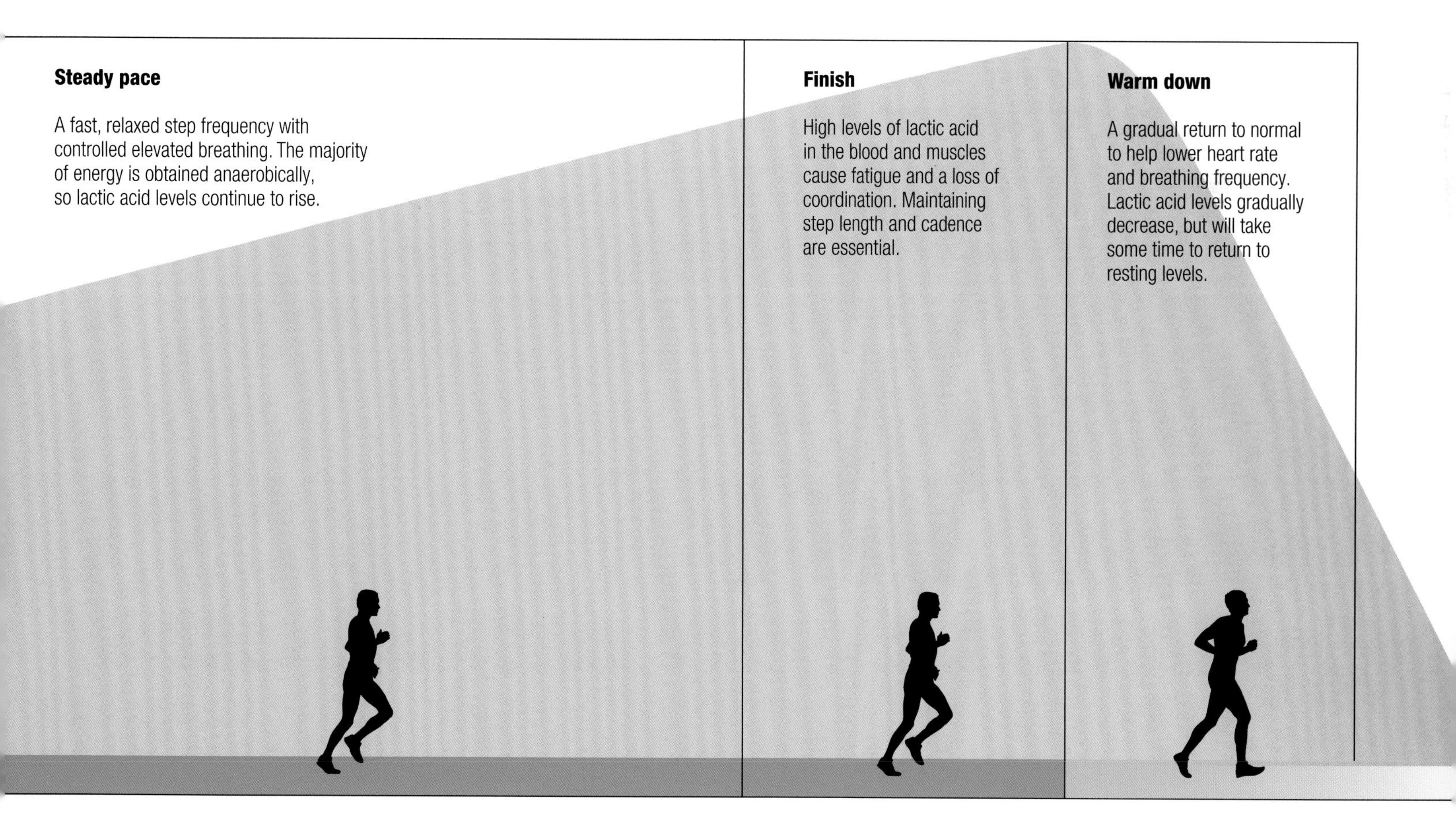

What are the main physiological changes that occur with aging?

Am I getting too old for this?

As with any organism, the human body deteriorates with age, a process that normally starts after adulthood is reached in the early or mid-20s. The underlying mechanisms are unclear, but it would seem likely that this is due to a combination of "wear and tear" and lifestyle. While aging is inevitable, the rate at which it occurs can be slowed through a combination of exercise and nutrition.

One of the principle organs of the body that becomes less efficient with age is the heart. Older cardiac muscle contracts more slowly, resulting in a smaller volume of blood being pumped around the bloodstream with each beat. By the age of 90, scientists have shown that the output of the heart is only 50% of that in someone in their mid-20s. This reduces

Records by age group and sex

▼ ***Running through the ages*** *The world best times for all running distances get longer with each age category from the mid-20s. Yet even with older runners these times are still very impressive, and demonstrate the capacity of the human body for high levels of human performance even with aging. Running is a sport that is suitable for individuals of all ages, and many elite older runners can produce times that are faster than those of runners who are many years younger.*

1,500 m

10,000 m

 under-20 female  under-20 male  20 to 24-year-old female  20 to 24-year-old male 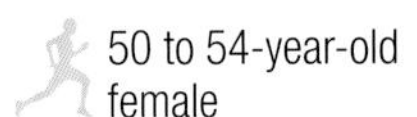 50 to 54-year-old female 50 to 54-year-old male

the capacity to deliver oxygen to the muscles. Concurrently, there is an age-related reduction in the muscles' ability to use the oxygen that they receive, mainly due to a drop in the density of mitochondria in the muscle cells—the "powerhouse" structures where energy is produced.

Strength peaks at around 25 years of age, but can decrease by 25% or more by the age of 65. This is due to a loss of protein from the muscle fibers, and a consequent decrease in their size and strength. As a result, powerful movements and speed decline with age. There is also a decrease in bone strength, resulting from a loss of calcium and reduction in bone density, an issue that is more prevalent in women than men.

Studies have shown that the age-related decline in physiological capacity can be reduced with exercise, resulting in aging that successfully maintains lifestyle and physical capabilities. For many, this will improve quality of life and longevity. Even in older people who have not exercised, targeted exercise programs have resulted in significant improvements in aerobic capacity and strength. While peak performance over shorter running distances where speed, strength, and power are critical tends to occur in the mid-20s, endurance runners often find they can sustain better performance and peak capacity much later in life.

Changes to the heart and arteries from aging

Younger man

Older man

Heart size
1 At rest
2 Start of beat
3 End of beat

Artery wall
A Adventitia
B Media
C Intima
D Arterial lumen

35:05.70 30:55.16 30:26.50 29:17.45 26:41.75 26:17.53

◀ ***Aging heart*** *As with all muscles, the heart becomes less powerful due to the aging process. As a result, in older runners it needs to beat at a faster rate to pump the equivalent volume of blood—known as cardiac output—when compared with younger runners. With aging, maximum heart rate decreases, and the heart rate required to sustain cardiac output increases.*

Is running performance dictated by nature or nurture?

Am I going to be a winner over 100 m or the marathon?

While training and coaching are fundamental to successful running performance, our genetic characteristics have a major impact on the events that we are good at. One factor that plays a large part in dictating whether a person is predisposed to either sprinting or endurance running is the composition of the muscles. Each muscle consists of many millions of fibers, which contract when they receive an electrical signal from the nerves. This contraction is fueled by the breakdown of molecules such as glycogen (the body's storage carbohydrate), and while some fibers are able to do this in the absence of oxygen, others are more readily able to contract when oxygen is present. Those fibers that can contract rapidly without oxygen are known as fast-twitch or type 2 fibers—they contract rapidly, but fatigue quickly. Those fibers that utilize oxygen to breakdown fuel, on the other hand, contract more slowly, but take much longer to fatigue. Consequently, they are known as slow-twitch or type 1 fibers. Not surprisingly, individuals with a high proportion of fast-twitch fibers are better endowed for sprinting, while those with a high proportion of slow-twitch fibers are better able to perform endurance running. The proportion of each muscle fiber type varies from muscle to muscle and is largely genetically determined, so the chances of being an elite runner will largely depend on genetic characteristics. Having said that, training and coaching can improve performance over any distance—indeed, although muscle fibers are either predominantly slow or fast twitch, there is evidence to suggest that endurance training can promote changes within fast-twitch fibers, and convert them into fibers that are more predisposed to endurance running.

Typical concentrations of different muscle fibers

Other characteristics that determine whether a runner will be good at sprinting include reaction time, strength, and flexibility. Endurance runners require a well-developed cardiovascular system and a high oxygen uptake capacity (VO_2 max). All of these can be developed through training, but the extent to which improvements can be achieved will still be dependent, to a degree, on genetic characteristics.

▶ ***Definitions*** *Each type of muscle fiber has characteristics that define its use. Rate of fatigue and speed of contraction are critical characteristics that determine whether a muscle fiber is best suited to sustained endurance activity, or to powerful, rapid contractions that produce high-speed running.*

▼ ***What's your type?*** *While muscle fiber type is not critical for recreational runners whose aim is to take part rather than to win, for elite runners, a predisposition for sprinters to have a majority of fast fibers, and for endurance runners to have a majority of slow-twitch fibers, is essential for success at the highest level.*

Properties of different muscle fibers

Type (twitch)	**1** (slow)	**2a** (intermediate)	**2b** (fast)
Respiration	Mainly aerobic	Aerobic and anaerobic	Anaerobic
Maximum duration of use	Hours	Up to 30 minutes	Under a minute
Capillary density	High	Intermediate	Low
Myoglobin concentration	High	Medium	Low
Glycolytic capacity	Low	High	High
Oxidative capacity	High	High	Low
Mitochondria density	High	Medium	Low
Fiber diameter	Small	Intermediate	Large
Contraction speed	Slow	Moderate	High
Force capacity	Low	Intermediate	High
Fatigue resistance	High	Intermediate	Low

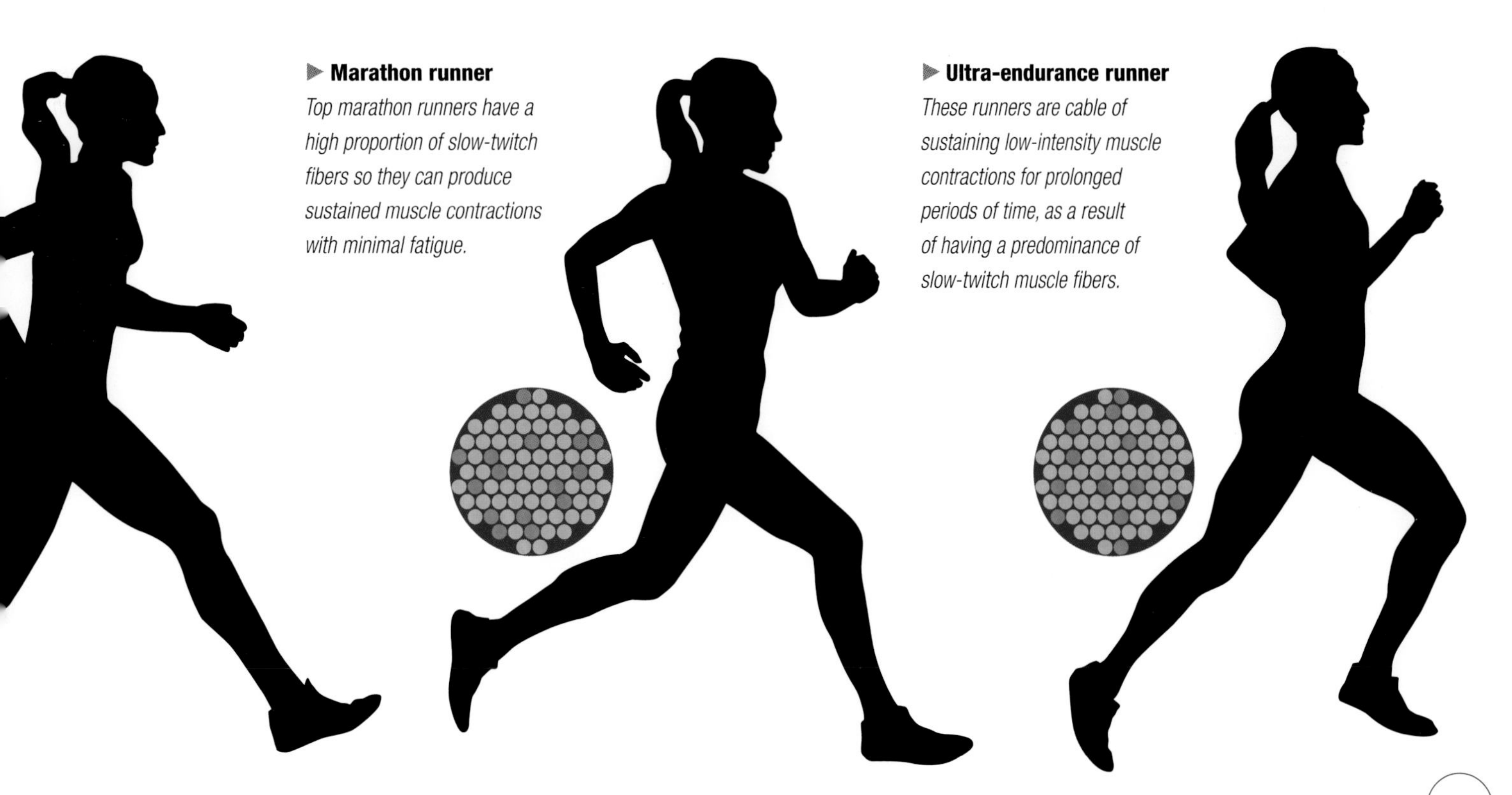

▶ **Marathon runner**
Top marathon runners have a high proportion of slow-twitch fibers so they can produce sustained muscle contractions with minimal fatigue.

▶ **Ultra-endurance runner**
These runners are cable of sustaining low-intensity muscle contractions for prolonged periods of time, as a result of having a predominance of slow-twitch muscle fibers.

What are the physiological adaptations of training?

How much fitter will I get from running?

The physiological consequences and adaptations of training are specific to the type of training that is undertaken. Speed training results in gains in strength and muscle size. This hypertrophy occurs because the muscle fibers increase in diameter as more protein develops within them. Endurance training results in many changes, most of which are related to the development of the cardiovascular system. The lungs become more efficient at extracting oxygen from air that enters them and at transporting this oxygen across thin membranes into the blood. This is mainly due to the opening up and development of tiny air sacs in the lungs called alveoli. The heart becomes a more efficient pump and expands in size, particularly the left ventricle, which is the chamber that pumps blood around the body, often reaching output values in elite athletes that are in excess of 30 liters per minute. Blood leaves the heart in arteries, then enters a broader network of smaller arterioles, and finally enters a very fine network of capillaries that surround organs and muscles, where the blood can release oxygen to support the provision of energy. Running and training help to increase the capillary network so that it is far easier for oxygen to disassociate from the blood and enter the muscles. Within the muscle cells, the "powerhouses" where the production of energy occurs are the mitochondria. With training, the number of mitochondria increases, making it easier to produce the energy that is required to support high-quality running.

Running also strengthens the tendons that join muscles to bones, and the ligaments that join one bone to another. The steady impact forces that go through the joints and muscles as a result of running also strengthen cartilage and bones, something that is particularly important as people get older.

In addition, running releases endorphins, chemicals produced by the body that create a natural "high." So as well as the physical and physiological benefits, running can also have positive psychological effects and enhance feelings of well-being and self-esteem.

Tendon and ligament strengthening

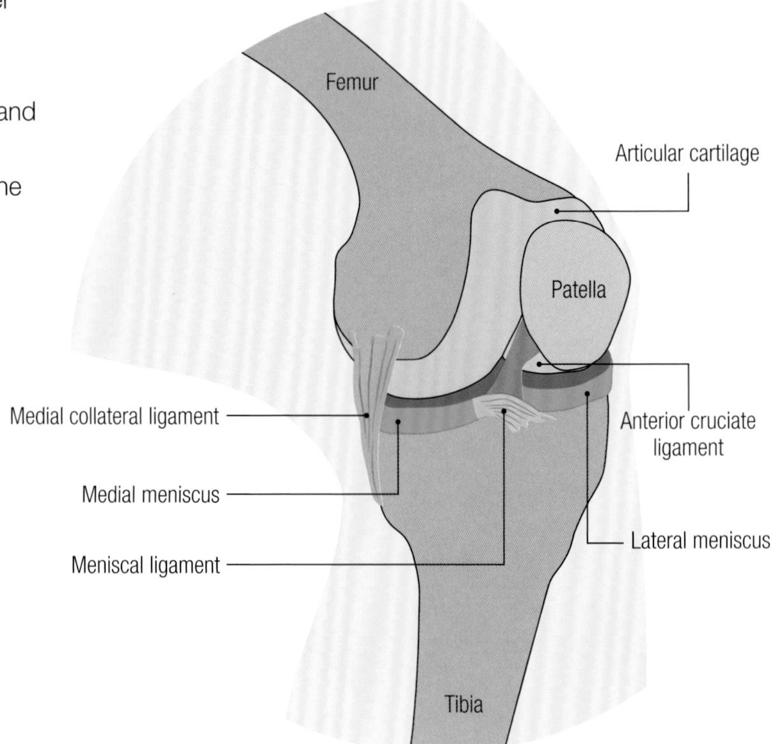

▶ ***Training benefits*** *Injury is an inevitable hazard and consequence of regular running. However, running can improve the strength and mobility of tendons and ligaments, which reduces the risk of injury, and helps the human body to sustain regular impacts and high forces. Regular training can make tendons and ligaments stronger and more flexible, thus reducing the risk of injury.*

Speed training

Normal muscle

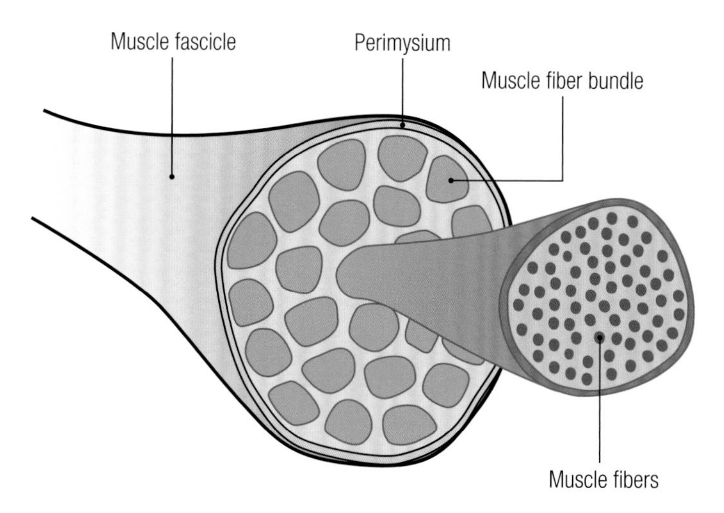

Bulked muscle

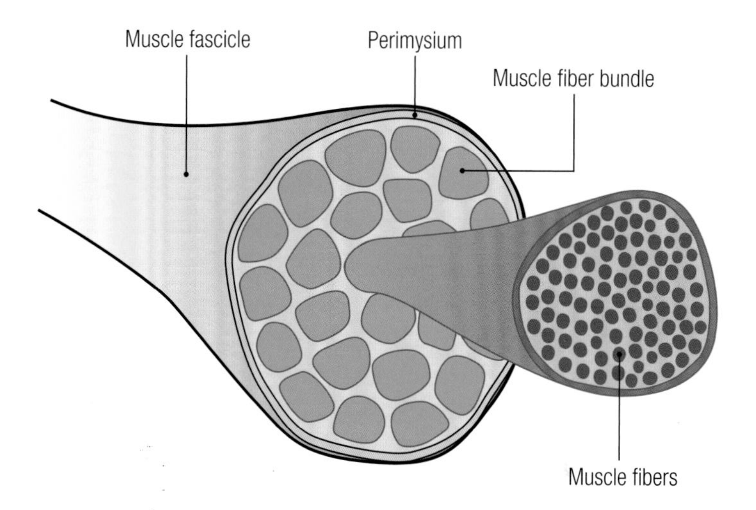

▲ ***Muscle bulk*** *Speed training will help to develop the anaerobic capacity of a human muscle fiber. It results in extra protein building up within a muscle fiber, making it larger and stronger, and more capable of producing powerful and rapid contractions. The fibers will increase in diameter and contain more proteins that work together to produce rapid, fast contractions that, in turn, result in high-speed running.*

Endurance training

Bronchus

Bronchioles

Alveoli

Normal lungs

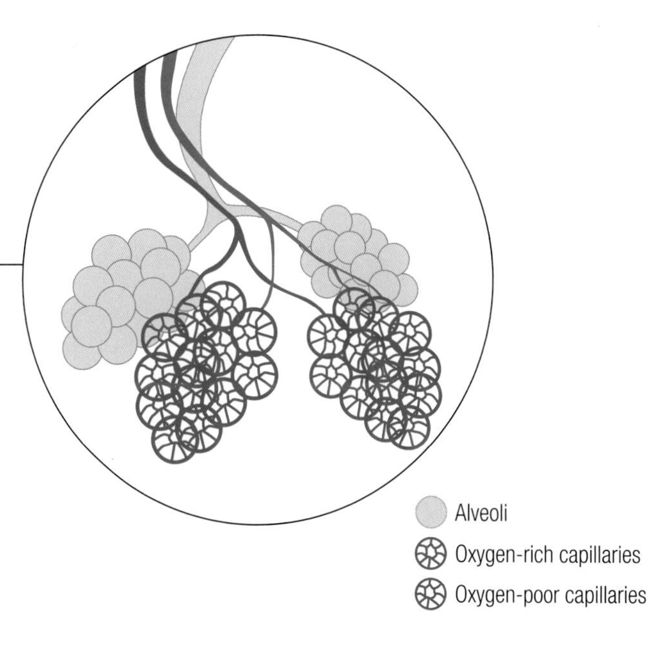

◀ ***Oxygen efficiency*** *Endurance training improves the body's ability to absorb oxygen from the air that enters the lungs. It increases the size of the lungs by opening up new branches and small sacs, called alveoli, so that oxygen can diffuse more easily across thin membranes and enter the bloodstream, before being transported to the muscles to help with energy production.*

SCIENCE IN ACTION

running in hot climates

Running places a challenge on the human body that significantly exceeds the physiological demands experienced at rest. But these demands can be even higher in challenging environmental conditions, and particularly when conditions of heat and humidity are encountered. Adapting to and dealing with hot and humid conditions is something that many runners are faced with at some point in their running careers, and failing to do so properly can lead to unpleasant and even fatal consequences.

Running raises body temperature, and the heat that is produced needs to be lost to prevent a dangerous overheating condition known as hypothermia. Sweating is the first way in which heat loss occurs, but this is less effective when environmental conditions are hot, or when moisture in the air—known as humidity—prevents sweat from evaporating efficiently. To counter this, the body subconsciously diverts blood from the core muscles to the skin so that hot blood can lose some of its heat into the external environment. To enable this to occur, the heart has to beat more rapidly because blood still needs to be sent to the muscles to help with energy production, so running feels—and is—much harder in environments that are hot and humid.

Many elite runners will adapt to hot conditions during their training, moving to warm environments to train, or use high-tech heat chambers to acclimatize to greater heat and humidity. For the majority of recreational runners, this is not practical, but it is still possible to use simple self-help measures to prepare for running in hot and humid conditions. Wearing extra layers as a means of replicating a hot and humid "microclimate" is a simple and low-cost means of preparing for races that have the potential to be held in such conditions. While the external conditions may not be hot and humid, the climate between the clothing and the skin is. This creates an environment of warmth and humidity that would otherwise not be possible, and, as a result, the body will start to adapt to running in these conditions. Training wearing extra layers in the build up to an important race, or one that is likely to be in hot or humid conditions, has real benefits and can also be of help even if conditions are normal, because running will feel easier with less heat stress after a period of acclimatization.

▶ ***Staying cool*** *Losing body heat is critical in hot conditions, and the evaporation of fluid from the skin—whether from sweat or an external source—can help to keep the body cool and prevent dangerous overheating.*

SOLE 2 SOUL
EXTREME
ADVENTURE
MÉXICO
1
adventurecorps
HAMMER
GEL

What is the optimum method for ventilation when running?

Should I breathe in through the nose or the mouth?

Getting air into the lungs is critical for runners, since it provides oxygen that is essential for energy production. The transport of air into and out of the lungs is known as ventilation, and ventilation rates of around 150 liters of air per minute are not uncommon when running at a high intensity.

Ventilation is controlled by the diaphragm, a layer of muscle underneath the lungs, and the muscles of the rib cage, known as the intercostals. As in any muscle, there is an energy cost to the contraction of the diaphragm and intercostal muscles, which increases as ventilation rate increases. Much of this energy cost is due to resistance that the air encounters on its way into and out of the lungs, which increases when narrow passageways are encountered.

Air can enter the lungs through the mouth, or through the nose. At low running intensities, where the ventilation rate is modest, either route works. However, studies have shown that the mouth offers the route of least resistance, and that when exercise intensity increases, the body uses both the mouth and nose to breathe.[1] This is known as oronasal ventilation. Ventilation rate is calculated from the breathing frequency and the volume of each breath—breathing frequency can increase from around twelve breaths per minute at rest to over sixty breaths per minute during exercise. The volume of each breath (known as tidal volume) also increases from rest, and could be double at maximum intensity effort.

In order to promote nasal breathing, some athletes have used nasal strips to widen the diameter of the nostrils and reduce nasal resistance. However, research has demonstrated that these strips have little or no effect on ventilation rates or performance, since the nose still provides greater resistance than the mouth for incoming and outgoing air.[2]

Need to know

Ventilation rate (V_e) can be calculated using the following equation:

$$V_e = V_t \times F_b$$

where:

V_t = Tidal volume

F_b = Breathing frequency

At rest, tidal volume for an adult is around 0.5 liters and breathing frequency is about 12 breaths per minute, so ventilation rate is normally around 6 liters per minute.

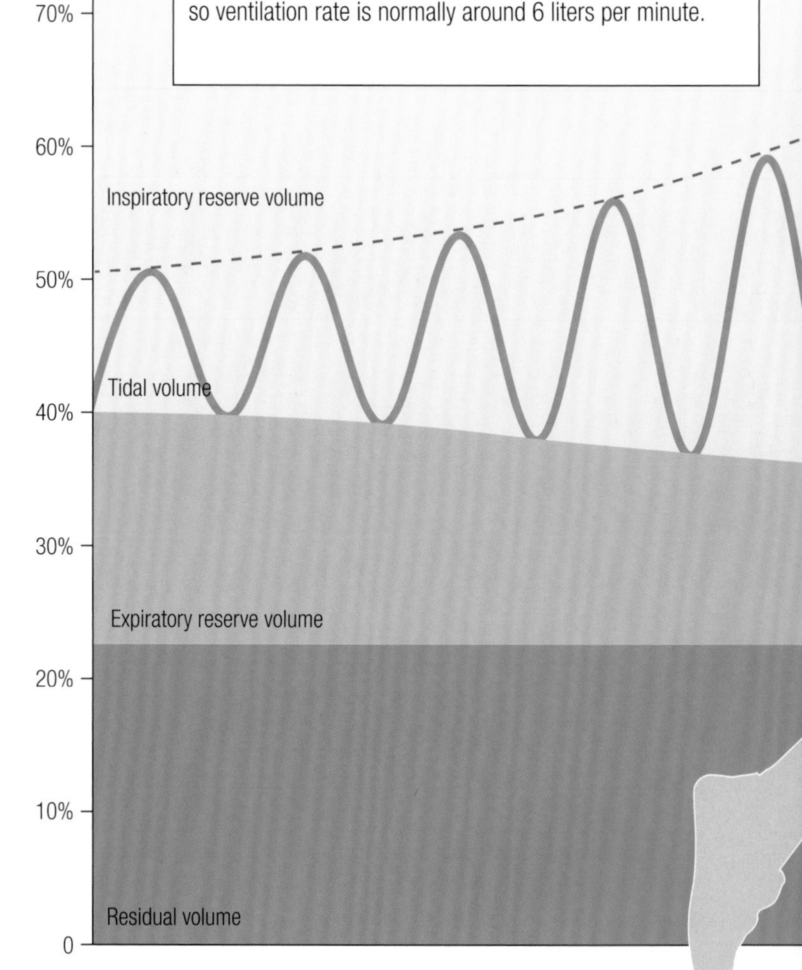

▼ ***Deep breaths*** *High volumes of air enter the lungs during exercise; this process becomes inefficient if too much resistance is encountered. Scientists have found that when ventilation rates exceed 40 liters of air per minute, the most efficient route, encountering least resistance, is through the mouth. The nose is used for breathing during exercise, but only when small volumes of air are required.*

Respiration during exercise

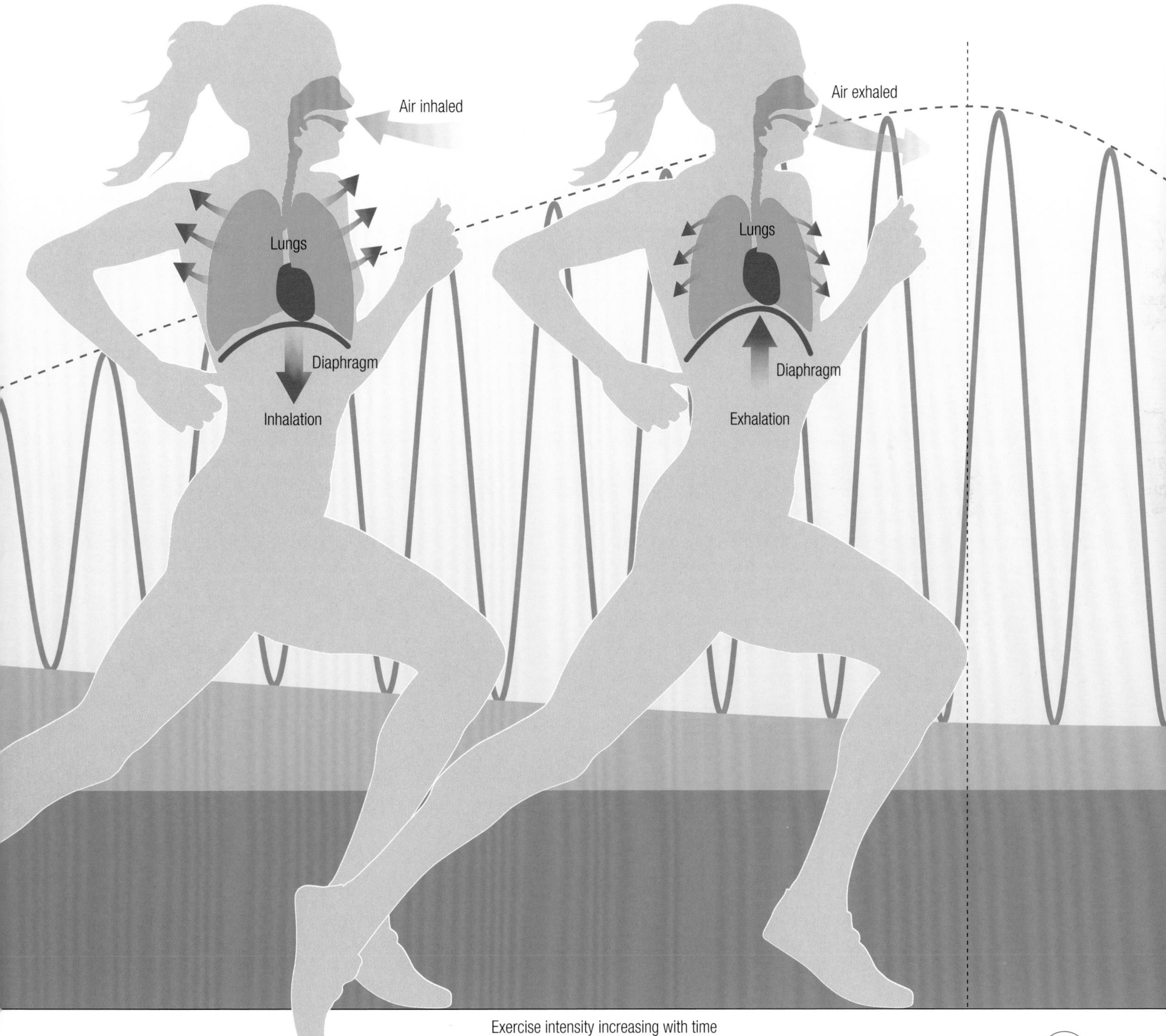

events:
sprinting

Speed, power, and—above all—explosive strength best summarize the art of sprinting. It may appear to involve nothing more than an all-out dash for the line, but in reality sprinting involves three distinct phases—drive, transition, and maintenance—and an understanding of effort every bit as important as for a long-distance event.

Phase one, drive, is about the first thirty meters of the sprint, even in events as long as 400 m. Reaction time, drive in the first three or four steps, and body position are all vital elements to get you up to speed as quickly as possible. From there, phase two involves smoothly transitioning to your normal running form, taking another twenty to thirty meters to do so; and thereafter, phase three is about maintaining form to the line.

Starting blocks are used in sprint races. The angle of each plate on the block can be changed and there are numerous papers discussing the optimum position to create maximum drive from power generated, but using blocks, more often than not, is much more about comfort and control than exact science. Some of the world's greatest athletes position their feet so they are off the ground at the start, others prefer to be in contact. Body size, strength, and mobility all have roles to play when creating the best position on the starting blocks.

Keep in mind biomechanics—those first few steps are all about driving forward at a 45° angle, before coming up into an upright position. Not surprisingly, core strength, gym work, intensity, and a superb range of mobility are all important elements sprinters must include in their training program. While muscle composition plays a role—sprinters need a greater proportion of fast-twitch fibers compared to slow-twitch fibers—step length and cadence are also significant. Improving your step length by just 50 mm (2 inches), with the correct training, may result in a two-meter improvement over 100 m.

When it comes to running, sprinters understand their event is anaerobic in that it involves the rapid use of a limited fuel supply—explosive energy. Longer sprints such as the 200 m and 400 m are predominantly anaerobic, with the demand for fuel outstripping supply, but with a small aerobic contribution as the race progresses.

Perhaps more than any other running event, sprinting is highly technical and responsive to body position, so working on drills to improve posture can prove every effective.

Article by **Paul Larkins**

▶ ***Holes*** *(pre-1938)*
Until the 1948 London Olympics, a trowel would be used to dig a hole in the track to create an area to drive off from. Of course, a hole in the ground can be unstable and destructive to the track surface, hence the evolution to the use of blocks. Nonetheless, the practice continued well into the 1970s and early 80s at events taking place on cinder or dirt tracks.

Development of starting block

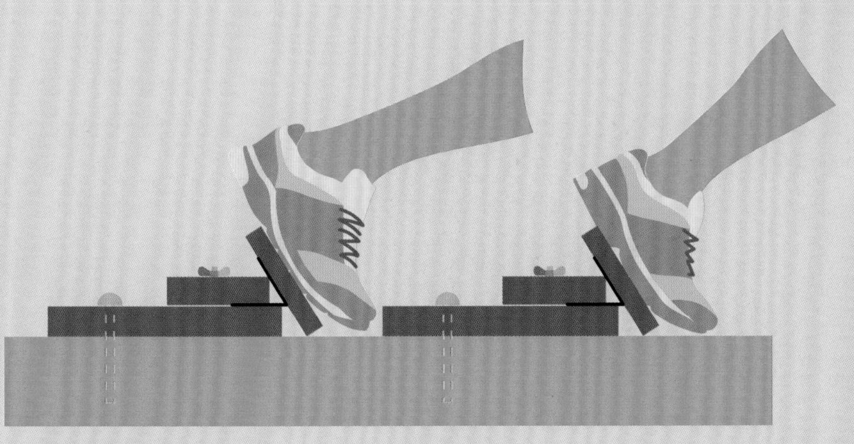

▲ ***Starting block patent*** *(1927)*
The original foot support patent used wood for the "blocks," which were hinged together to create the base for a runner to drive off from. Before long, blocks were constructed from steel, which formed a more stable base.

▼ ***First Olympic Blocks*** *(1948)*
Starting blocks were approved by the International Association of Athletics Federation (IAAF) in 1937, but the outbreak of the Second World War meant they were first used in the Olympics in 1948.

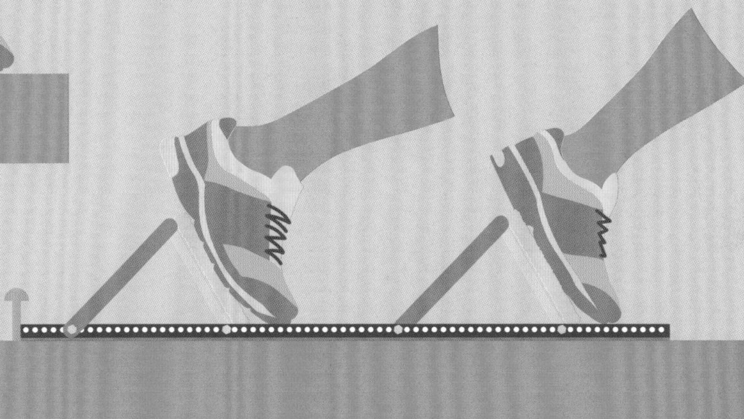

▼ ***Tall blocks*** *(2016)*
Modern blocks used at major meetings have speakers so that all the runners hear the sound of the gun at the same time. They also have sensor plates fitted that warn if the athlete reacts quicker than 0.1 second, a time based on tests that show athletes cannot hear and process the gun's sound any faster than this.

▲ ***Short/low blocks*** *(1996)*
Block design has actually evolved very slowly, with only minor alterations over the years. The size of the plate from the which the athlete drives away from has been the primary area of change, from large to small for most of the latter part of the twentieth century and currently back to large. Large footplates allow the athlete to drive back hard and tend to be favored by power athletes, while smaller footplates throw the athlete onto their toes very quickly. It's very much a case of personal preference.

This chapter introduces some of the common aspects of biomechanics and how they explain human running performance. Biomechanics is a much misunderstood and maligned area of sport science, and it is often seen as too confusing or abstract to be of use to coaches or runners. However, any coach who does not understand the basics of biomechanics is missing an important knowledge area. Biomechanics is defined as "the study of motion in biological systems." If we assume that one of the most important jobs of a running coach is to improve movement efficiency in a human biological system, a working knowledge of biomechanics is vital in order to understand the fundamental principles behind human running performance. Athletic performance is principally governed by how much and how quickly an athlete can apply force in an appropriate direction to cause the intended outcome; biomechanics explains why certain movement techniques achieve this in an efficient way, helping to maximize running performance.

chapter two

perfect motion

Iain Fletcher and Laura Charalambous

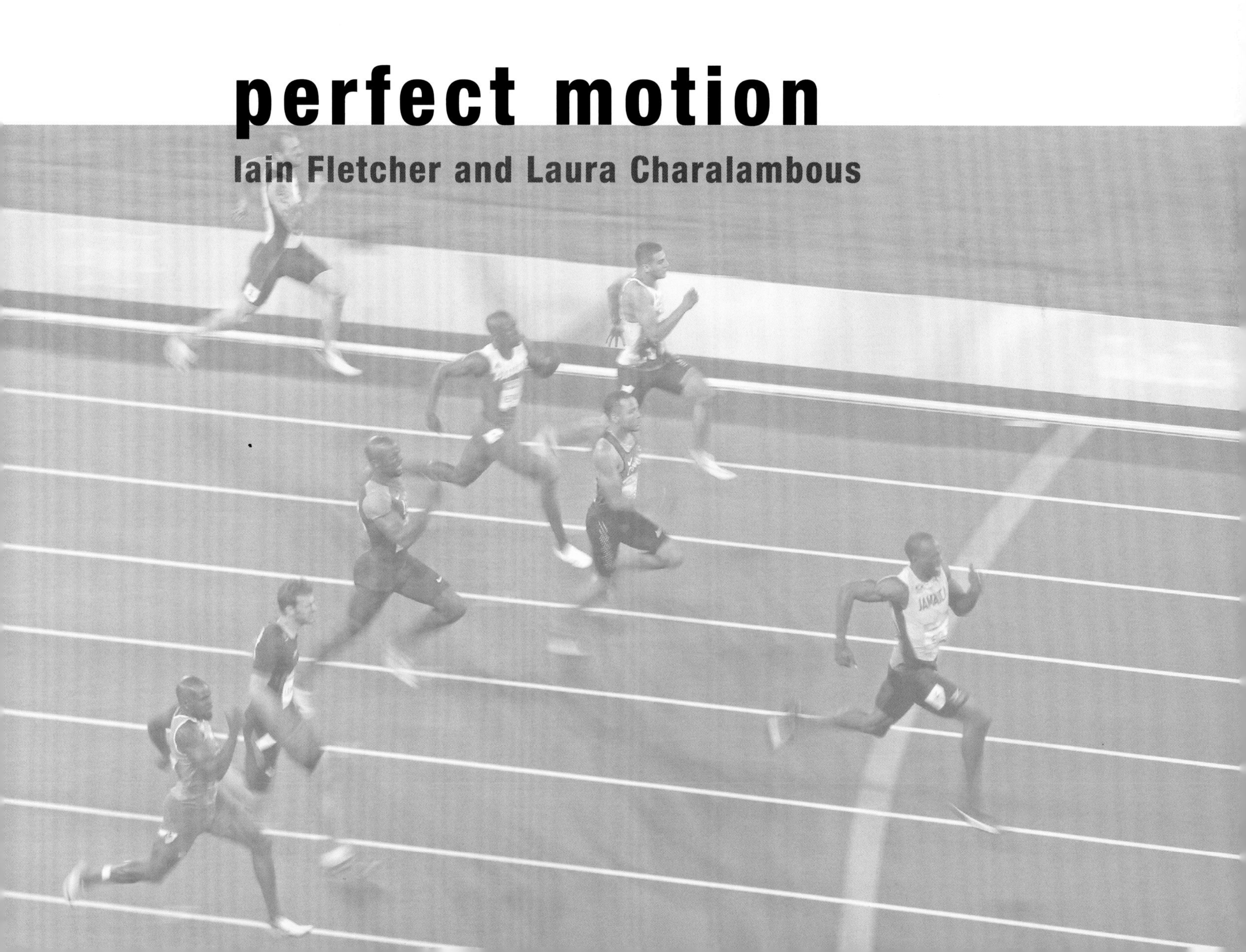

Is there an optimum sprint start position?

How do I get off to a good start?

Perhaps surprisingly, there is no single optimum start position for sprint events, due to anthropometric and strength differences between athletes. But there are some important considerations to be made when finding the optimal position for a specific athlete. Ultimately, the athlete aims to exit the blocks with the greatest horizontal acceleration possible, and in a position to continue this acceleration. So the optimal sprint start position is one that best orientates the athlete to produce maximal horizontal force, in the fastest time. This optimal set positioning of both the blocks and the athlete has been the focus of much research.

From the set position, the athlete extends both legs to drive from the blocks and the hands are lifted. The rear leg exits the block and is pulled forward as fast and as low as possible. The front leg continues to extend as the arms swing and the head should remain in a natural line with the trunk. When near full extension of the front leg is achieved, the front toes leave the blocks and block propulsion is complete.

The optimum starting position is specific to the individual, and depends on the athlete's strength and anthropometric characteristics. The overall aim is to position the center of mass (CM) over the legs, as far forward as possible while maintaining balance.[1, 2] This means the knee and hips should not be too flexed, ensuring the first movements do not create excessive vertical displacement.

Need to know

Impulse–momentum relationship

force × time = impulse

impulse = change in momentum

momentum = mass × velocity

The aim of the sprint start is achieve maximum horizontal velocity in minimal time. From the impulse–momentum relationship above, increasing either force applied or the duration of force application will increase the impulse and therefore the block exit velocity. Since time is obviously constrained, however, increasing the force applied is more effective for enhancing sprint start performance.

Finding the optimum start position

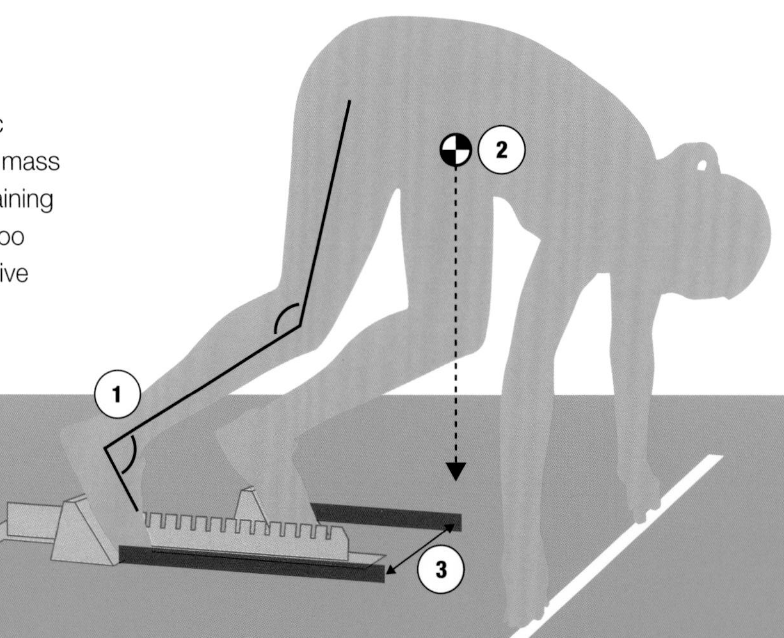

▶ ***The perfect start*** *Key aspects of the block and athlete positioning should be considered when working to optimize sprint start performance.*

1 Lower leg angles

Studies suggest that there is no single optimal position in the blocks and that a large range of set positions exist, even within a group of elite athletes.[3] For example, no significant correlations were found between body position in the blocks and block performance in a group of 16 elite to well-trained 100 m sprinters, whose personal best times ranged from 9.98 to 11.60 seconds.[4]

2 Center of mass

Positioning the center of mass (CM) above the legs enhances pretension in the leg muscles.[5] Additionally, if the CM is too low or positioned behind the feet, horizontal propulsion will be impinged and excessive vertical displacement will occur. An excessively forward CM will also be detrimental to horizontal force production.

3 Mediolateral block spacing

Studies manipulating mediolateral spacing (the distance between the left and right blocks) found no effect on performance when adjustments were made between 0.24 and 0.52 m or between 0.25 and 0.45 m.[6, 7] Advice to athletes should therefore be to ensure they are in a position that allows maximum contact area between the feet and block faces.

4 Anterior–posterior block spacing

The larger the distance between the front and back blocks, the greater the horizontal impulse—due to greater forces exerted by the rear foot, for a longer time. As a result, block exit velocity is greatest with a longer block spacing. However, an elongated position means more time is spent in the blocks, which is clearly a negative effect. Consequently, a medium block spacing is widely accepted as the most beneficial as it allows sprinters to generate relatively large forces without spending a detrimental amount of time doing so.[8, 9, 10]

5 Block face angles

The angle of the front block surface relative to the track affects the start velocity, with a more acute or smaller angle resulting in greater force production at the ankle.[11] This has been attributed to the more dorsiflexed ankle allowing pretension in the calf muscles and Achilles tendon, and indeed further research has confirmed this mechanism, showing that a 40° front block angle resulted in longer muscle–tendon lengths of the gastrocnemius and soleus (calf muscles), greater ankle joint moment and power, and consequently greater block exit velocity than a 65° angle.[12] The angle of the back block should also be steep to maximize block exit velocity, though a little less so than the front—for example, 30° front and 50° rear.[13] However, studies have also found that different block face angles have no effect on the performance or movement patterns when exiting the block or by the first stance touchdown, suggesting the effects of block orientation are temporary.[14]

6 Distance of the front block from the start line

You might think that the blocks should be very close to the start line, so that the athlete's center of mass starts as far forward as possible, reducing the total distance to be covered during the race. But in fact it has been found that when the blocks are positioned further from the line, the sprinter produces greater horizontal impulse and thus achieves a higher block exit velocity.[15]

If the front block is too close to the starting line, much of the body mass will rest on the front leg and the knee angle will be less than 90°, detrimental to force production. A compromise is required to ensure the athlete's mass is distributed over both legs while not being a detrimental distance from the start line.

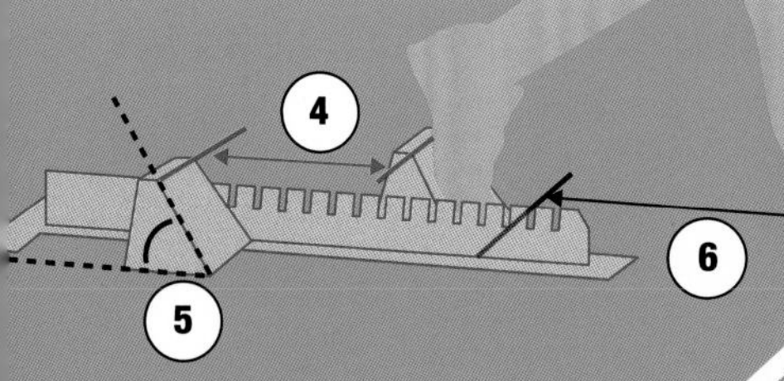

How does running surface influence an athlete's technique?

What's the difference between running on concrete and on grass?

The aim of running is to achieve and maintain a high horizontal velocity of the athlete's center of mass, propelling the body forward, while limiting the vertical movement between steps. Running can be considered as a "bouncing gait"—the actions of the body's many musculoskeletal structures are integrated together so that the overall system behaves like a vertical, single spring.[1] The spring is compressed during the first stage of the ground contact, with the system storing energy, before lengthening during the latter phase and releasing energy. The stiffness of this spring represents the average overall stiffness of the musculoskeletal system during the ground contact, which is termed "leg stiffness." An ideal balance in stiffness is required to optimize the kinematics and kinetics of the body's interaction with the ground,[2] maximizing performance[3] while minimizing the risk of injury.[4] Researchers have shown that an athlete can manipulate the level of stiffness depending on the type and velocity of movement,[5, 6] ground surface,[7] and footwear worn.[8]

The level of stiffness of a surface is described by its elastic modulus. For example, a pavement or road is a stiffer "spring," with a higher elastic modulus, than grass or a mud trail. Studies have shown that runners undergo kinematic (movement) changes in order to adjust leg stiffness between steps, according to the anticipated stiffness of the surface they

The relationship between leg and surface stiffness

are about to contact.[9] In this way, they are able to keep the overall stiffness of the leg–surface–footwear spring system consistent, while also maintaining both a vertically level center of mass and similar impact forces whether running on hard or softer surfaces.[10]

Road runners who are frequently injured are often advised to run on a softer surface. However, softer surfaces can also predispose runners to injury. For example, running on a harder surface, and using a lower leg stiffness to compensate for this, involves greater flexion. The muscles and tendons of the leg absorb more of the impact, and may therefore be at greater risk of injury. Running on a softer surface with a higher leg stiffness, on the other hand, has been associated with higher rates of plantar fasciitis,[11] tibial stress fracture injuries,[12] and Achilles tendon injuries.[13] A reasonable approach could be to vary your running surface, thus not over exerting either the bone or soft tissue structures. One certain advantage to running on softer surfaces is that they tend to be more irregular. If you run on a road, every step is about the same as the last, just as on a treadmill, However, if you run on a trail there are small variations in each step, leading to a greater adaptation to exercise and a more robust musculoskeletal system.

▼ ***Adaptation*** *When running on different surfaces, athletes will adapt their leg stiffness to account for changes in surface stiffness (quantified by the elastic modulus of the surface), maintaining a consistent stiffness of the leg–surface–footwear spring system. The athlete will pretune their leg stiffness before impact by combining all available information to anticipate the stiffness of the upcoming surface. This will include proprioceptive feedback from the previous contact, external feedback such as visual clues, and their past experiences of contact with the surface. Leg stiffness is calculated as the vertical ground reaction force divided by the change in leg length of the athlete. One way the athlete adapts their leg stiffness is by changing the level of pretension in their muscles before ground contact. This often-increasing flexion at the knee will result in a decrease in leg stiffness.*

What is the optimum relationship between step length and frequency? → Should I take fewer, longer steps or more, shorter steps?

Running time is ultimately dependent upon the average step velocity for a given distance. Individual step velocities are the product of the step's length and frequency. Consequently, step length and step frequency are considered the higher-order variables of running technique, evident in the hierarchical models developed of running and sprinting.[1]

Studies have found that, as velocity increases, step length and step frequency do not each increase in linear proportion with it.[2] At lower velocities, step length increases more steeply as the runner speeds up,[3,4] while step frequency appears to be more important in achieving higher velocities.[5,6] Near maximum velocity, an increase in either step length or frequency may be accompanied by a decrease in the other, because of the way the two factors are inherently linked.[7] Consequently, an optimum balance between step length and frequency is required to maximize sprint velocity.

Research has investigated how step characteristics interact within individual sprinters, and compared results between individuals, often with conflicting results. It has been suggested that between individuals, longer step length produces greater top speed, but within individuals, increases in step frequency are more effective in enhancing velocity.[8] So it seems unclear whether training to develop step length or frequency is more important in achieving an optimal step characteristic relationship.

Step lengths and frequency for record times

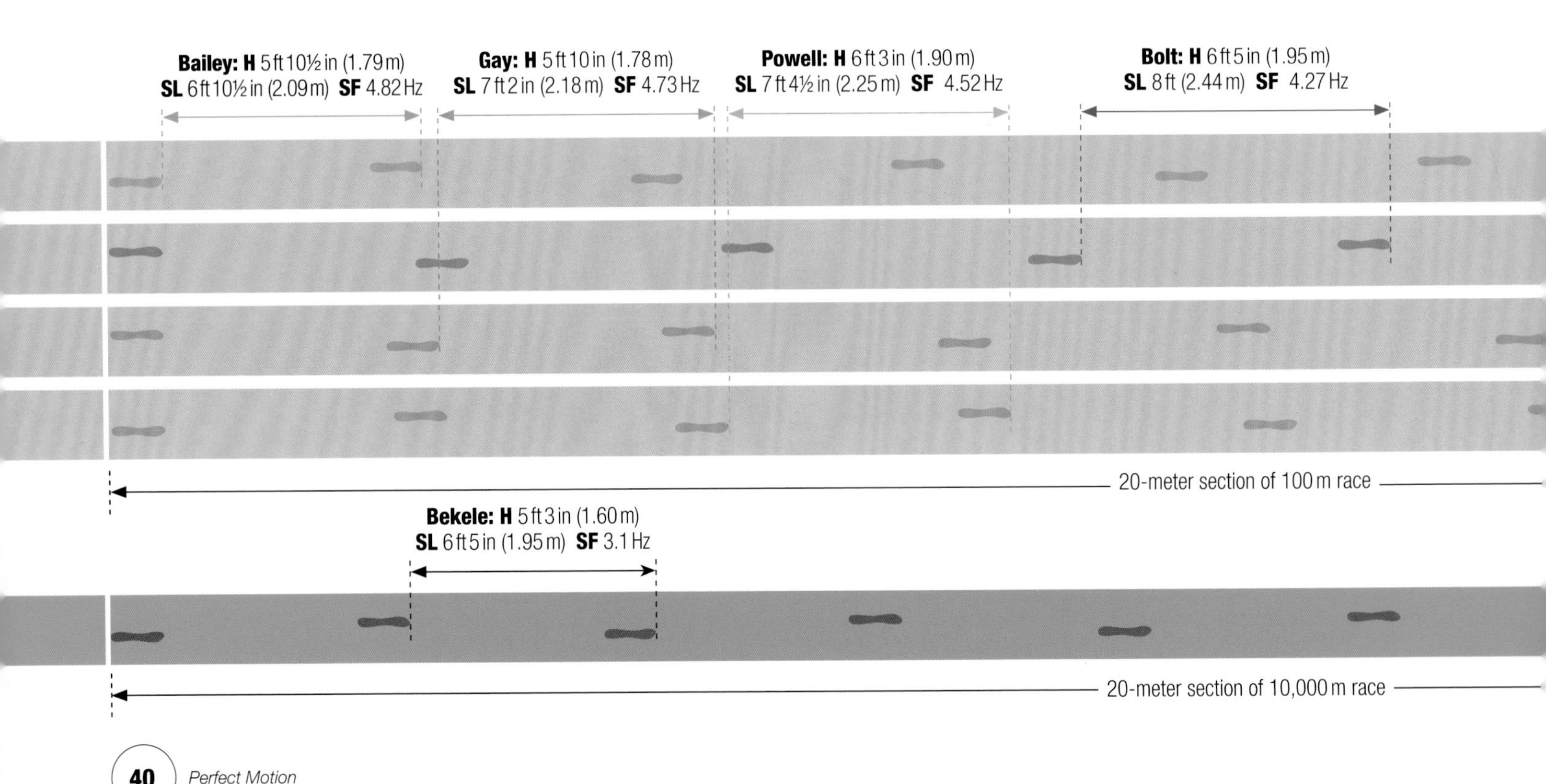

A study of eleven top sprinters' competitive performances[9] revealed that different athletes can perform at an elite level with either step length or step frequency being the characteristic with the greatest effect on velocity—similar results have been found in recreational athletes,[10] and in both males and females.[11] In terms of training, step-frequency reliant athletes should focus on neural activation to allow quick turnover of the legs, precompetition. Meanwhile, step-length-reliant athletes should maintain strength and flexibility, ensuring they can optimize their step length and therefore their velocity.

In summary, step length usually naturally increases with step velocity at lower running velocities, but towards maximum velocities (that is, sprinting) it can be either step length or step frequency that is key to improved performance, depending on the athlete.

Need to know

$$SV = SL \times SF$$

where:

SV = step velocity (the average horizontal velocity from one foot contact to the next contralateral contact)

SL = step length (the average horizontal distance between the touchdown of one foot and the touchdown of the next contralateral contact)

SF = step frequency (the number of steps per second)

▼ ***Step up*** *Data from the 2009 World Athletics Championships men's 100 m final, when Usain Bolt achieved a world-record time of 9.58 seconds, reveals some interesting trends in step length and frequency. Even when normalized to height, Bolt has the longest step length—does this mean it is better to run with longer steps? Not if you consider the second-place athlete (Tyson Gay) has the second-shortest step length and the second-highest step frequency, revealing how athletes can excel with longer steps, more steps, or sometimes a good value of both.[12] In comparison, a world-class 10,000 m runner, Kenenisa Bekele, achieves his velocity by combining a step length almost as long as the sprinters (normalized for height and it's the third longest) with a lower step frequency.[13]*

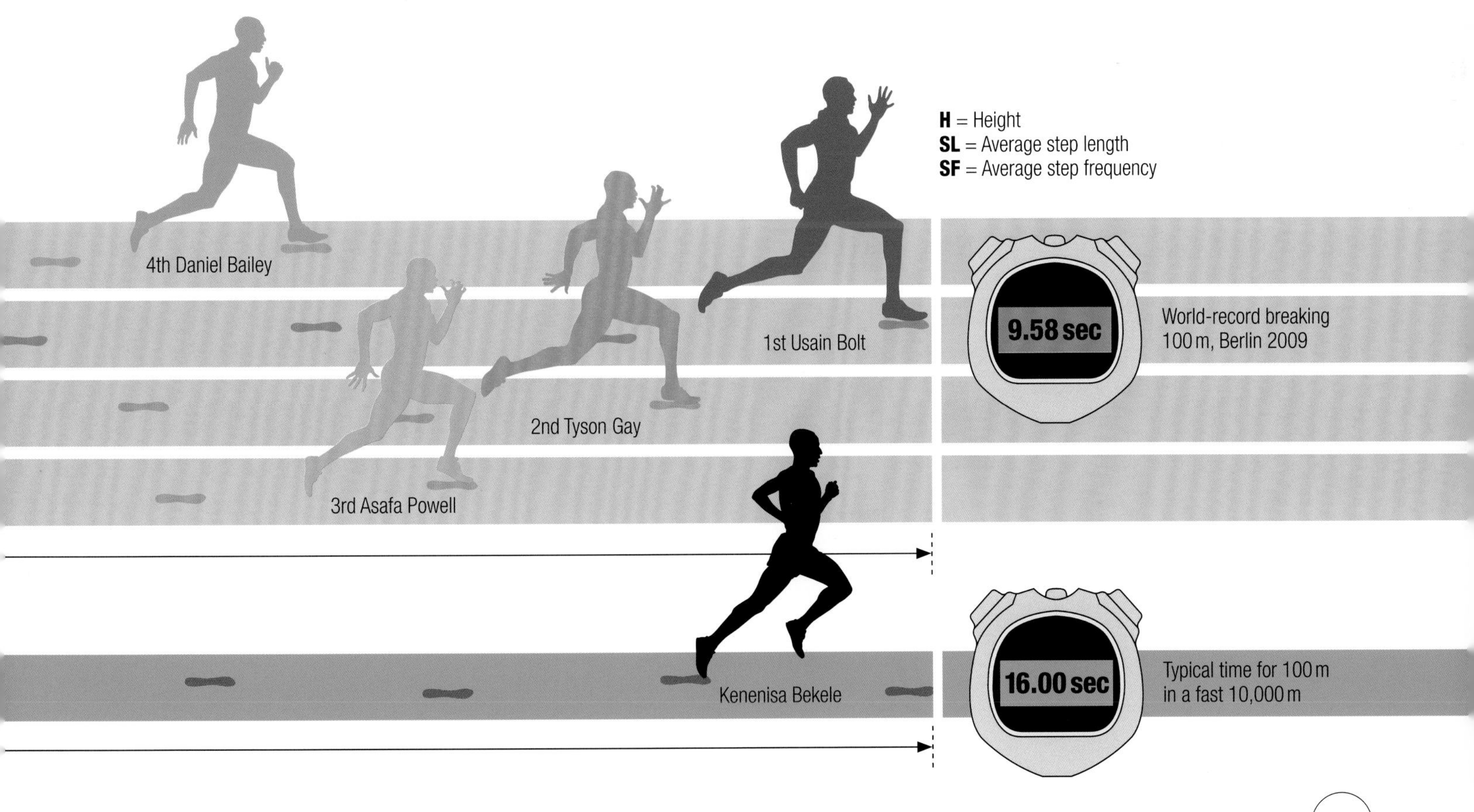

What is the most efficient endurance running technique?

Can I become a better runner by changing my style?

What makes a good endurance runner? Running performance depends not only on a large maximum rate of oxygen uptake (VO_2 max), but also on the percentage of VO_2 max at which an athlete can perform. Elite marathon runners are reported to run at 80–90% of VO_2 max and 10,000 m runners at up to 95% of VO_2 max.[1] At this top level, there is very little difference between athletes' oxygen capacity or uptake, so what explains differences in elite performers? It seems that the critical factor in differentiating elite athletes is running economy—the oxygen cost at a given running velocity (below the ventilator, or anaerobic, threshold).[2] Efficient runners win. But what do we mean by "efficient" running?

Efficient running technique involves making the most advantageous use of muscle mechanics, with the use of the stretch–shortening cycle (SSC) being key. One SSC is the stretch of a muscle followed by its immediate shortening. The muscle works to maintain a fixed contractile component,[3] usually found at a muscle's resting length, and involves the active component of the musculotendinous unit, or MTU (the bundles of muscle fibers, called fascicles), contracting isometrically—that is, without changing length—while the passive structures of the MTU stretch on ground contact, storing elastic energy, and recoil to release this energy into the propulsive stage of running. Using this elastic energy helps to minimize the metabolic cost of running—that is, the cost of producing energy from the chemical breakdown of substrate molecules during respiration.

In practical terms, a running technique that allows the ankle to be around neutral on ground contact, but dorsiflexed in the late swing phase to cause a stretch on the gastrocnemius, triggering pre-activation of the calf, will allow a strong isometric contraction of the muscle fibers. This needs to be combined with a forefoot strike—that is, touchdown with the ball of the foot—allowing the passive part of the MTU (particularly the Achilles tendon) to stretch and recoil, gaining energy to enhance propulsion.

Forefoot-strike runners therefore have an advantage over heel strikers in terms of running efficiency and liberation of elastic energy via the Achilles tendon. With this in mind, athletes might consider working to change their running technique toward a forefoot-strike style, but the potential costs in terms of injuries associated with such a change must be taken into account. Forefoot-strike runners have greater ankle joint moments (angular force) than heel strikers,[4] which helps running efficiency, but could put more stress on the calf MTU and foot muscles, leading to an increase in Achilles tendinopathy and plantar fasciitis.[5] It seems that forefoot-strike running is more efficient, but if you are going to change running style it should be done gradually, and include calf MTU and foot-musculature strengthening as part of the process to mitigate any potential injury problems from a change in running style.

Shod heel-strike running technique

▶ ***Heel strike*** *For this experiment, force was measured via a force plate embedded in the floor. There is a force spike on heel strike that is linked to increase stress-related injuries. The direction of the lower limb positions the foot strike in front of the body's center of mass, creating a breaking force that will need to be overcome to enable forward propulsion, causing an increase in energy cost.*

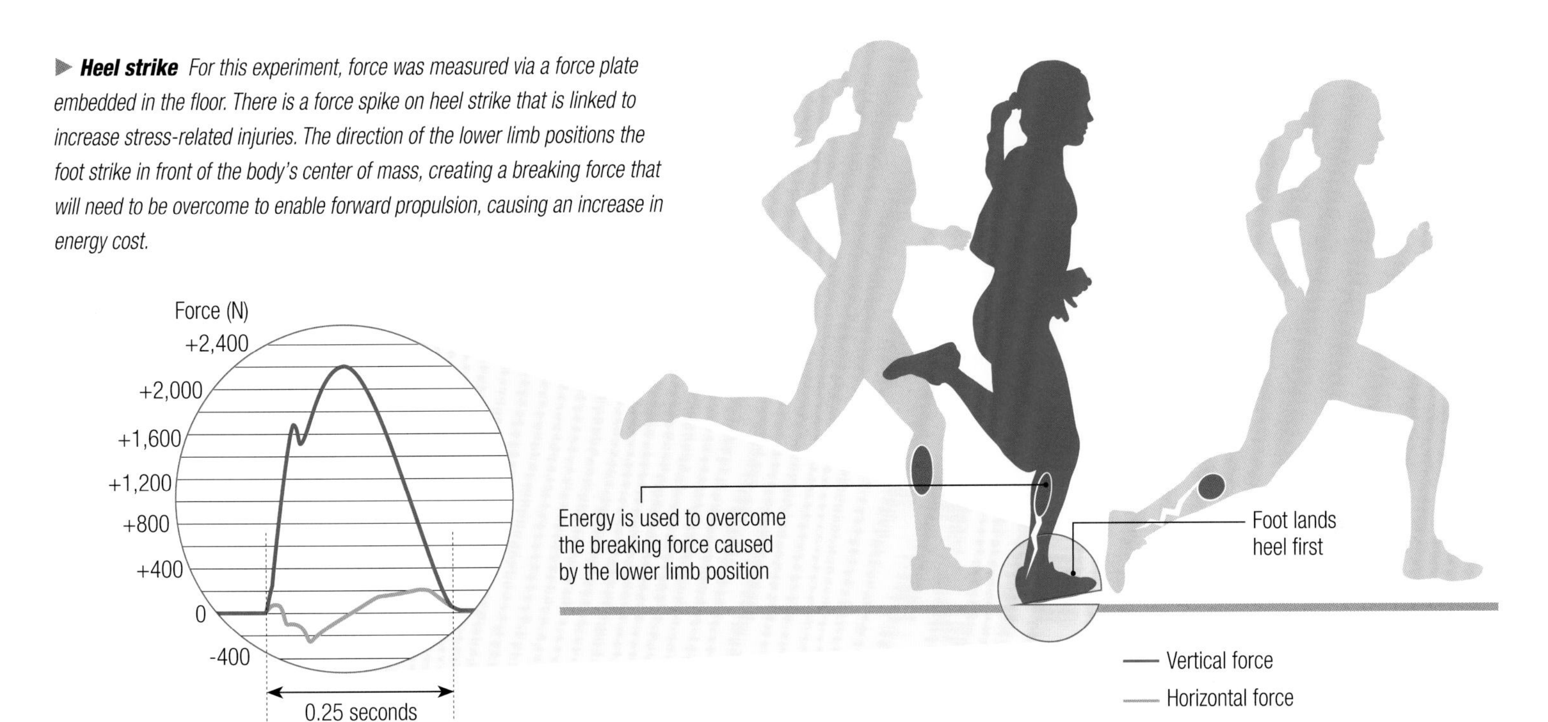

Shod ball-strike running technique

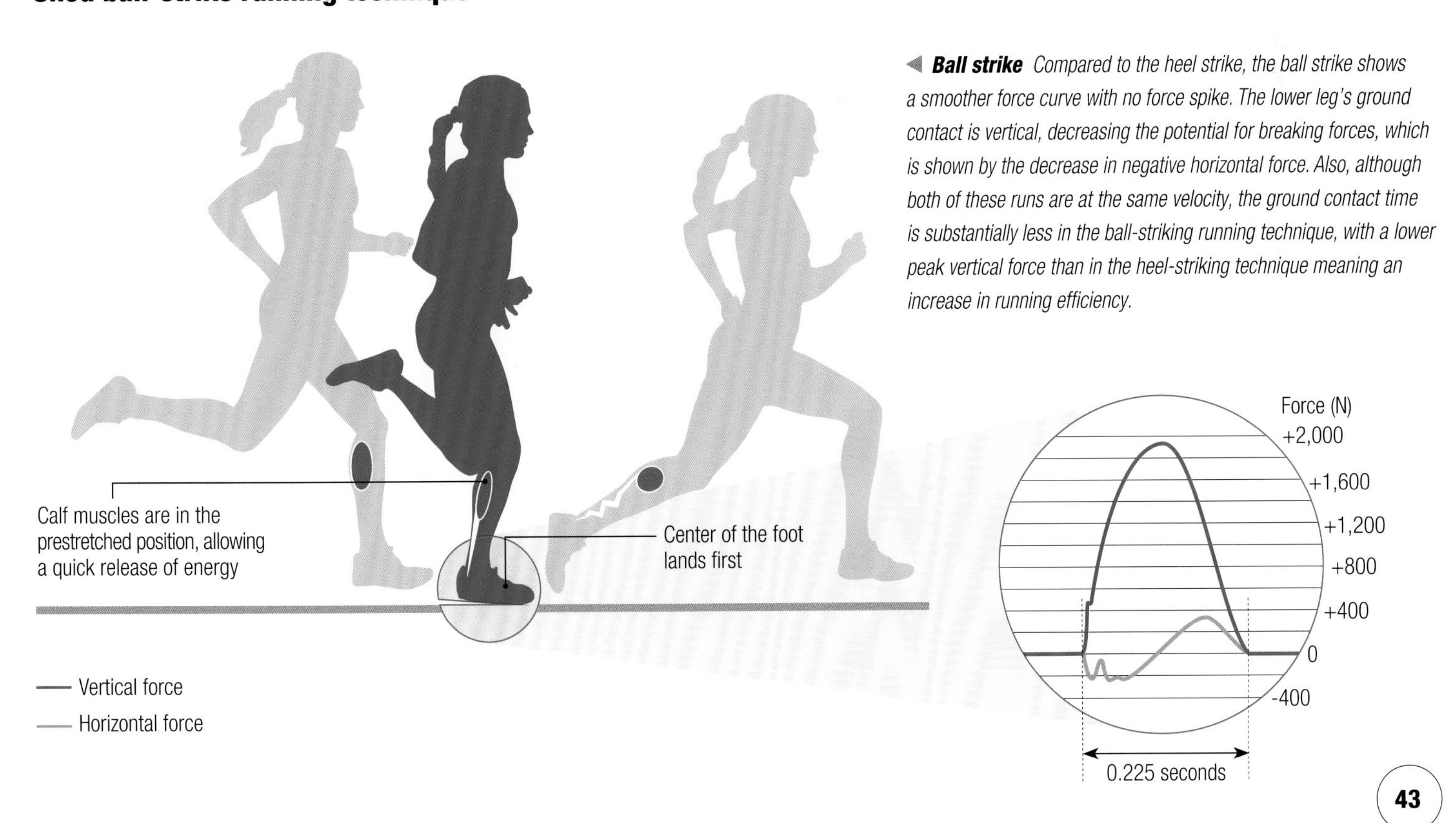

◀ ***Ball strike*** *Compared to the heel strike, the ball strike shows a smoother force curve with no force spike. The lower leg's ground contact is vertical, decreasing the potential for breaking forces, which is shown by the decrease in negative horizontal force. Also, although both of these runs are at the same velocity, the ground contact time is substantially less in the ball-striking running technique, with a lower peak vertical force than in the heel-striking technique meaning an increase in running efficiency.*

SCIENCE IN ACTION

a magic ingredient?

East Africans have dominated distance running for the past 20 years, as exemplified by Denis Kimetto achieving a new world record at the Berlin Marathon, Germany, in 2014. His 2:02:57 time equates to an average of 4 minutes, 41.5 seconds per mile—that is less than 70 seconds for every 400 m. So why does this part of the world dominate these events? What is their magic ingredient?

No marked difference in physiology has been identified in athletes from East Africa compared to European, Asian, or North American athletes. It seems that VO_2 max, percentage of VO_2 max at race pace, and biochemical parameters in elite endurance performers are similar regardless of racial background.[1] At this level, the most efficient runner wins, and this may be where East Africans gain an advantage.

Studies have found that elite East African runners do indeed exhibit superior running economy than their Caucasian counterparts,[2] perhaps because these athletes have a unique biomechanical makeup of the triceps surae musculotendinous unit (MTU) in the calf. East African endurance runners tend to be forefoot runners and therefore use their calf and ankle musculature to its optimum. They have a longer gastrocnemius MTU and greater Achilles tendon cross-sectional area, compared to Caucasian and Japanese runners,[3] which gives them a mechanical advantage. For a given running velocity, East Africans tend to have a smaller tendon stretch and recoil during the ground contact phase[4] and less fascicle length change, allowing a more effective isometric contraction of the active muscle component. The gastrocnemius therefore requires less muscle activation during ground contact,[5] which means lower energy consumption for a given running velocity. This means East Africans can run faster than other athletes, for the same amount of MTU fatigue caused by the propulsive phase of the gait cycle.

So forefoot running may sound like the best way forward. But remember, before you try to change to this more efficient running style, that many East African athletes from impoverished backgrounds have had years of preconditioning in minimal or no footwear as children and adolescents. This leads to stronger lower limb musculature with better proprioceptive efficiency and a habituated forefoot running style, while their lighter frames (male East African runners average 132 lb, or 60 kg) cause smaller ground reaction forces to be applied to their bodies than for heavier athletes.

▶ ***Berlin Marathon*** *East African runners dominating the medal positions is a sight common around the world in male and female distance races. This success is not confined to one country, but the whole of Africa's east coast, and it is not confined to athletes born and bred in East Africa—we see athletes with East African lineage being brought up in the West achieving incredible success. How can these athletes from different countries with different upbringings and training regimes achieve such global domination?*

BMW
adidas
TATA
CONSULTANCY
SERVICES
AOK
Die Gesundheitskasse
ERDINGER
ALKOHOLFREI
Holiday Inn
TIMEX

How does treadmill running affect ground-based performance?

Is running on a treadmill the same as running outside?

Broadly speaking, treadmill and ground-based running are similar in their kinetic and kinematic parameters, with the same VO_2 max achieved,[1] as long as a 1% treadmill incline is utilized to take into account the lack of air resistance on a treadmill. But there are subtle differences any runner needs to know before deciding whether treadmill running is a useful adjunct to training.

In general terms, when compared to ground-based running, treadmill running decreases knee flexion, hip flexion, and ankle dorsiflexion range.[2] It decreases the velocity of ankle rotation in ground contact, while increasing eversion (turning out) of the heel and ankle.[3] Specifically, treadmill running at slower velocities (6.7–8.9 mph or 3–4 m/s) tends to cause a shorter flight phase, with decreased step length and increased cadence compared to ground-based running,[4,5] while at faster (sprint) velocities of around 19 mph (8.5 m/s), step kinematics and support and flight times are similar. However, the support shank (lower leg) tends to be less erect, with a greater range of motion and increased angular velocity during treadmill running.[6] This tends to cause a flatter foot strike and increased forward lean at faster running speeds.[7]

Importantly, the kinetics of running can also be altered in treadmill running. The peak propulsive and medial-lateral forces are reduced on a treadmill, probably due to the trampoline effect of the treadmill belt providing some elastic energy to help the athlete's propulsion. This causes a decrease in vertical force and knee moment forces, while the flatter foot strike causes greater ankle moments compared to ground-based running.[8]

Treadmill and ground-based running positions

▶ ***Different angles*** *Runners generally exhibit a slightly flatter foot contact on the treadmill than on the ground. The treadmill runner's support limb is also positioned slightly in front of the athlete's center of mass. This will cause some breaking force, which will decelerate the athlete and need to be overcome to maintain running velocity.*

So training on a treadmill means employing a slightly different running style, with increased forward lean and flatter foot contact, which is exacerbated as the incline increases. In addition, it means runners miss out on the improved dynamic balance and joint stability gained through ground-based running on a varied terrain, which helps condition the musculoskeletal system to the impact forces associated with ground contact. Runners also have no opportunity to develop turning or downhill running techniques on a treadmill.

However, treadmill training can still be of value. It can produce significant cardiovascular gains and, importantly—as ground reaction forces are decreased on a treadmill—it may be useful for athletes suffering or recovering from chronic stress-related injuries. Interestingly, because running cadence is higher on a treadmill for a given velocity, some athletes use it as a form of over-speed training, but no research has yet explored whether this impacts performance on returning to ground-based running.

Effects of increasing velocity on the treadmill

▶ ***Foot strike and shank angle*** *As the treadmill velocity increases, the athlete's running technique changes. The foot strike becomes increasingly flat and the support limb starts to get in front of the runner, creating a negative shin angle, associated with breaking forces.*

7.5 mph (3.35 m/s)

10 mph (4.47 m/s)

12.5 mph (5.59 m/s)

How does force output vary from acceleration to steady state running?

How should my style change from starting to running flat-out?

Acceleration and maximum velocity running are fundamentally different with regard to technique and how forces are applied to the ground. Both need to generate impulse (force × time) in order to change the athlete's momentum, but the direction in which this impulse is applied is different.

When you start to accelerate, you are trying to create as much horizontal impulse as possible in order to change velocity in the desired direction. (In fact, the largest impulse is in the vertical direction—gravity still dominates and forces are needed to overcome its effects—but in order to accelerate forward, a large horizontal impulse is also vital.) Acceleration can be divided into four distinct phases: toe off, when the athlete leans forward to help generate horizontal propulsive force; mid-flight, when the rear leg drives forward and the front leg moves down and back forcefully toward the ground; ground contact, in which the foot hits the surface below the athlete's center of mass, ideally causing as little breaking force as possible; and finally, the ankle cross phase, during which the front leg drives back and the rear leg pulls forward, so the ankles cross at a low level.

Although maximum velocity sprint technique can similarly be analyzed in four phases, it contrasts with acceleration technique in a number of ways. In maximum velocity (or steady state) running, the athlete is concerned with creating enough vertical force to overcome gravity in as short a time as possible, in order to maintain forward velocity. Perhaps counterintuitively, horizontal forces become less important than in acceleration and in fact need to be minimized in some respects. If, in the ground contact phase, the foot strikes in front of the body's center of mass, a horizontal braking force is exerted, which then needs to be countered by some propulsive impulse. Faster sprinters tend to be able to minimize this braking horizontal impulse.[1]

We differences in step length and step frequency between acceleration and steady state running. Both increase with running velocity, but step frequency becomes more dominant at maximum running velocities.[2] At the same time, ground contact time becomes shorter from acceleration to maximum velocity (reducing from 200 ms down to below 100 ms), and vertical center of mass displacement decreases,[3] indicating a stiffer system in maximum velocity than in acceleration, in which there is less leg flexion and extension during ground contact.

▶ ***Accelerating*** *In order to gain the necessary impulse to accelerate or to run at maximum velocity, an athlete's technique needs to change. At toe off (**A**), the athlete's torso should lean forward with a straight line through the rear leg to the ground in a powerful triple extension action (the more horizontal this line, the more efficient the generation of horizontal forces). The front-leg shin angle should mirror the torso and rear leg angle, with a high, punched-thigh position during mid-flight (**B**). The rear leg starts to drive forward with the front leg driven down and backward forcefully toward the ground (**C**), where the foot should strike below the athlete's center of gravity in a position to generate enough vertical force to overcome gravity, while maximizing horizontal force to accelerate forward. It should be noted if the athlete's foot strike is in front of the center of mass, then this will cause a braking force and too much vertical impulse, resulting in the athlete "popping up" and failing to generate enough horizontal impulse to accelerate quickly. Lastly, as the athlete drives their front leg backward, the rear leg will be pulled through in a low position causing the ankle cross (**D**) to be below the stance leg's knee.*

▶ ***Steady state running*** *At toe off (**A**), the rear leg does not extend fully because the main requirement is for vertical force; once the support leg has got beyond the athlete's center of mass it is no longer producing force in the required direction. During mid-flight (**B**), the athlete positions their front leg with a shin angle that is vertical; this means when the leg is driven into the ground (**C**), forces will be produced in the required vertical direction. This ground contact should be as close to the body's center of gravity line as possible. However, it should be noted that some braking force may be required for an efficient instigation of the lead leg's stretch-shortening cycle.[4] The rear leg will swing forward in a much more flexed knee position than during acceleration, with good technique exemplified by the ground contact and swing leg's thigh being level with one another at the point of ground contact. The swing leg is driven far higher than during acceleration (**D**), showing an ankle cross at or above the stance leg's knee.*

Forces when accelerating

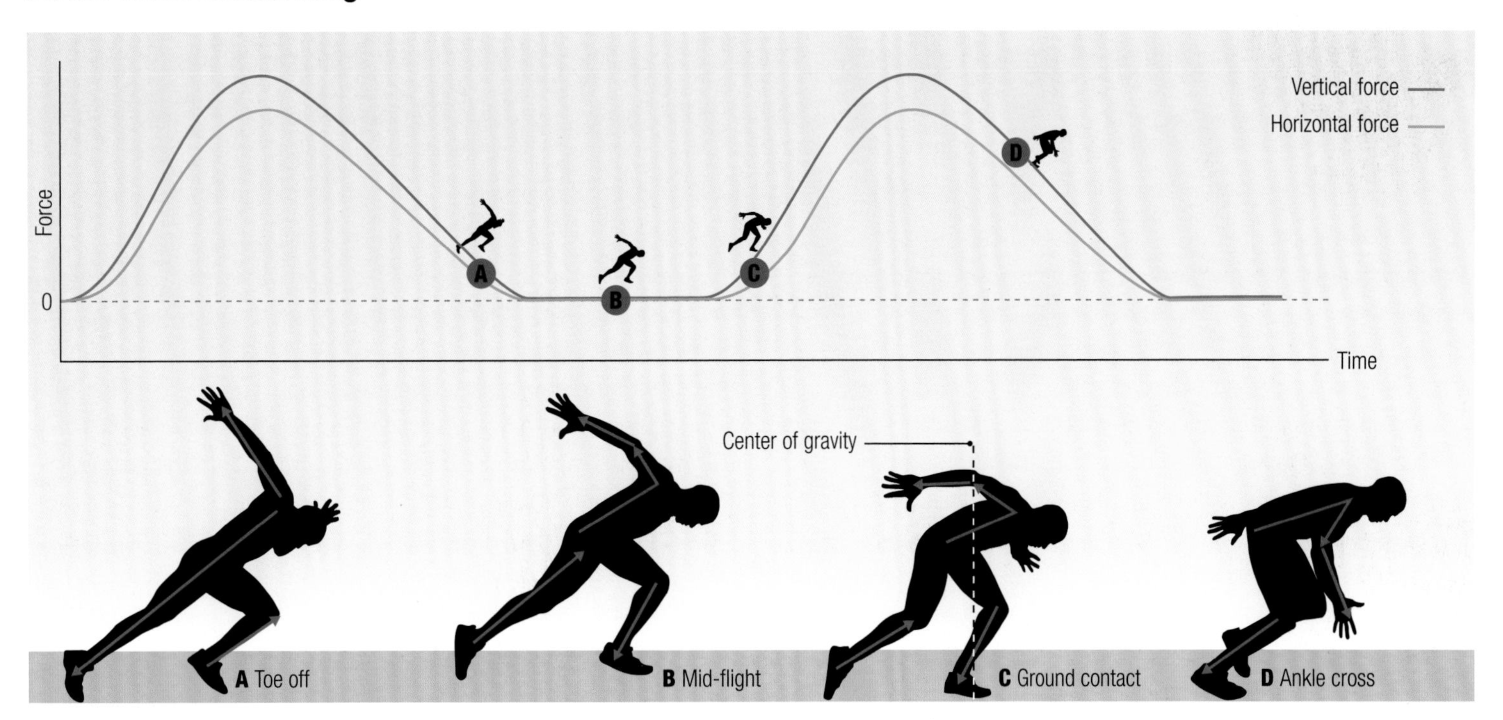

Forces in steady state running

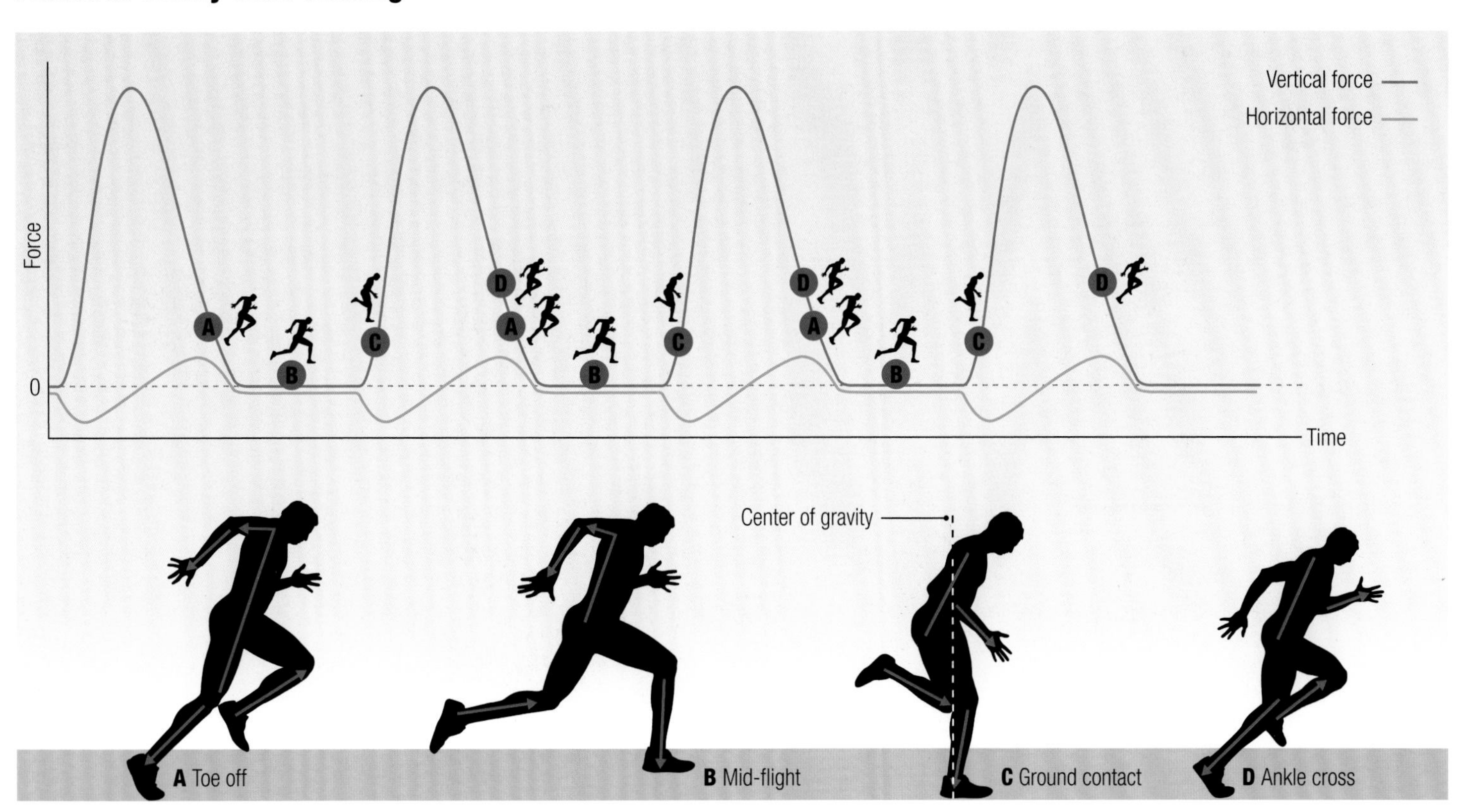

How is arm action linked with leg movement and performance?

Will pumping my arms harder make me run faster?

Contrary to many people's belief, pumping your arms harder will not help you run faster. Nevertheless, arm action is vital for running performance. The basic principle is linked to Newton's Third Law, which states that every action has an equal and opposite reaction—the arms are used to counteract the angular momentum of the legs, causing net angular momentum to be zero.[1] The arm swing in steady state running is a passive process, responding to forces exerted on the torso by the swinging legs. So, as running velocity increases we see greater leg accelerations, leading to greater arm accelerations, which act as a mass-dampening system preventing extreme trunk rotation.[2] If the arms are prevented from passively countering the leg momentum, there is a large increase in oxygen consumption as the athlete has to expend energy in actively trying to dampen rotation of the torso.[3]

During running, when the right leg descends into the ground contact phase, it rotates down and back, with the mass of the leg distributed a long way from the center of rotation (the hip). This large angular momentum will cause the right side of the body to start to rotate backward unless the left arm counteracts it. So the left arm must rotate backward, extending at the elbow

Endurance runner's arm action and spinal movement

to shift the arm's mass further from the center of rotation (the shoulder) and also accelerating the hand, to increase the arm's angular momentum in line with the right leg. At the same time, the left leg folds under the body, aligning its mass closer to its center of rotation (hip) and decreasing its angular momentum. This means the right arm, canceling out the left leg's momentum, rotates to the front, flexing at the elbow to decrease its angular momentum. This contralateral arm–leg movement prevents the torso from rotating excessively, avoiding inefficient movements and providing stability around the longitudinal axis.

The legs and arms are, of course, linked by the torso, which works in conjunction with the limbs to store and release energy for efficient running. The theory of the "spinal engine"[4] describes how efficient energy transfer up and down the kinetic chain between the lower limbs, torso, and upper limbs is helped by spinal movement, with the lumber spine rotating the pelvis while the thoracic spine rotates the shoulder girdle in the opposite direction. This constant winding and unwinding of the torso through a countermovement action stores and releases energy through a stretch–shortening cycle, facilitating energy transfer into the lower and upper limbs and making running more efficient and faster.

▼ ***Sprinters and endurance runners*** *The endurance runner and sprinter have very different arm actions because they are coping with very different leg momentums. The sprint athlete has a more pronounced action because of the need to counter the greater leg momentum caused by a faster leg velocity and a more extended rear leg position. As we run faster, the arm action should naturally become more pronounced. The arm action is backward and forward, with little to no movement across the body. The "spinal engine" action also differs. In endurance running, the movement of the spine constantly winding and unwinding helps to control the lower limbs' momentum, decreasing the need for excessive arm swing to do this job and therefore saving energy from unnecessary arm movement. In sprinting, this spinal movement is reduced, with a much stiffer torso presented, therefore arm swing needs to be more vigorous to counter the effects of the legs' momentum.*

Sprinter's arm action and spinal movement

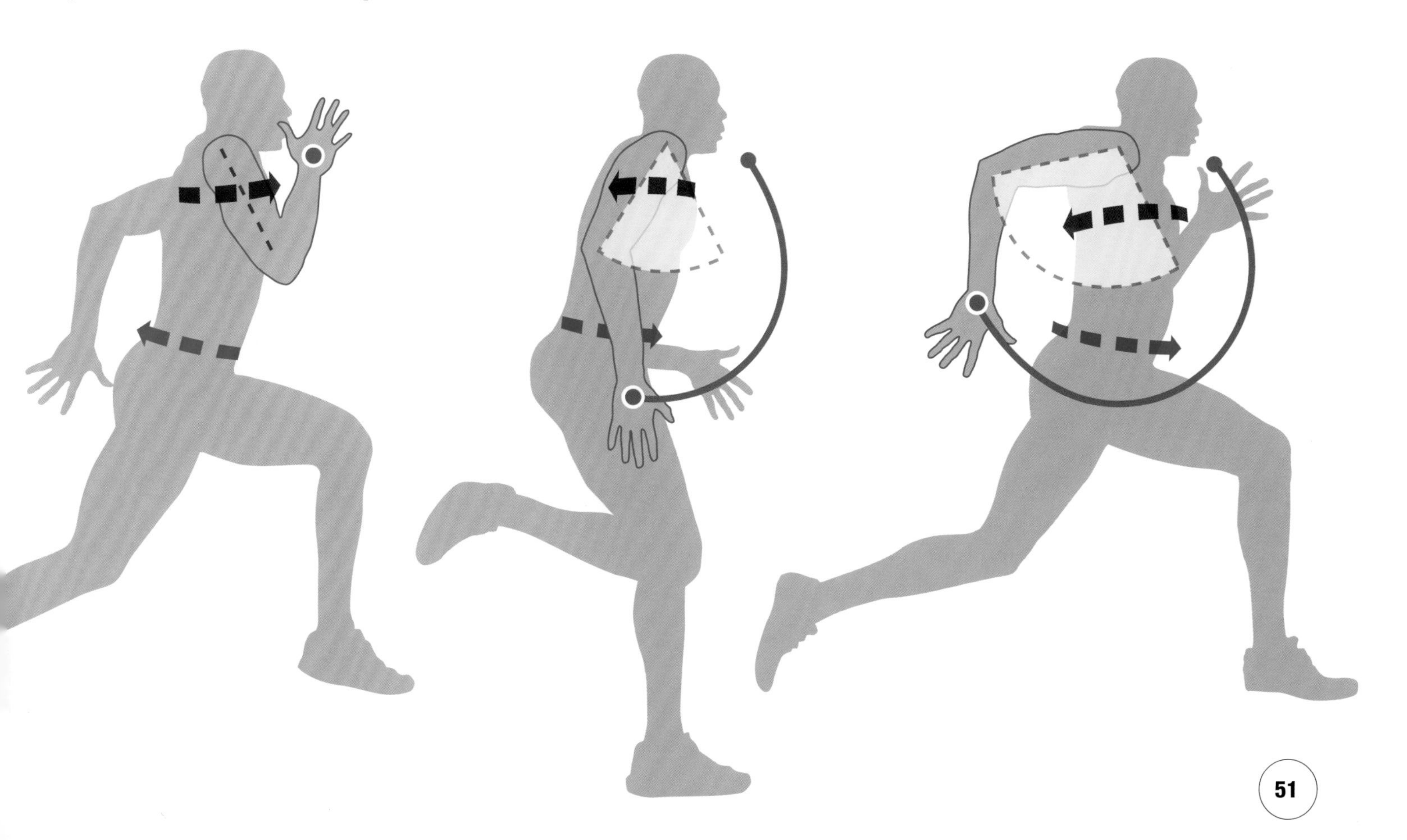

Does a forward lean help performance and reduce the risk of injury?

Should I learn to lean into my run?

The posture of the trunk during running is an important technical aspect for both enhancing performance (running economy) and minimizing injury risk. In the sagittal plane (front to back), trunk posture can be described by the amount of forward or backward lean of the trunk. Runners who lean forward to a greater extent are more economical (run faster for a given oxygen usage).[1] The forward lean needs to start at the ankles, not the waist, and promote alignment of the body in a straight line, all the way up to the head. Researchers and practitioners have been working to identify the mechanisms explaining why leaning forward is advantageous, and to determine the optimal lean angle beyond which further lean increases become detrimental. From their results, it appears that the performance benefits can be attributed to greater activation of the hip extensors (hamstrings and gluteals) during the driving phase of the running stance.

In addition to enhanced performance, a forward lean is also beneficial because it reduces the risk of injury. Around half of injuries reported in runners are knee injuries, and patellofemoral joint (PFJ) pain is the most common of these. A recent study highlighted the value of running with a slight forward trunk lean.[2] Injury-free athletes were asked to run under three posture conditions: "self-selected trunk lean," in which they ran with their normal trunk posture (average 7.3° lean during the stance phase of a running step); "reduced trunk lean," in which they ran leaning further back than usual (average 4.0° lean); and "increased trunk lean," in which the runners leaned further forward than normal (average 14.1° lean). In the "self-selected" condition, those with a more forward lean (more flexed trunk) exhibited less PFJ stress. Following changes to their posture, the increased trunk lean resulted in a significant decrease in peak PFJ stress, while the reduced lean resulted in a significant increase in peak PFJ stress (when compared to the levels experienced under runners' self-selected posture). The forward lean did not increase ankle stress, but did increase hip stress, supporting the notion that changes in technique only move the load to other structures. However, for many runners, reducing knee stress and engaging the large hip extensor muscles is a positive change. Maintaining a forward lean without losing the straight alignment over long distances requires a certain level of torso strength, which is why strength and mobility exercises are fundamental in improving running performance and reducing injury risk.

Leaning from the hip or ankle

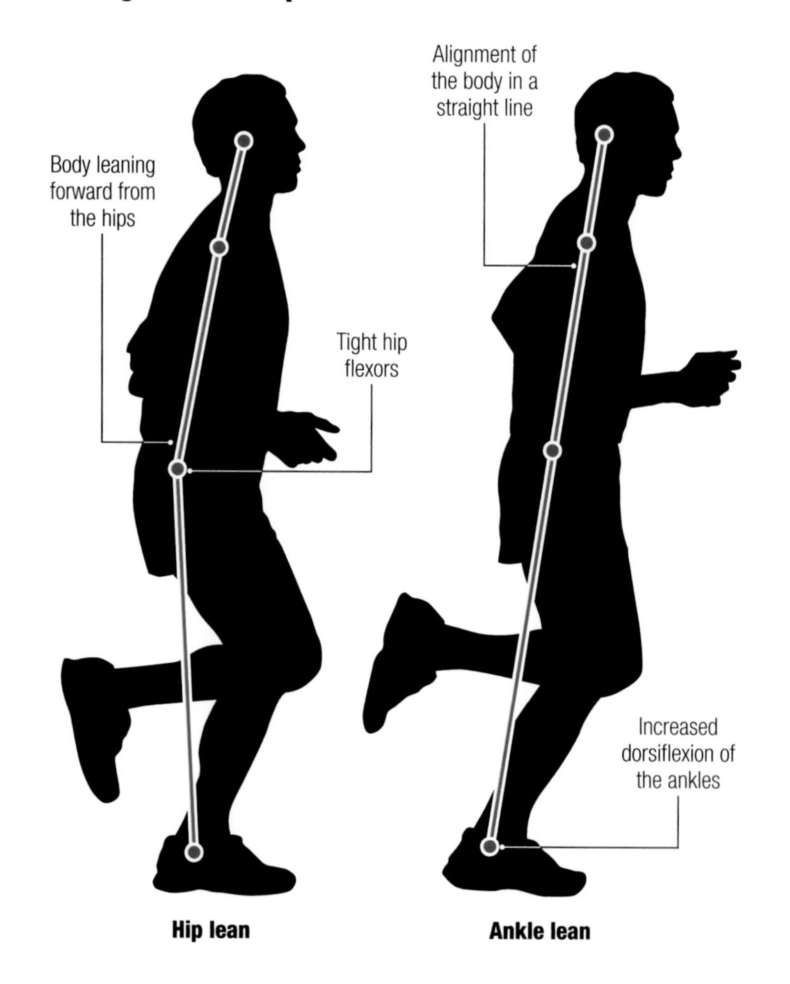

Different leaning angles

◀ ***Slight forward lean*** *The amount of forward lean an athlete exhibits affects both performance (through running economy) and injury risk. An ideal amount would be a "slight" forward lean, considered to be between 10° and 16° from the vertical.*

◀ ***Limited forward lean*** *A lean of less than 10° can lead to more hip flexor and less hip extensor activity, causing overstriding. A slight backward lean can cause ground contact in front of the center of mass, causing excessive ground contact forces and a braking effect.*

◀ ***Excessive forward lean*** *A lean of over 16° limits the use of hip flexors and can put strain on the runner's lower back.*

◀ ***Angle from the ankles*** *A forward lean from the ankles enables a straight body alignment and more efficient transfer of force from the extension of the hip, knee, and ankle into forward movement without excessive upward movement. Leaning from the hips or waist will increase knee movements (and therefore stress) and limit hip extensor activity.*

events:
middle-distance

Achieving the perfect symmetry of four minutes, four laps, and the mile may be more than sixty years old now, but there's still a certain romanticism about middle-distance running. It's an event that combines thought—planning the perfect tactics—with speed and endurance. The greatest middle-distance runners are international standard 400 m sprinters and world-class long-distance runners, boasting 47-second 400 m pace along with 28-minute 10 km strength.

With that in mind, the best training needs to include work that caters for five different types of paces and energy systems; two below and two above the race distance pace for a mile. That involves conquering the technique and power of the short, sharp sprints, and the aerobic requirements of mid-distance speed such as the 400 m, as well as being able to cope with distance running and the aerobic nature of long, easy efforts over 20 miles.

Perhaps it's for this reason that middle-distance runners were highly regarded in ancient Greece as far back as 776 BCE, where races from around 400 m to 4 km were hugely popular. Modern-day middle-distance athletes are held in similar regard, with the 1,500 m tagged "the Blue Riband" event of the Olympic Games.

Nutrition and recovery play a significant role in a middle-distance athlete's week because carbohydrate for fuel, fats to rebuild broken-down muscle, and adequate hydration all work together to allow a runner to achieve all the work required. It's not uncommon for a middle-distance runner to log up to 120 miles a week or more during the winter preparation phase while at the same time fitting in hours in the gym. In the summer, mileage is replaced by running intensity and gym work along with mobility and sprint technique.

▶ ***Cinder track***

Cinder tracks tend to be slightly slower than synthetic versions—suggestions vary between one and two seconds per 400 meter lap. Harvard University conducted an experiment in the 1970s, comparing times run on their new synthetic track compared to ones run away from home on older, cinder surfaces, and concluded there was a 2.9% improvement in times on the new surface.

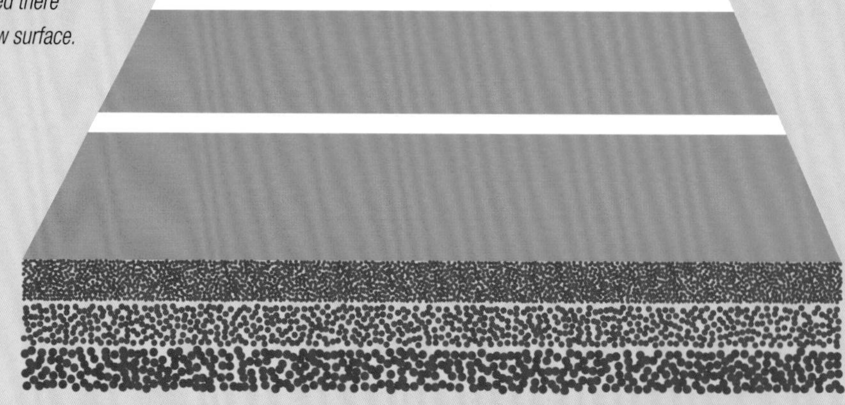

Track development

► ***Synthetic track***

The composition of the rubber on synthetic tracks has an impact on different distances. Sprinters tend to prefer a much harder surface, which allows them to generate more speed and run quicker, while long-distance runners like a softer track because harder surfaces create too much muscle damage. However, it's a fine line, and early experiments quickly concluded that too much cushioning soaked up a runner's energy, while too little resulted in the athlete bouncing off the track too quickly, before any stored energy could be released.

◄ ***Grass track***

A well-kept grass track is surprisingly fast; science has shown it returns the same amount of energy as a road surface, although being softer means that this is at an energy cost. Grass tracks are great for training and building leg power.

Experienced runners are in tune with their bodies—muscles, joints, heart, lungs, and stomach—and are constantly on the lookout for negative feedback that could mean impending fatigue or injury. For runners who train and race hard, keeping their bodies well fueled and hydrated goes a long way in helping achieve personal bests by prolonging the onset of fatigue and minimizing the risk of injury. Muscles are the engines that drive our bodies forward, ideally establishing a pace consistent with our fitness, a pace that can only be maintained if our muscles have ample fuel and coolant in the form of carbohydrates, water, and electrolytes. Between training sessions, the consumption of a varied, well-balanced diet provides the proteins, fats, vitamins, minerals, and other micronutrients that aid recovery and promote the cellular adaptations required for faster running.

chapter three

fuel and fluid

Bob Murray and Daniel Craighead

How can muscle glycogen stores be optimized?

Do I have to load up on carbs to be a good endurance runner?

Carbohydrate is the primary fuel for muscle contraction. The carbohydrate stored in muscle—glycogen—is used progressively during running, as it is broken down to provide the glucose molecules that are metabolized by muscle cells to create the required energy. After a hard training session or race, muscle glycogen stores are lower than at the start. The only way to restock is to consume ample carbohydrates—in the form of vegetables, fruits, grains, and simple sugars—and allow at least 24 hours for them to be assimilated into muscle glycogen.[1]

To ensure that muscle glycogen stores are "topped off" prior to an important race, athletes can "carbo load." When carbo loading was originally devised in the 1960s, athletes were instructed to train hard for three to four days on a low-carbohydrate diet to deplete muscle glycogen stores, then cut back on training and increase carbohydrate intake over the three to four days prior to competition. Research showed that muscles responded to the glycogen deprivation by super-compensating—storing even more glycogen than usual, a good thing for race day. Later research demonstrated that the same results were more easily achieved just by cutting back on training intensity and mileage while consuming a high-carbohydrate diet for two to three days before a race.[2]

Traditional and modern carb loading

Some coaches and runners advocate the notion of "train low, race high"—the concept of training with low muscle-glycogen levels in order to stimulate muscle cells to store even more glycogen when they have the chance. There is some evidence that such stimulation may occur, but "training low" is a high price to pay. Runners often train with low or significantly reduced muscle-glycogen levels simply because they fail to eat enough carbohydrate on some training days, and most are aware of how awful they feel when that happens. Those days take both a physical and a psychological toll. Perhaps early in a season it might be a toll worth paying on occasion but, because train-low sessions feel so dreadful and also make rapid recovery difficult, they probably aren't a great idea during race season. Most importantly, there is not a lot of good scientific evidence that "train low, race high" is any better for performance than just training hard and eating ample carbs throughout. There is no need to go carb crazy every day, but you do have to consume enough carbohydrates in your daily diet to fully replace those your muscles use in training and racing. Sometimes that requires loading up, other times not.[3]

Calorie content for different exercise durations

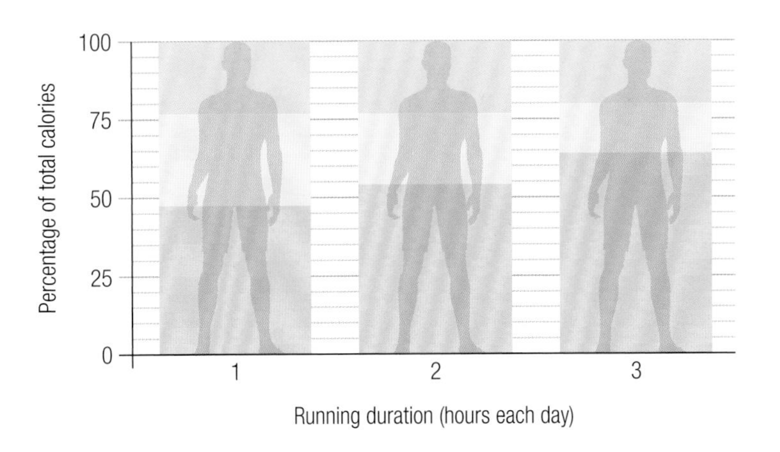

▲ ***Changing needs*** *The carbohydrate, fat, and protein content of a runner's diet varies with training intensity and duration. This figure illustrates that as training time increases, so does the need for dietary carbohydrate. The same is true on days of very high-intensity training that can quickly lower muscle glycogen stores.*

Glycogen resynthesis

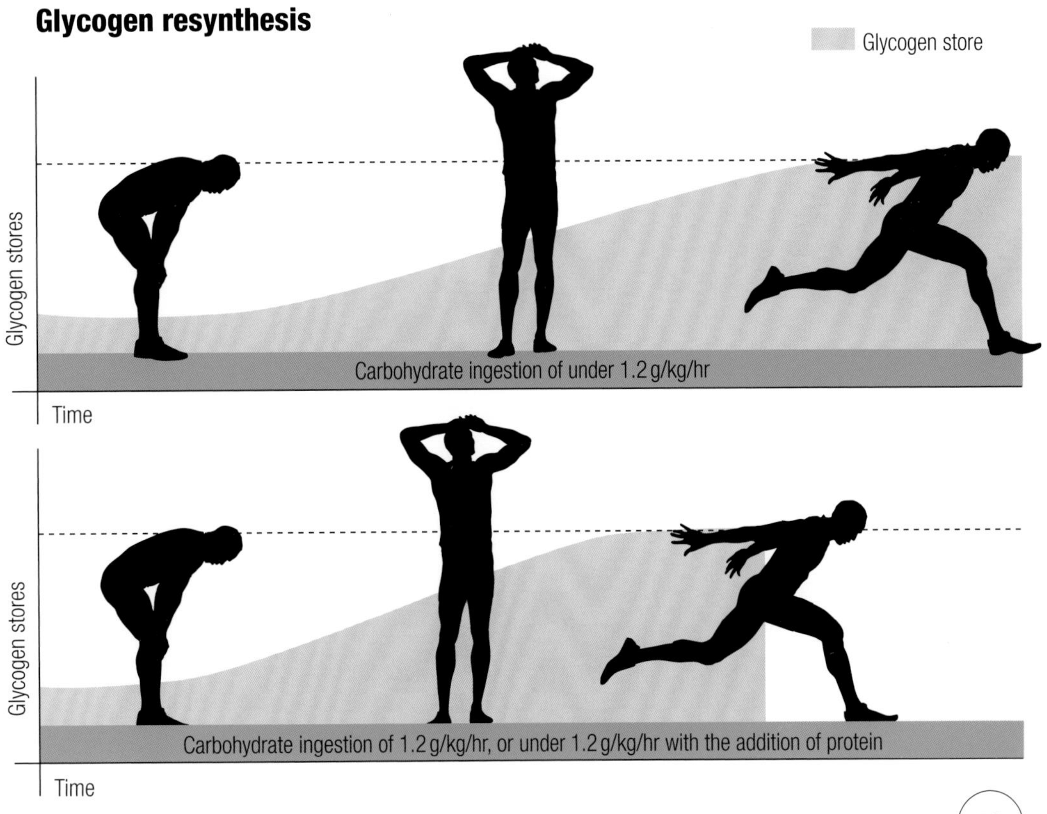

◀ ***Different methods*** ***(A)*** *The traditional method of carbo loading required a glycogen depletion phase of three to four days, followed by a high-carb diet. The trouble with traditional carbo loading is that training is extremely difficult during the depletion phase.* ***(B)*** *The current advice for carbo loading is an easier way to achieve the same results. Simply cutting back on training load (primarily a reduction in training distance rather than intensity) for a few days prior to a race will result in muscle glycogen supercompensation provided the runner consumes a high-carb diet.*

▶ ***Efficient refueling*** *While refueling guidelines suggest an optimum carbohydrate intake of 1.2 g/kg/hr, in some circumstances the same rate of glycogen synthesis can be achieved with a lower carbohydrate intake through the addition of protein.[4] A lower intake means that it takes longer to replenish glycogen stores.*

SCIENCE IN ACTION

why muscles love carbs

In scientific terms, the capacity to establish and maintain a fast running pace depends upon the capacity to establish and maintain a high power output. Factors contributing to a high power output include proper running mechanics and high running economy, a large aerobic capacity (VO_2 max), and a high anaerobic threshold—the percentage of VO_2 max that can be sustained over time. Effective training programs incorporate periods of exercise stress along with appropriate rest and nutrition to promote the many adaptations required to sustain a high running speed.

During high-intensity running, muscles rely on carbohydrate in the form of blood glucose and muscle glycogen to produce the ATP (adenosine triphosphate) needed to sustain a fast rate of muscle contraction. This is because proteins and fats cannot be broken down (oxidized) rapidly enough to produce ATP at the rate required to sustain running paces above roughly 50% VO_2 max. The faster the running speed, the greater the reliance on carbohydrate as fuel.

It's no wonder that muscle glycogen falls during training and racing—muscle cells are reliant upon the glucose molecules supplied by glycogen. Proper training along with a diet high in carbohydrates results in increased muscle glycogen stores, one of many adaptations that allow for faster running. Some of the ATP needed to sustain fast running speeds does come from the oxidation of fat (fatty acids, to be more precise). Fat oxidation is a continuous process at rest and during exercise, but the rate at which fatty acids can be oxidized to produce ATP is much lower than for glucose. In addition, fatty-acid oxidation requires more oxygen per ATP produced, another limitation of fat oxidation during exercise.

There have been a number of efforts to further improve the fat-burning capacity of muscle. Eating a very low-carbohydrate (ketogenic) diet is one such attempt to force the body to increase its capacity to oxidize fatty acids, particularly during exercise. In fact, research shows that such diets do increase fat oxidation and spare muscle glycogen. Unfortunately, endurance performance is not improved at race-pace intensities. It turns out that under these circumstances, glycogen-sparing becomes carbohydrate-impairing: the reason that glycogen is spared is because the muscle loses some of its capacity to oxidize carbohydrate, a big negative during high-intensity efforts. Other attempts to train or sleep with low muscle glycogen have produced inconsistent effects on performance.[1]

▶ ***Fuel for success*** *While training does improve muscle's ability to oxidize fatty acids, during high-intensity exercise, muscle glycogen and blood glucose remain the predominant fuels.*

UBS
THOMPSON
Zürich
UBS
SCHIPPERS
Zürich
5

What is the optimum amount of protein for a runner to consume daily?

Why do I need protein?

Compared to sedentary people, athletes benefit from consuming more protein each day. Protein macromolecules in the diet—regardless of whether they are animal or vegetable proteins—are broken down during digestion into individual amino acid molecules, which are absorbed from the small intestine into the bloodstream. The amino acids are taken up by cells, and used to create new proteins. In muscle cells, the amino acids supplied by the diet are used to repair damaged proteins and to synthesize new proteins. Most of the proteins in muscle cells are contractile proteins—long strands that enable muscles to contract and relax. But, like all cells, muscle cells also need to synthesize other protein molecules such as enzymes, signaling molecules, structural proteins, and proteins in mitochondria—the energy factories inside most cells.

Every cell in the body is constantly breaking down and building up proteins. Some do this rapidly—liver cells, for example, turn over a lot of protein. Muscle cells are not as active in protein metabolism, but because there are so many, much of the protein in the diet is used by muscles. Non-athletes require about 0.4 g of protein per pound of body weight each day (0.8 g/kg/day) to meet their protein needs. In other words, a 220 lb (100 kg) person needs less than 3 oz (only 80 g) of protein to meet daily requirements. Most people consume far more protein than this.[1]

Muscle protein synthesis

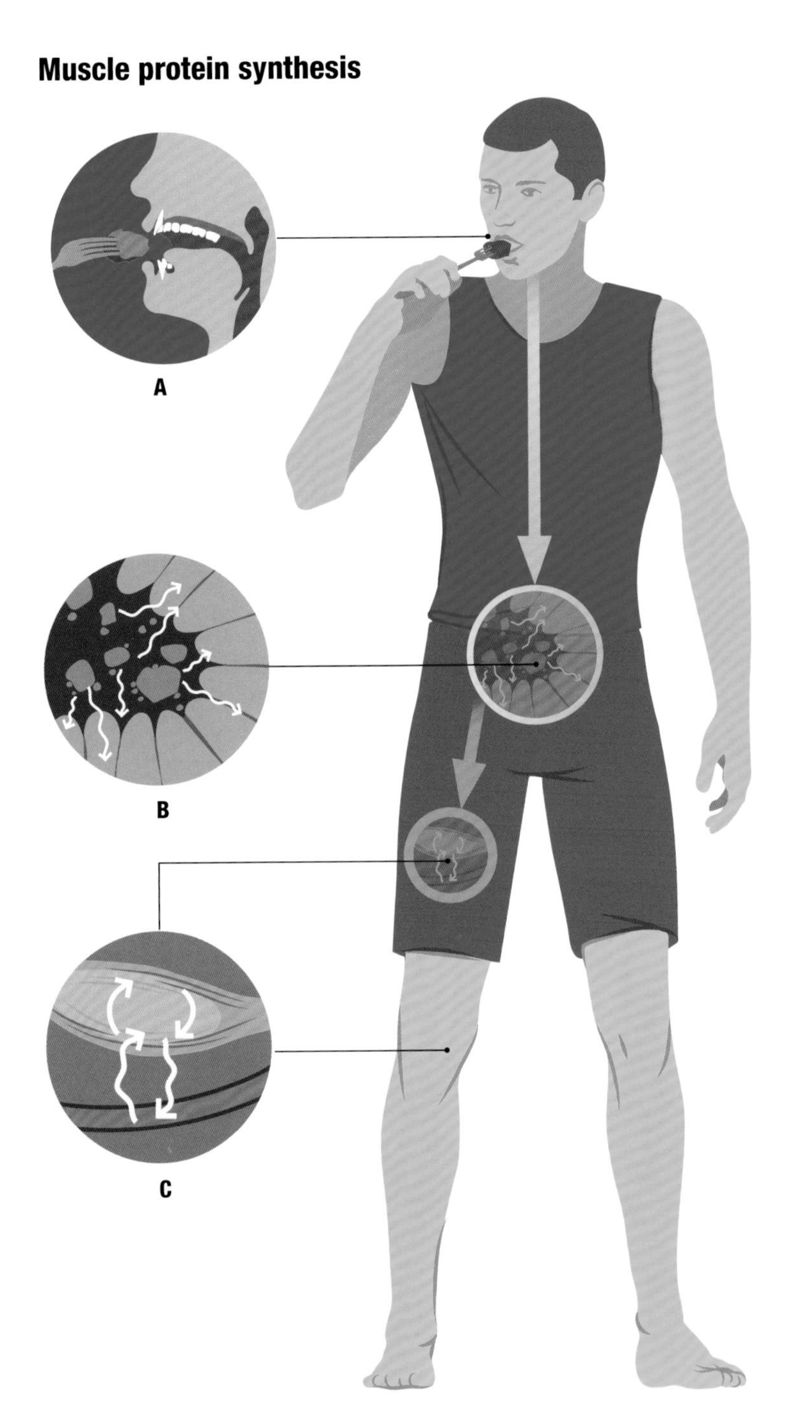

▶ ***Amino acids*** *Proteins in foods such as milk, meat, eggs, beans, and grains (**A**) are broken down in the stomach and small intestine and absorbed into the bloodstream as individual amino acids (**B**). Those amino acids are transported in the bloodstream and delivered to muscles and other tissues (**C**). Muscles take up the amino acids and use them for repair and growth of the contractile, structural, regulatory, and mitochondrial proteins inside muscle cells.*

Research has shown that athletes have elevated protein needs: 0.5–0.7 g or more per pound of body weight per day (1.2–1.6 g/kg/day). This amount is also easily achieved, especially with typical Western diets—a 220 lb (100 kg) athlete might need 120–160 g of protein each day, substantially more than a sedentary person, but this is still only 4–6 oz daily, the equivalent of 12–24 oz of meat and fish. Protein deficiencies are extremely rare among athletes except in extreme cases of malnutrition resulting from very restricted energy (calorie) intake.

Protein intake is particularly important to support recovery from training. Runners' muscles literally take a beating during each training session and the damaged cells and proteins have to be repaired or replaced. Equally important is that most adaptations that allow for faster running rely upon the production of new proteins. Immediately after a hard training session, runners should consume proteins and carbohydrates to spark recovery. Scientists have shown that muscle protein synthesis is maximized when around 1 oz—that is, 20–30 g—of protein are consumed after exercise. Dairy proteins seem to be particularly effective, but meat, fish, egg, soy, and other protein foods also boost muscle protein synthesis.[2]

▼ ***Boosting MPS*** *Muscle protein synthesis (MPS) naturally rises and falls throughout each day. MPS falls slightly during training sessions and then rises after exercise, especially when protein is consumed. This increase in MPS is sustained for 24–48 hours, gradually falling back toward normal baseline levels. This is one reason why protein intake after training and throughout each day helps promote muscle recovery, repair, and growth.*

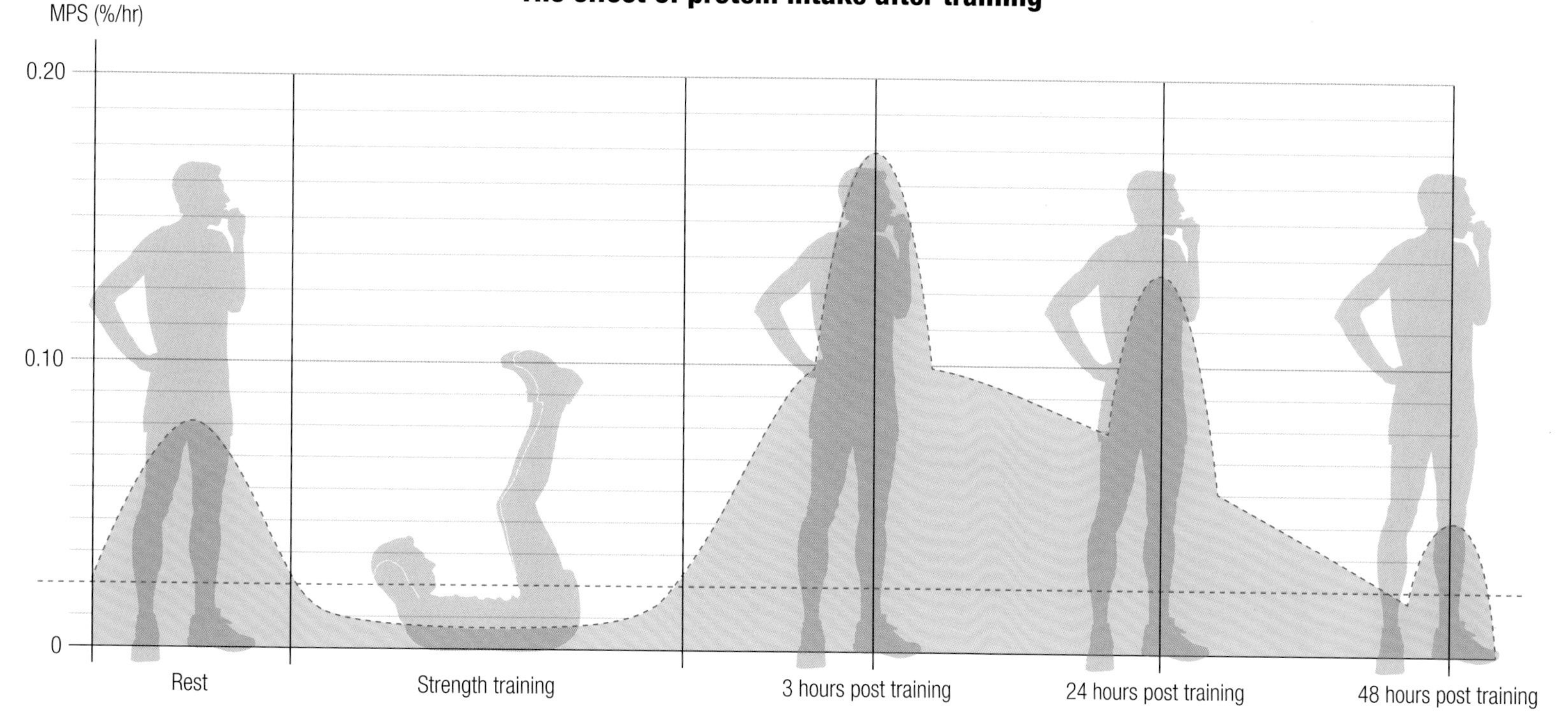

Do female runners need more vitamins and minerals than male runners?

Will supplements improve my running?

The recommendations for daily consumption of vitamins and minerals vary to some extent between males and females, as well as with age and pregnancy. For a few nutrients, female athletes may benefit from higher intakes than males—for example, the US Institute of Medicine recommendations for iron intake are 8 mg/day in adult males, 18 mg/day in adult females, and 27 mg/day in pregnant females. For most vitamins and minerals, however, the requirements are broadly the same for both sexes.

▼ ***Mineral requirements*** *The recommended average daily requirements for minerals are similar for males and females. Female runners should pay attention to calcium and iron intake because intake of those two minerals are often less than ideal. Both male and female runners will likely require more sodium on training days to replace the sodium lost from sweating.*[4]

Runners usually eat more than sedentary people of equal size and so consume more vitamins and minerals. If runners consume a balanced, varied diet, their larger appetites help ensure adequate vitamin and mineral intake. But for some female athletes, even that may not be enough to supply enough iron and calcium to match daily losses of those minerals. Iron, for example, is lost from the body through gastrointestinal bleeding, by the breakdown of the oxygen-carrying pigments myoglobin (in muscle cells) and hemoglobin (in red blood cells), and in females through monthly menstruation. For these reasons, some female runners develop iron-deficiency anemia, a performance-sapping situation that requires iron supplementation to reverse.[1]

Female athletes often have difficulty ingesting adequate calcium each day because they often avoid foods high in calcium such

Recommended daily mineral requirements for adults under 50 years of age

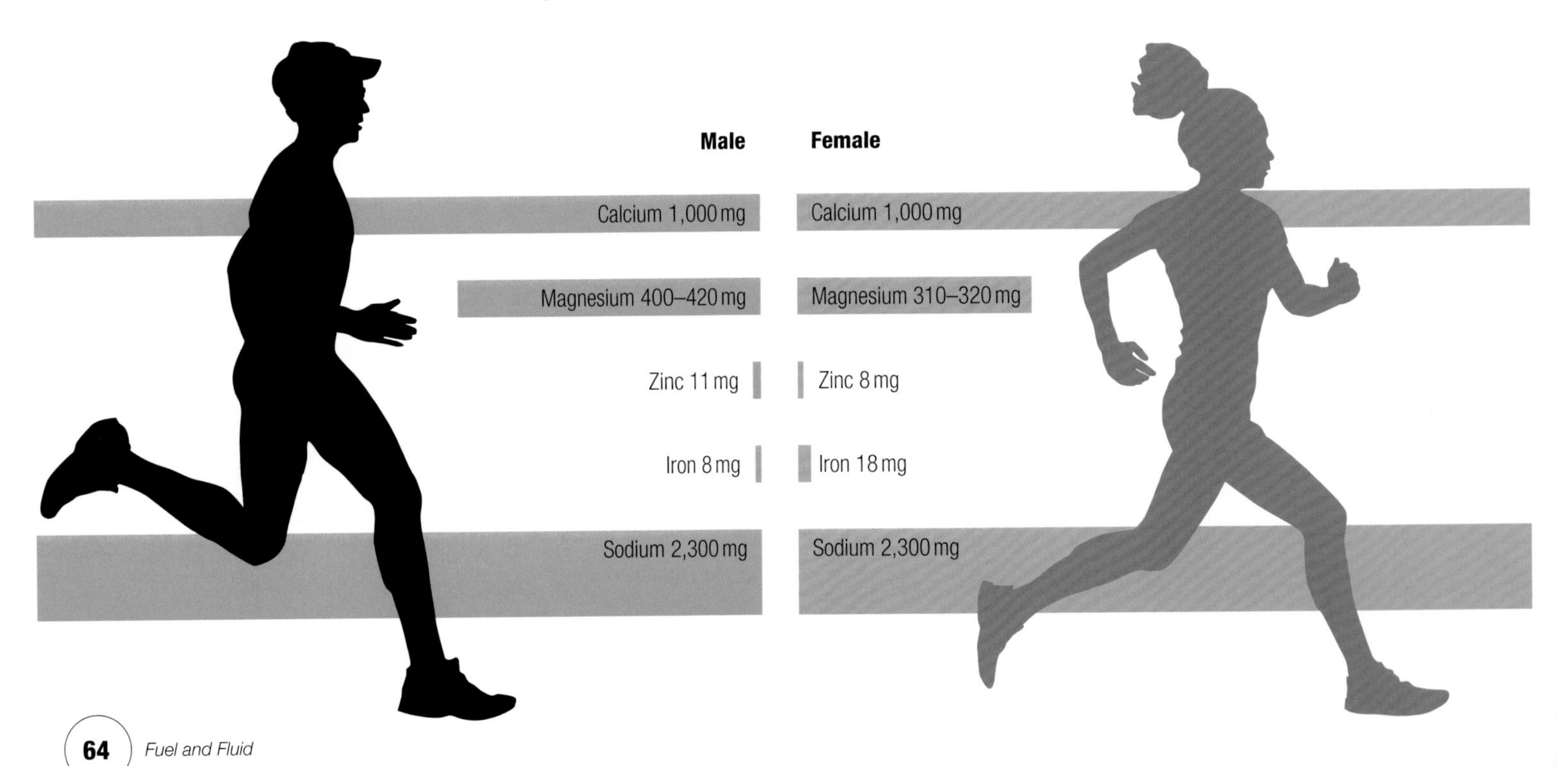

as dairy products, dark leafy greens, and enriched breads. The recommended calcium intake is 1,000 mg/day for both males and females (1,300 mg/day in adolescents). Calcium is lost in small amounts in sweat and those losses can further compromise calcium status in runners who do not ingest the recommended amount.[2]

Any runner who sweats on a daily basis also loses sodium and chloride—salt—sometimes in large amounts, which has to be replaced by salt in the diet and in sports drinks. Five tablespoons or more of salt can be lost in the sweat of athletes who train twice a day in warm environments. Sodium and chloride are vital in maintaining blood volume and in many other functions in the body, so replacing what is lost in sweat is important. For most athletes, that goal is easily accomplished because salt is found in so many foods. Runners who are on a sodium-restricted diet should take extra care to replace the salt lost in sweat.[3]

▼ ***Sweating it out*** *Sweat contains small amounts of a variety of different minerals. Sodium (and chloride) are lost in the greatest quantities in sweat, along with lower amounts of potassium, magnesium, and calcium.*

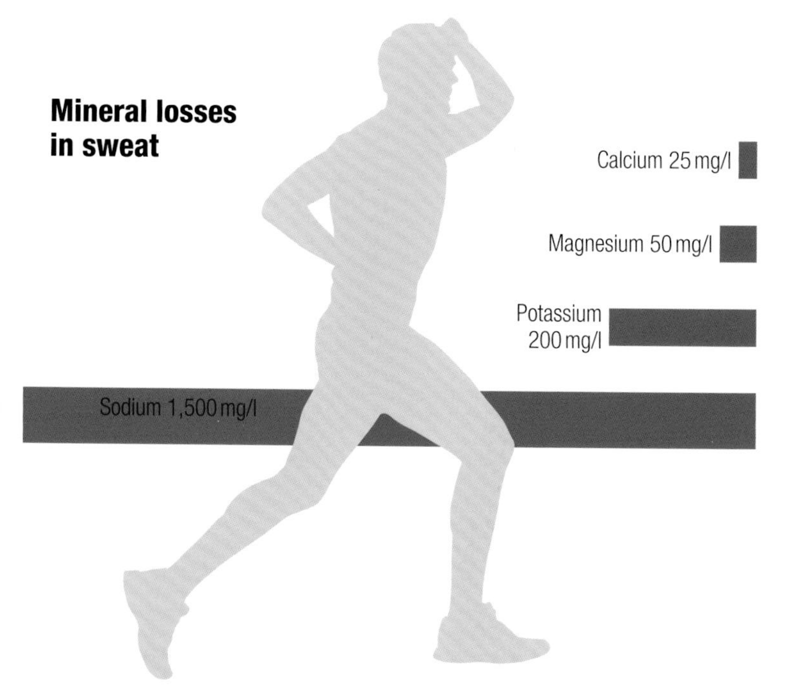

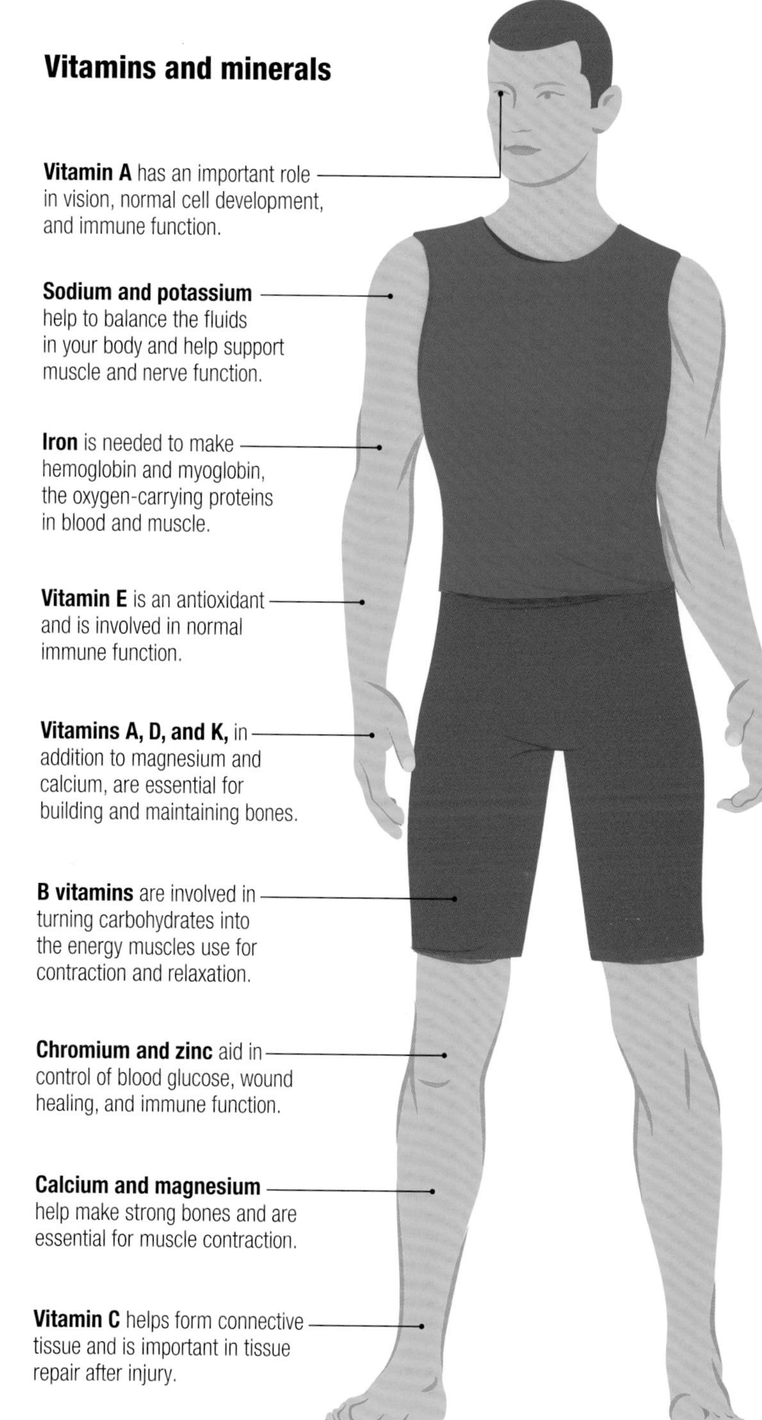

▲ ***Healthy body*** *Vitamins and minerals are involved in thousands of different functions throughout the body, examples of which are noted here. Runners who consume well-balanced diets will usually ingest adequate amounts of vitamins and minerals. A low-dose vitamin and mineral tablet (one that provides more than 100% of daily requirements) can be taken to ensure adequate intake in runners who are concerned that their diets are not always adequate.*

How does dehydration affect running performance?

Why should I keep drinking during endurance running?

Dehydration takes a toll on performance because it triggers dozens of physiological responses, each contributing to impaired function. The most important response involves blood volume. A major benefit of run training is an increase in circulating blood volume, which contributes to a higher maximal oxygen uptake and improved endurance because more blood allows more oxygen and nutrients to be delivered to working muscles. In addition, a greater blood volume enables the body to direct ample blood both to active muscles and to the skin to prevent overheating—blood removes heat from active muscles and transports it to the skin where it is lost to the environment, primarily through the evaporation of sweat.[1]

Sweating is essential for maintaining a safe internal body temperature, but also leads to dehydration unless we drink enough during exercise. The more we sweat, the greater the loss of water from the blood. As a result, blood volume progressively falls and the blood becomes thicker, which makes it more difficult for the heart to pump, necessitating an increase in heart rate. To make matters worse, less blood flows to the skin, so heat loss is reduced and body temperature rises. As the body becomes hotter, the brain sends inhibitory signals to muscles to prevent internal temperature from climbing to dangerous levels. This is why running in the heat feels much more difficult than holding the same pace in a cool environment. Dehydration adds to that feeling of difficulty, causing most runners to slow their pace and thereby prevent heat exhaustion or heatstroke.[2]

▶ ***Stay hydrated*** *When it comes to running performance, being well hydrated is always better than being dehydrated. Dehydration during running impairs physical and mental functions, especially during runs in warm or hot weather. The greater the dehydration, the greater the negative impact on performance.*

Although dehydration does not always mean running pace is slowed, it does make it much more difficult to maintain pace, and, in terms of body function, staying well hydrated is always better. So it's important that runners develop individual hydration plans—some runners sweat a lot and therefore need to drink a lot, others under similar conditions will sweat considerably less and so do not need to drink as much.

To gauge your hydration needs, weigh yourself before and after training runs. Weight loss indicates dehydration and the need to

The effects of dehydration

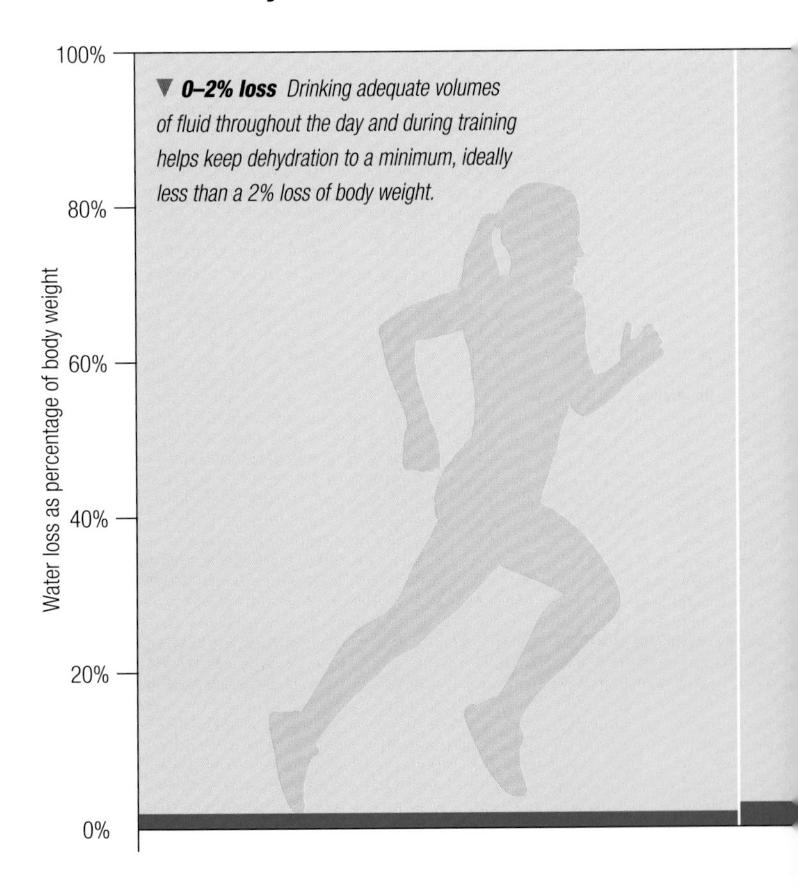

▼ ***0–2% loss*** *Drinking adequate volumes of fluid throughout the day and during training helps keep dehydration to a minimum, ideally less than a 2% loss of body weight.*

▶ ***Heat rising*** *Body temperature naturally rises whenever we run because muscles generate heat with each contraction. Under most circumstances, body temperature plateaus once a consistent running pace (exercise intensity) has been established and then gradually returns to normal after exercise. But when the environment is too hot or humid or if a runner is dehydrated, body temperature can climb to dangerous levels, increasing the risk of life-threatening heatstroke.*

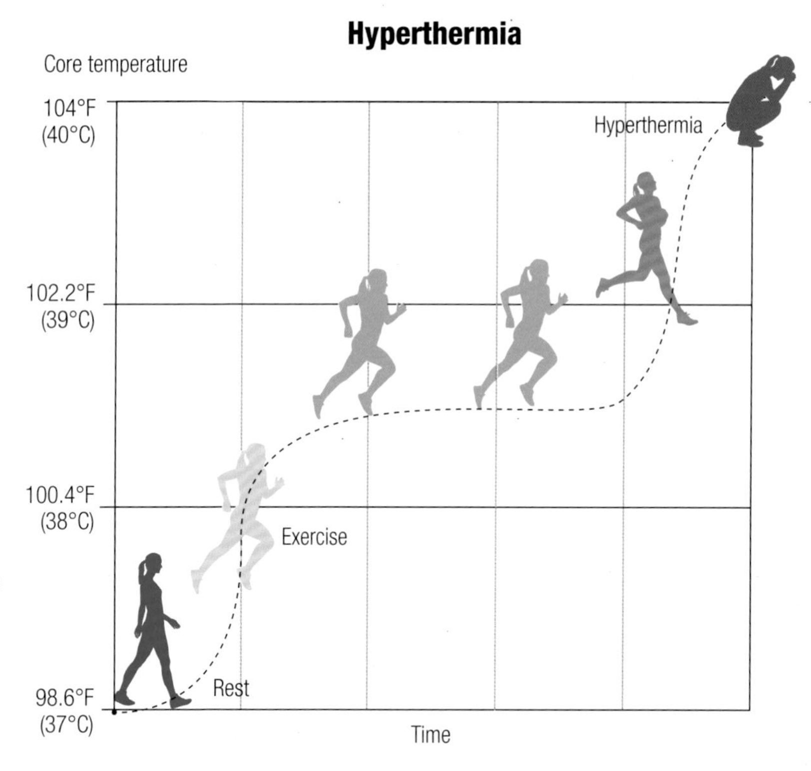

increase fluid intake during future runs. If you gain weight, you have consumed too much fluid and can cut back in the future. Aim to complete training runs of less than two hours weighing close to your prerun weight. A small loss is fine, but you should try to drink enough to limit weight loss to less than 2% of body weight. However, excessive fluid intake can dilute the concentration of fluid in the cells and tissues, reducing the concentration of electrolytes and impairing the function of nerves and muscles. This condition is known as hyponatremia and if left unchecked it can cause nausea, collapse, and potentially death.

▼ ***2–4% loss*** *When dehydration rises above a loss of 2% of body weight, it becomes more difficult to maintain running pace.*

▼ ***4–6% loss*** *As dehydration worsens, both physical and mental functions are severely impaired. Control of body temperature is compromised, adding to the risk of heat illness.*

▼ ***More than 6% loss*** *Severe dehydration can be extremely dangerous because of drastically reduced blood flow to the brain, kidneys, and other vital organs.*

What is the energy cost of endurance running?

How many calories do I burn when I run a mile?

Male runners of average weight burn roughly 125 kilocalories per mile (78 kcal/km), while female runners of average weight burn 105 kcal/mile (66 kcal/km). Much of the difference between men and women is due to the fact that men are generally heavier and it requires more energy to move heavier bodies. The fact that these values are averages means that they might not apply to you. On the other hand, they are pretty good approximations of the calories (kilocalories, more accurately) that you can expect to lose with each mile you run. Although it's not quite accurate, as an easy-to-remember rule of thumb, 100 kcal/mile allows you to calculate a decent estimate of how many calories you expend during a run. For instance, at 100 kcal/mile, the calorie cost of a 10-mile run is about 1,000 kcal, although the actual cost will be closer to 1,250 kcal for men and 1,050 kcal for women. To complicate things a bit more, the calorie cost for heavier runners will be greater than for lighter runners.

You might have heard that we burn the same number of calories when we walk a mile compared to when we run a mile. This is incorrect. In fact, running burns about 50% more calories than walking. When men walk a mile, they burn 90 kcal on average; for women, the calorie cost is about 75 kcal.

The fairly large difference in calorie cost between running and walking is because running involves more upper-body muscles and also places larger demands on the leg muscles to compensate for the greater and more frequent vertical movements during running.

In other articles and books, you may see different values for the calorie cost of running (or walking) a mile. That's because calorie cost can be expressed as total calorie cost or as net

Kilocalories burned per mile

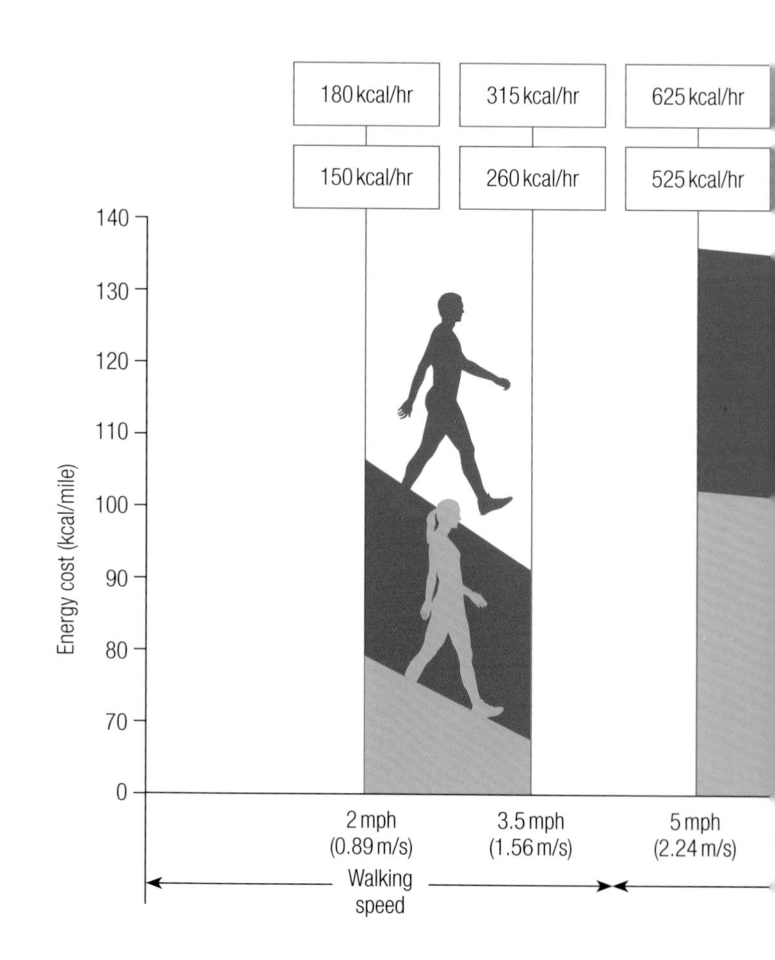

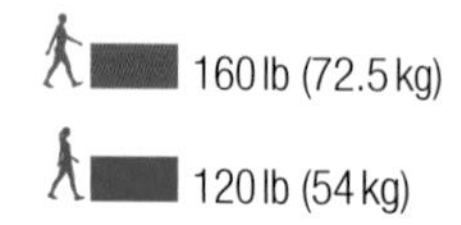

▶ ***Energy expenditure*** *This graph depicts the energy cost of running and walking at different paces. Note that the estimated values for calories expended rise with running speed and are greater for heavier runners than for lighter runners.*

*▶ **Walk or run?** At low speeds, the energy cost (calories expended) of walking is slightly less than that of jogging for equal periods of time, but as walking and running speeds increase, it becomes more economical (fewer calories expended) to run rather than walk. These differences are related to changes in the biomechanics of walking and running.*[1]

calorie cost. Total calorie cost is just what it seems: the measured calorie cost of the activity. Net calorie cost is a lower value because scientists subtract out the calorie cost associated with rest. In other words, there is a calorie cost for just sitting on the couch and if that cost is subtracted out, what remains is the net cost of running. The values here are total calorie costs because, after all, that represents what most people are interested in: the total number of calories expended.

Energy cost of running and walking

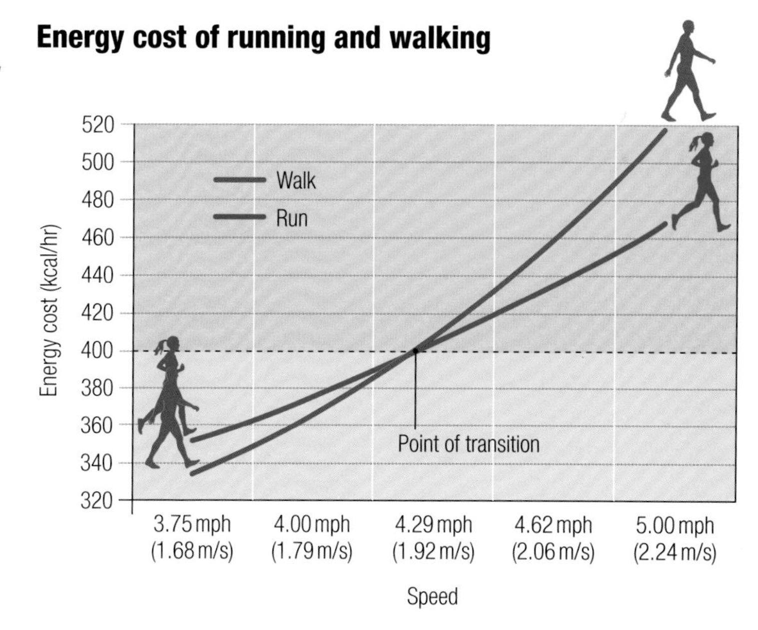

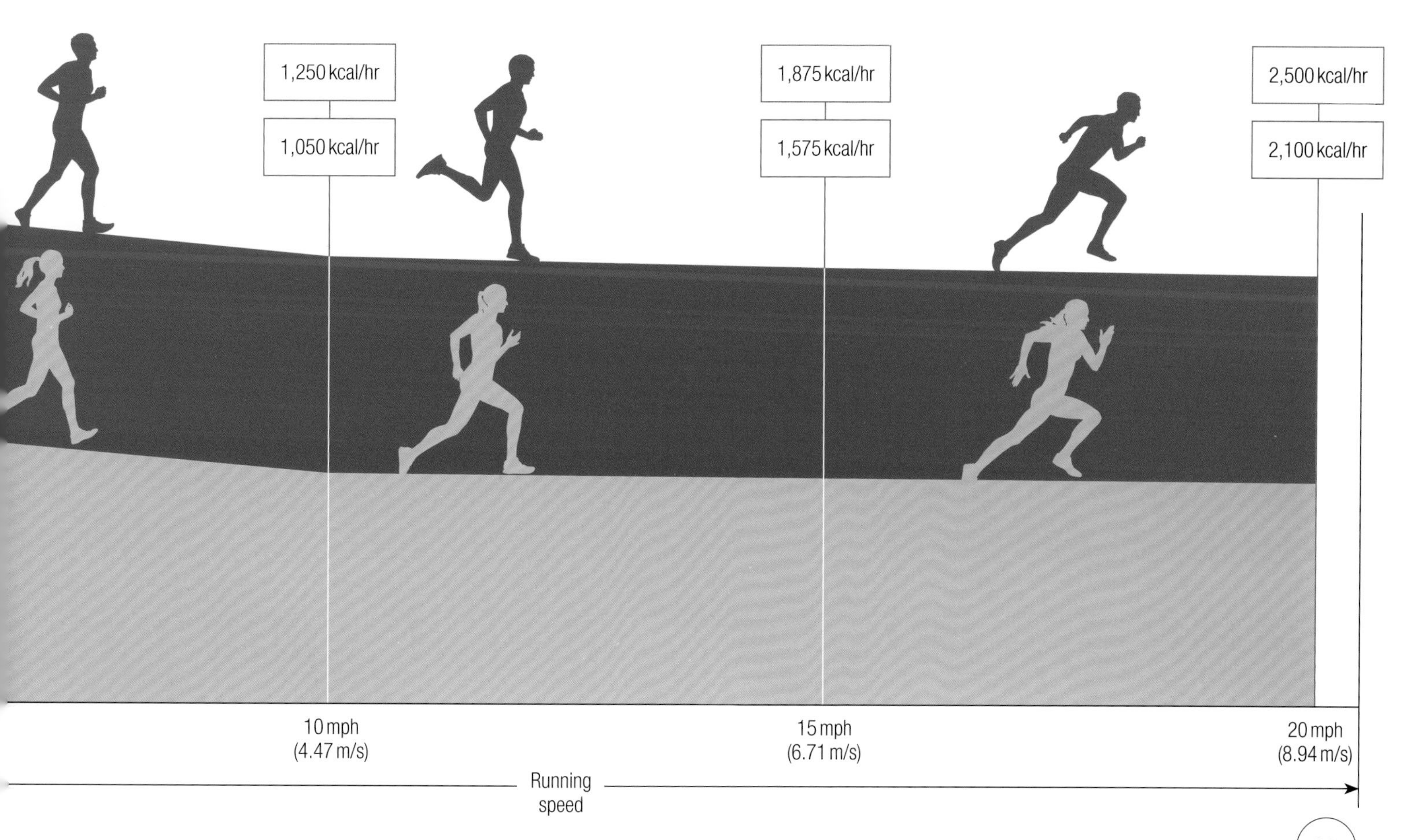

How does caffeine affect running performance?

Will a cup of coffee help me run better?

Caffeine is a stimulant that has ergogenic—work-enhancing—properties. Of course, as with all pharmacological agents, the dose of caffeine and the timing of its intake affect the way the body responds. Too little caffeine is insufficient to improve performance; too much caffeine risks negative side effects such as anxiety, nervous shaking, and a racing heart rate.

Caffeine is a non-specific adenosine-receptor antagonist, a fancy way of saying that caffeine counteracts the effects of the molecule adenosine. Adenosine is produced by all body cells and is important in energy metabolism and many other cell functions. When levels of adenosine begin to rise in the brain, drowsiness sets in as the adenosine molecules interact with receptors on brain cell membranes. Caffeine molecules bind with the same receptors, preventing adenosine from attaching and thereby reducing drowsiness. Taken before or during exercise, caffeine intake is associated with reduced feelings of discomfort and exertion (running feels easier), increased alertness, and perhaps even an increase in the number of muscle cells that are recruited during running.

Studies have demonstrated that consuming 2.27 milligrams of caffeine per pound of body weight (5 mg/kg) 45 to 60 minutes before exercise is usually enough to improve performance in endurance and shorter-duration activities by 3–5%. For a 160-lb (72.5-kg) runner, a caffeine dose of 2.27 mg/lb (5 mg/kg) is 363 mg, a large amount of caffeine. By comparison, an energy drink may have 111 mg per 8.4-oz (250-ml) bottle, while coffee has around 136 mg in a 12-oz (355-ml) mug.

Runners interested in using caffeine to improve performance should experiment during training to determine the amount of caffeine that produces the best results. Those who regularly consume caffeine may need a larger dose compared to caffeine-naive runners, who may require 1 mg of caffeine per pound of body weight (2 mg/kg).[1]

Servings of different drinks to achieve 3 mg/kg concentrations

Energy shot in a 1.9-oz container (82.5-mg caffeine/oz)

Energy drink in an 8.4-oz container (8.2-mg caffeine/oz)

Instant coffee in a 12-oz mug (7.4-mg caffeine/oz)

Ground coffee in a 12-oz mug (11.3-mg caffeine/oz)

Espresso in a 1-oz shot (60.1-mg caffeine/oz)

Tea in a 6-oz cup (5.7-mg caffeine/oz)

Cola in a 12-oz can (2.6-mg caffeine/oz)

160-lb (72.5-kg) runner

Molecular structures

Caffeine
$C_8H_{10}N_4O_2$

Adenosine
$C_{10}H_{16}N_5O_{13}P_3$

▲ ***Similar structures*** *The similarities in the structures of caffeine and adenosine make it easy to understand how caffeine can block adenosine receptors in the brain and elsewhere.*

▶ ***Nothing instant*** *While the results are mixed, studies of the effects of ingesting low quantities of caffeine have tended to find little or no benefit for sprinters, but 1–2% time improvements for middle- and long-distance runners.*

> **Need to know**
> Although caffeine is not banned by the World Anti-Doping Agency (WADA), it is on the organization's monitoring program, which tracks the potential misuse of substances that are not prohibited. The U.S. Department of Health and Human Services and U.S. Department of Agriculture's latest dietary advice is to ingest no more than 400 mg of caffeine per day—approximately three mugs of coffee—and not to incorporate it into your diet if you don't consume it already.

▼ ***Caffeine rush*** *This figure depicts the volume of various drinks that would be required to consume 3 g of caffeine per kilogram of body weight based on typical container or serving sizes.*[2]

Performance improvements with low doses of caffeine

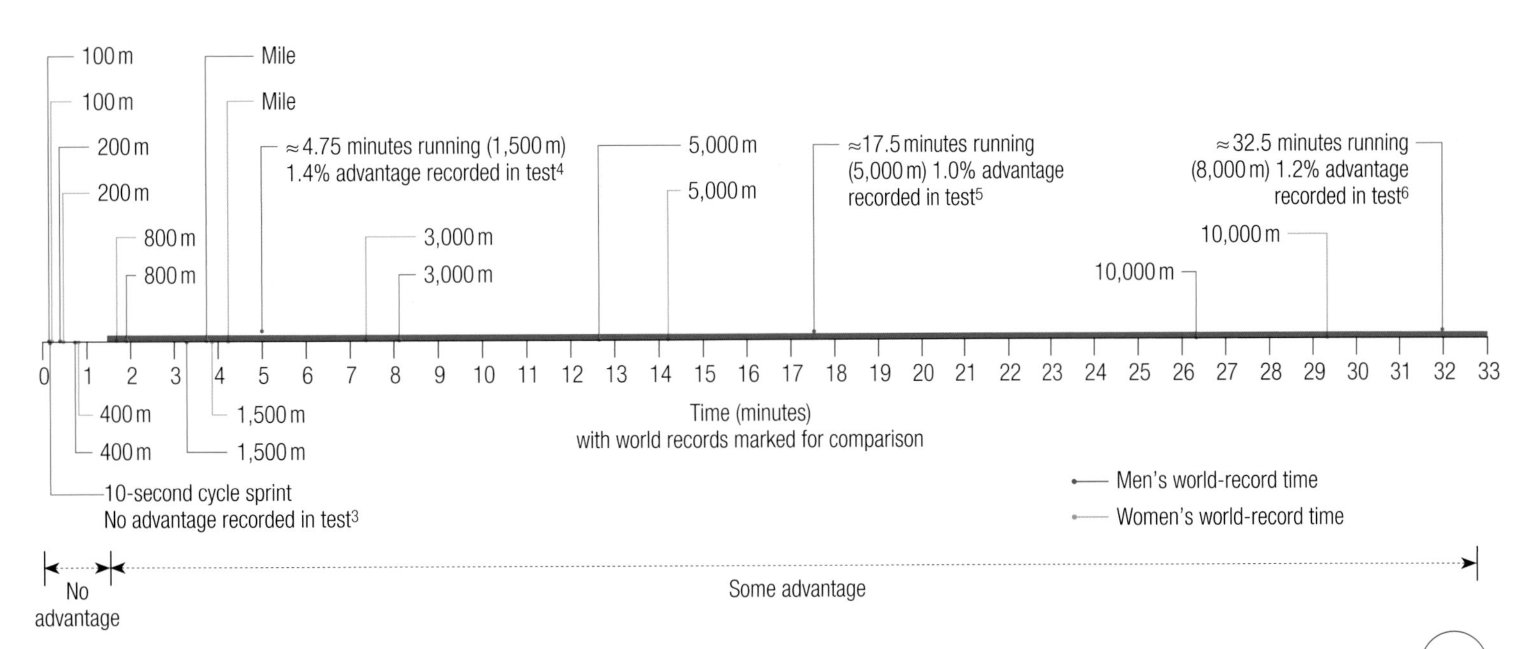

How much does body weight affect running performance?

Will losing weight make me a better runner?

A quick look at the bodies of top runners strongly suggests that leanness is an important characteristic for peak performance. Top sprinters and endurance runners, female and male, are all very lean, in large part because of their training volume, with some genetic predisposition mixed in. Extra body weight in the form of fat or muscle is extra baggage and runners pay an energy penalty for lugging around extra weight.

For example, maximal oxygen consumption (VO_2 max) is one important characteristic for endurance runners and is most often expressed as the volume of oxygen consumed per kilogram of body weight per minute (ml/kg/min). Proper training increases VO_2 max and top runners typically have values well in excess of 70 ml/kg/min. If a 128-lb (58-kg) runner has a VO_2 max of 72 ml/kg/min and manages to lose 4.5 lb (2 kg) of body fat, VO_2 max increases to almost 75 ml/kg/min, a big jump for any runner.

Sprinters are generally larger and more muscular than endurance runners, with much greater upper-body musculature. All runners have to expend energy to move their bodies forward and the heavier the body, the greater the energy cost involved. In sprinting events, the upper body is much more engaged in the explosive start, acceleration, and maintenance of running speed, so some extra muscle comes in handy, although it is still weight that must be moved. As a silly example, if a world-class sprinter could maintain the muscle mass in his legs but shrink his upper body to the size of a marathon runner's torso, he would theoretically run much faster because he has less weight to

The effect of excess weight

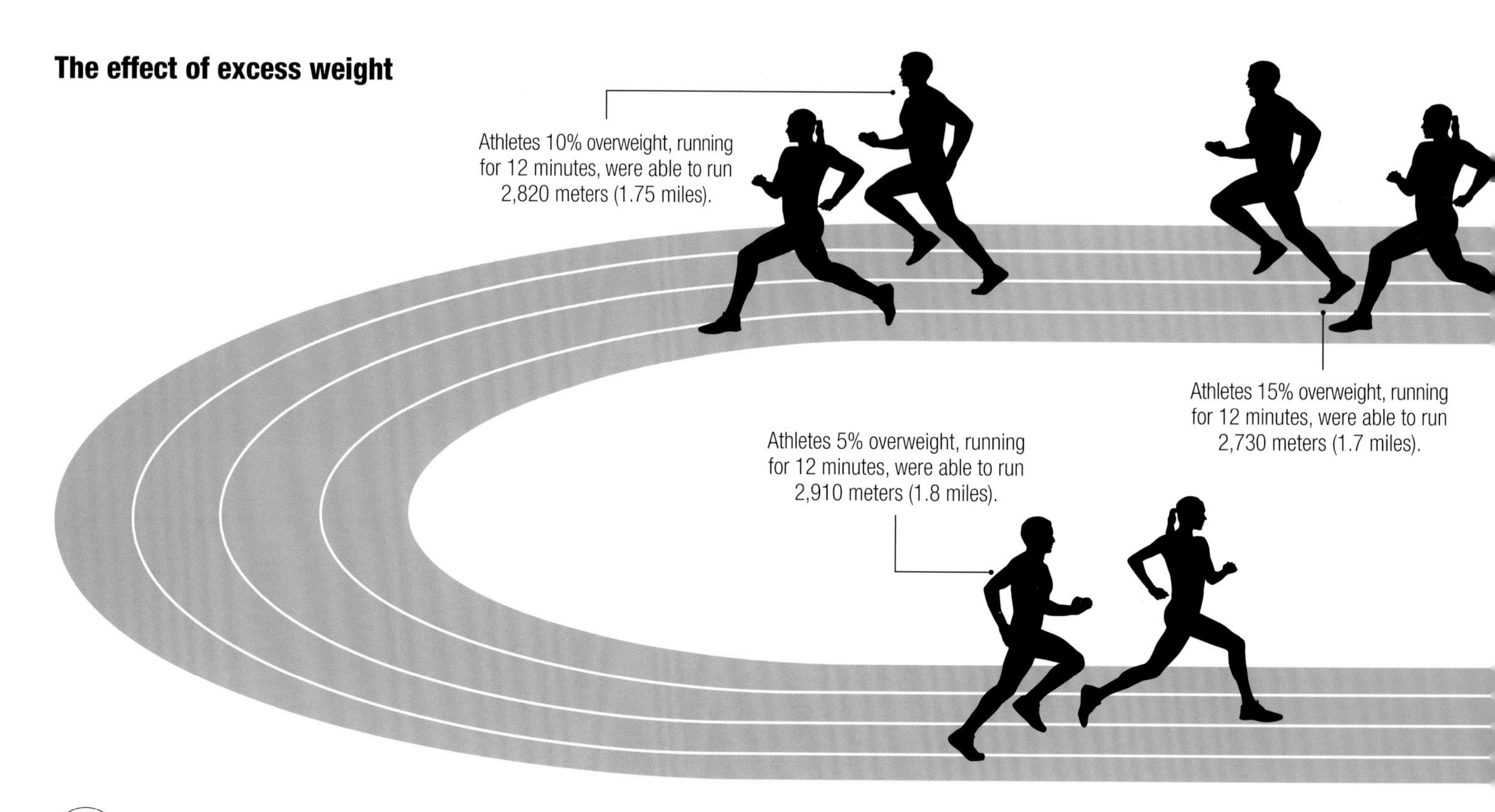

Typical Body Mass Index (BMI) for runners competing at different distances

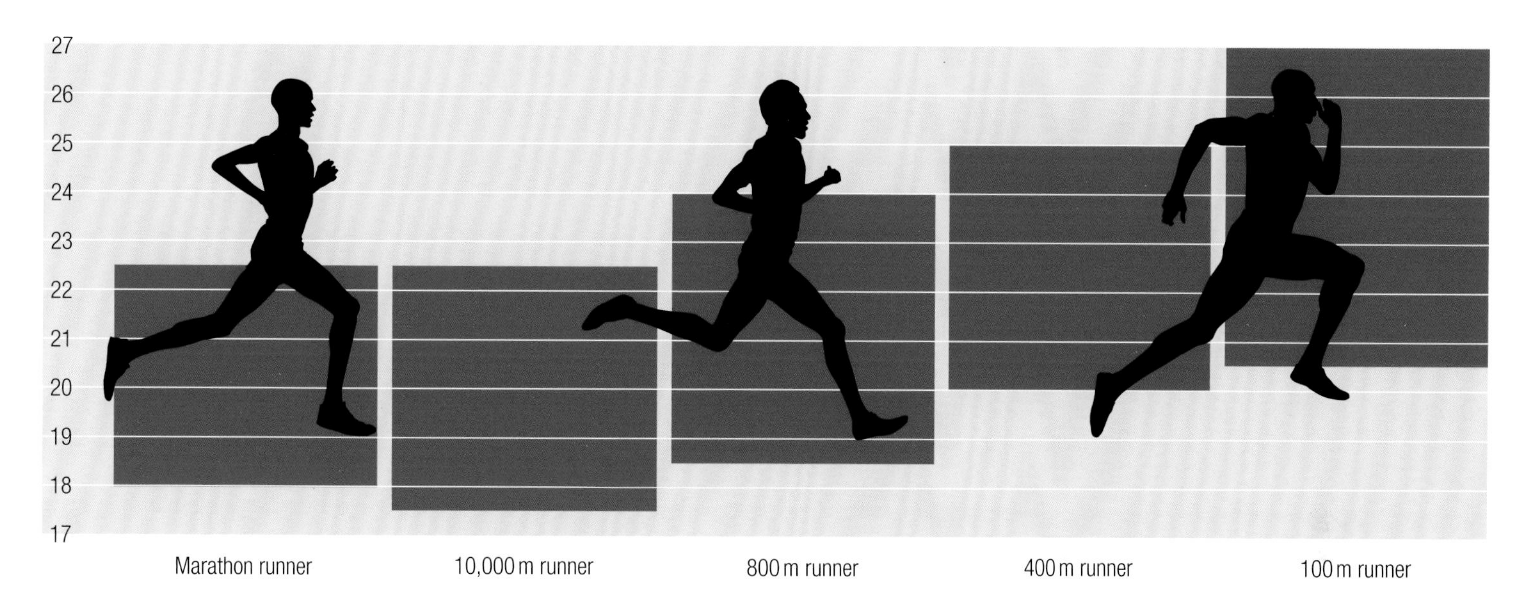

move forward. Of course, runners' upper and lower bodies are usually proportional, so some extra muscle comes in handy. Fortunately, successful runners come in all shapes and sizes, but being lean—in a healthy way—is a common characteristic.

▲ ***Different shapes*** *In general, top international distance runners have similar body types, as reflected by the narrow range of BMI values. On the other hand, the body types of international-class sprinters tend to be increasingly more varied the shorter the event, as demonstrated by the wider range of BMI values.*[2]

▼ ***Excess baggage*** *Researchers tested six subjects by having them run for twelve minutes at their natural body weight and three additional runs with increasing added weight. In addition to slowing the athletes, the added weights also reduced their* VO_2 *max relative to their body weight.*[1]

To what extent can sports drinks support peak performance?

Are sports drinks good for me?

If your goal is to train as hard as possible or run as fast as possible, a well-formulated sports drink can help. In any training session or race where you expect to sweat, a sports drink is superior to plain water at keeping the body hydrated and fueled. The carbohydrates and electrolytes in sports drinks help sustain the desire to drink, stimulate rapid fluid absorption, supply energy to working muscles, and maintain blood volume. All of these responses help support peak performance.[1]

When carbohydrates are ingested during exercise and absorbed into the bloodstream, they are quickly transported throughout the body, exposing all cells to an increased supply of glucose. Active muscle cells readily take up glucose molecules and quickly metabolize them to produce the energy (ATP) required to support continued muscle contraction and relaxation.[2]

Even fairly small amounts of carbohydrate can increase the use of glucose by muscles—a good thing for performance because glucose is so quickly metabolized—enabling muscles to maintain a high exercise intensity. For runners participating in races that last over one hour, consuming 30–60 g of carbohydrate per hour is enough to improve performance. For events lasting two hours or more, runners capable of maintaining a strong pace can benefit from ingesting 60–90 g each hour. Research shows that ingesting more than 90 g per hour overwhelms the body's capacity for carb absorption and metabolism.[3] Excessive carbohydrate inake can also lead to gastrointestinal discomfort.

For training and races that last less than an hour, runners can consume a sports drink prior to the start, in order to begin exercise with a steady supply of fluid, carbohydrates, and electrolytes entering the bloodstream. Interestingly, studies have demonstrated that performance in events lasting less than 1 hour can be improved simply by swishing a sugar solution around the mouth. The brain senses sugar in the mouth as incoming fuel and that is enough to temporarily reduce feelings of fatigue and maintain a high exercise intensity.[4]

Carbohydrate intakes for different exercise durations

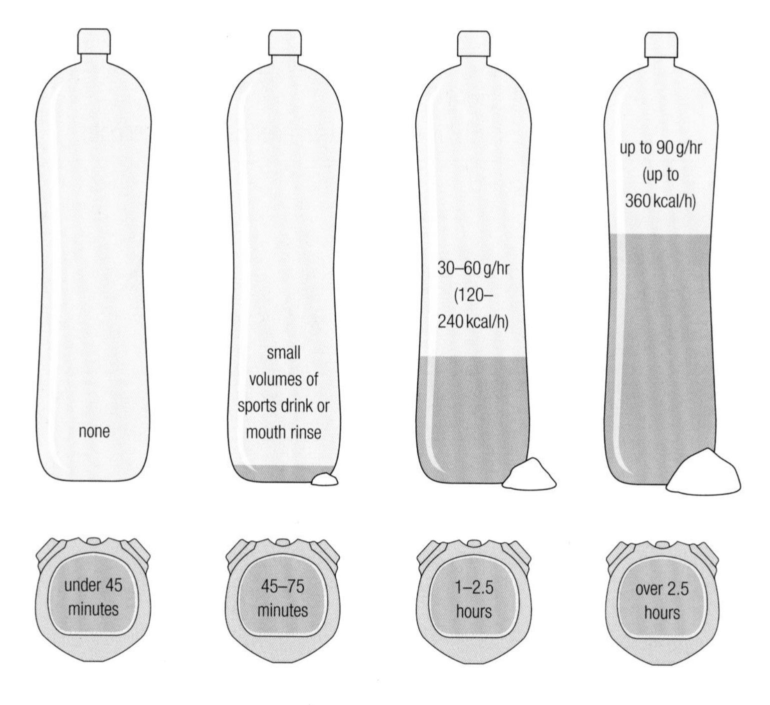

◀ ***Fuel intake*** *Sports drinks can be a primary source of fluid, carbohydrates, and electrolytes; runners in longer events may also choose to consume additional carbohydrates from gels or bars.*

Effects of different carbohydrates

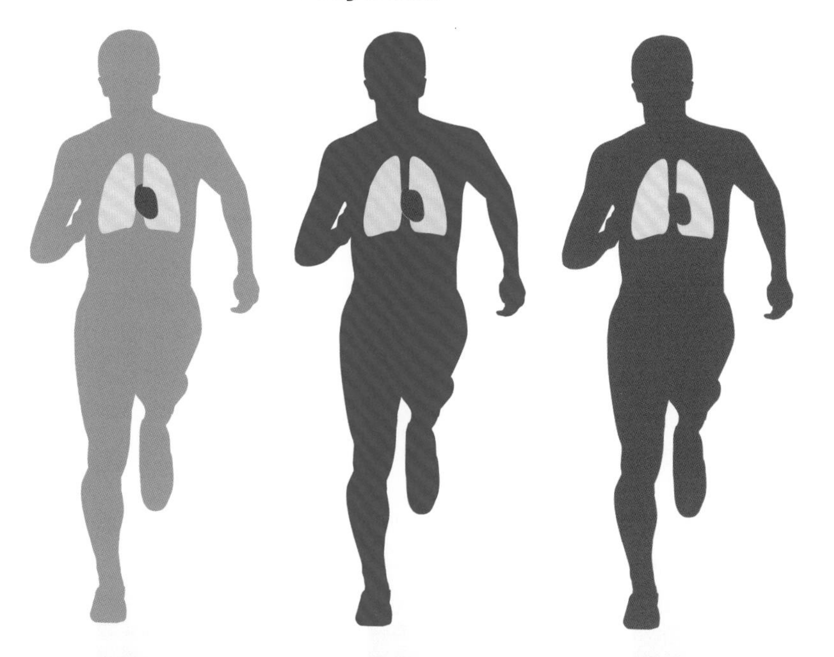

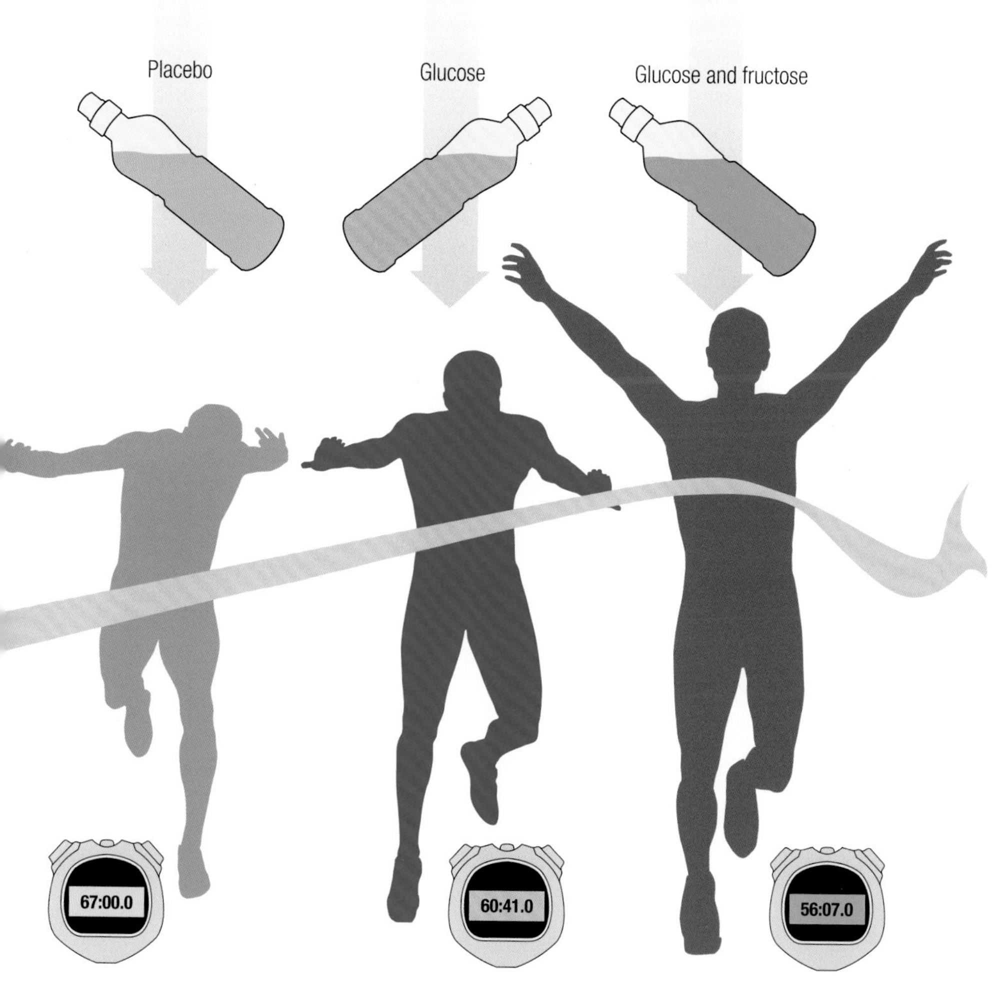

Energy source for muscle contraction

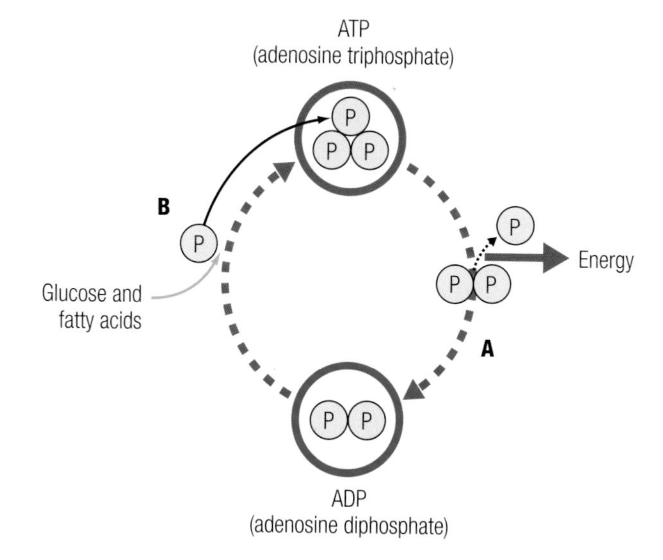

▲ ***Energy production*** *(**A**) ATP (adenosine triphosphate) contains three phosphate molecules. Energy for muscle contraction is produced when the bond of the last phosphate molecule is broken, resulting in ADP (diphosphate) and P (phosphate). (**B**) Glucose and fatty acids are oxidized inside muscle cells to provide the energy needed to turn ADP and P back into ATP.*

Need to know

ATP is shorthand for adenosine triphosphate, the molecule that all cells use as energy to support cell functions. Muscle cells use glucose and fatty acids to produce the ATP required during running. Glucose is the primary fuel for muscles during most running events.

◀ ***Winning combination*** *A combination of glucose and fructose—the carbohydrates found in most sports drinks—promotes better performance because of increased carbohydrate use by muscles. Carbohydrates can be oxidized by muscle cells faster than fatty acids and less oxygen is used in the process, making carbs the preferred fuel for intense exercise. The times listed are for a simulated endurance event after two hours of exercise at 40% of VO_2 max.* [5]

What nutritional intake best aids postexercise recovery?

What should I eat and drink after a run?

After a hard training session or race, you will have lost water and electrolytes in sweat, your body will have burned through a lot of muscle and liver glycogen, muscle cells will have been damaged from the repeated pounding of each foot strike, and cells throughout your body will be calling for the nutrients needed for adaptation and growth. It's time to eat and drink, and the sooner, the better.

Water, electrolytes, carbohydrates, and proteins are all required to speed recovery and enhance the adaptations to training. Water and electrolytes help ensure that muscle cells rehydrate quickly. A dehydrated muscle cell is a catabolic muscle cell—molecules such as glycogen and proteins are broken down to help attract water molecules into the cell. A hydrated muscle cell is an anabolic muscle cell—glycogen and proteins are synthesized to help maintain the cell's desired hydration state. In the case of dehydration, prevention is the best solution. Your goal should be to drink enough during training and racing to avoid dehydration altogether.

Depleted muscle and liver glycogen stores respond quickly to ingested carbohydrates, so eat what you can as soon as you can. Try to consume at least 0.45 g of carbohydrate per pound of body weight (1 g/kg) soon after your run to spark glycogen synthesis so that by the time you are ready for a full meal, some glycogen will already have been replaced.

Sample nutrition plan for training

▶ ***Daily diet*** *Runners in training require carbohydrates, proteins, and fluids to replenish lost stores and stimulate the desired adaptations to training. Protein intake should be fairly constant at each meal to optimize muscle protein synthesis throughout the day. Daily carbohydrate intake will vary depending upon the intensity and duration of training. In this example, the consumption of carbs at mealtime and in snacks would provide enough carbohydrate to meet the needs of moderate-intensity training. Consuming fluids throughout the day is required to replace urine losses, and drinking during training should keep dehydration to a minimum. Runners should rely on the consumption of nutritious whole foods and beverages to meet their daily needs.*

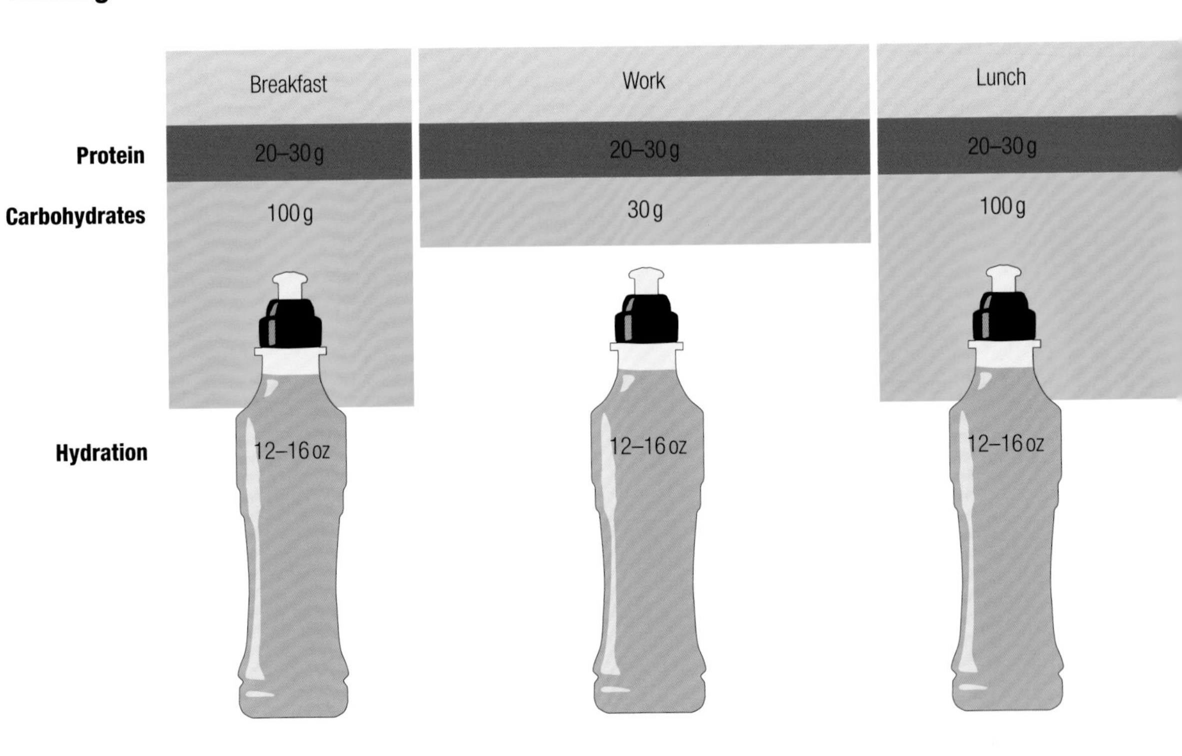

Hydrated and dehydrated cells[1]

▼ ***Exercise and recovery*** *Dehydration not only impairs performance during exercise, it also impedes recovery because dehydrated muscle cells don't recover as quickly as well-hydrated muscle cells.*

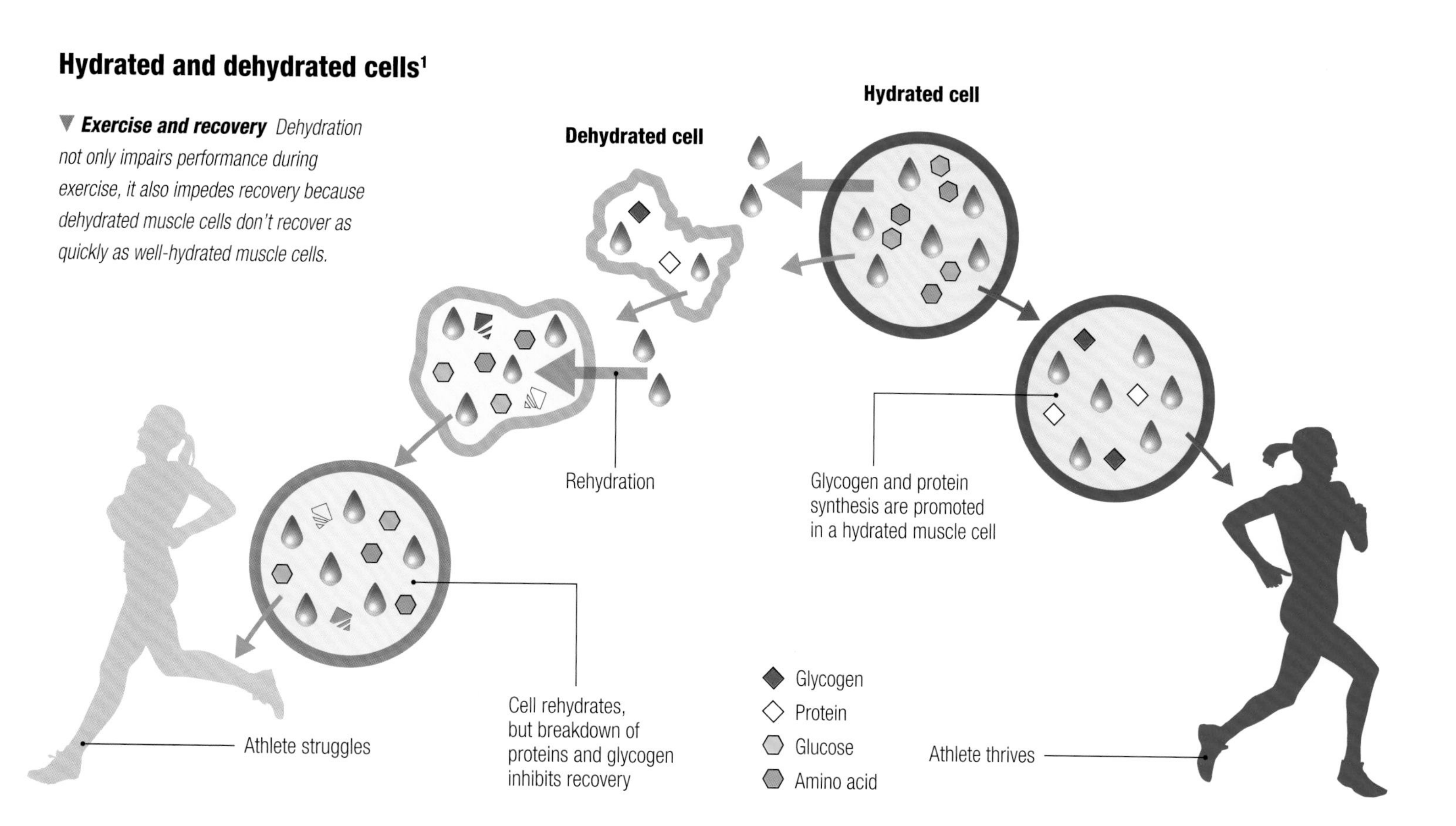

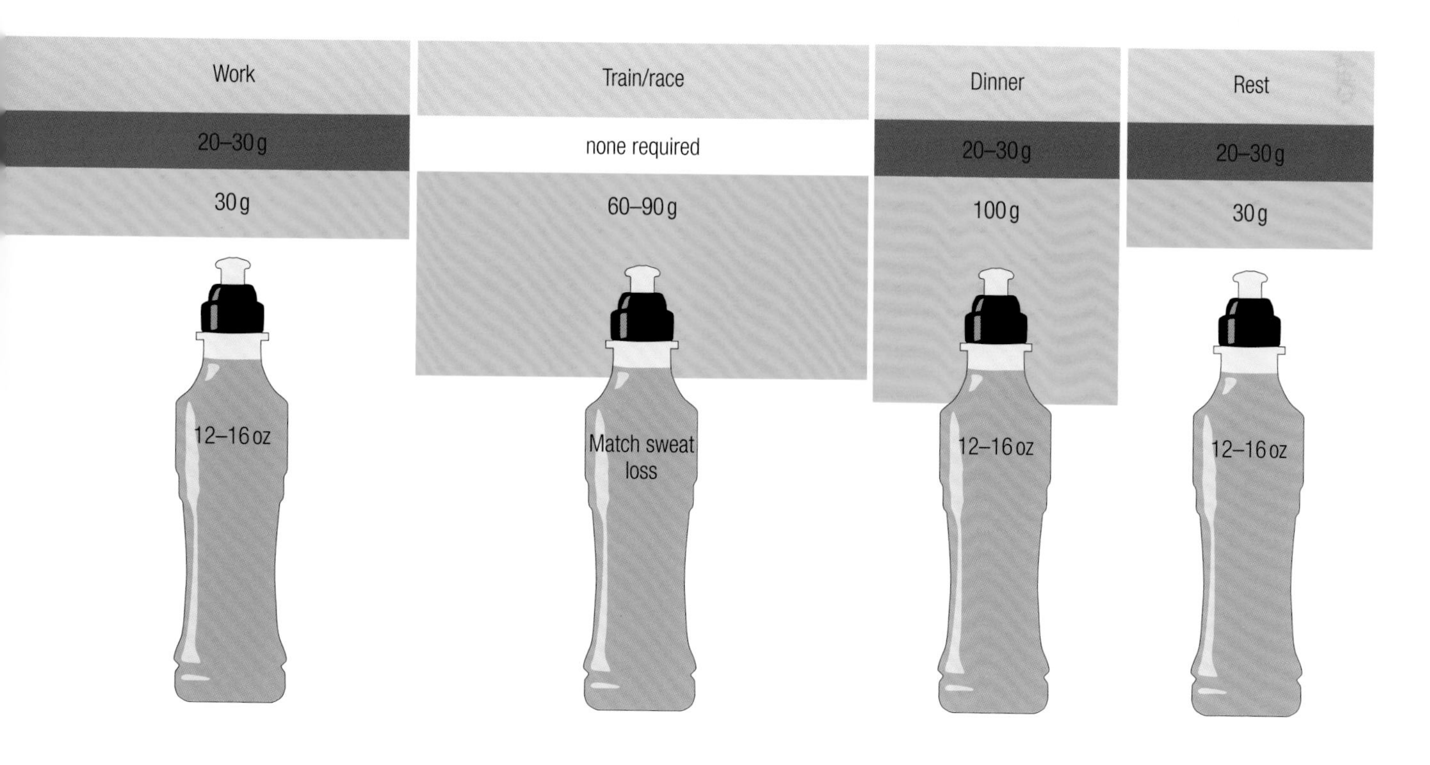

events:
long-distance

Traditionally, longer distances such as the 5,000 m and 10,000 m on the track were the domain of athletes who lacked speed but were hugely strong. However, such is the changing face of competitive running that books from a decade or two ago describing the physiology of long-distance runners no longer seem relevant. Endurance is still a major priority, but the current world records for both men and women's 5,000 m and 10,000 m perfectly illustrate what else is required. The modern long-distance runner may not possess the out-and-out speed of a sprinter, but they certainly pack a competitive punch when it comes to running fast. For example, run a 62-second opening 400 m in a men's 5,000 m race—close to four-minute mile pace—and you're already off the world-record speed in an event that's more than three miles long. Similarly, the women's 10,000 m now sees runners clocking 4 minute 30 second-miles mid-race.

As a result, budding long-distance runners need to not only think about getting the miles in, but they also need to include a decent amount of fast-paced work to cope with the unrelenting maintenance of speed required to produce your best 10,000 m. Pace, aerobic capability, and fuel are all vital when it comes to distance running. It's a well used-phrase but Rome wasn't built in a day—distance running is as much about an understanding that establishing a solid endurance background takes time—months, even years—and that it's about routine, dedication, and real understanding of how recovery is as important as the actual hard work when it comes to improving performance. So sleep, hydration, nutrition, and the use of foam rollers to improve circulation and help mobility are all elements today's long-distance runner must include in their training week.

Distance running is all about utilizing oxygen as efficiently as possible, and athletes who live and train at altitude or simulated altitude can do this by naturally encouraging new red blood

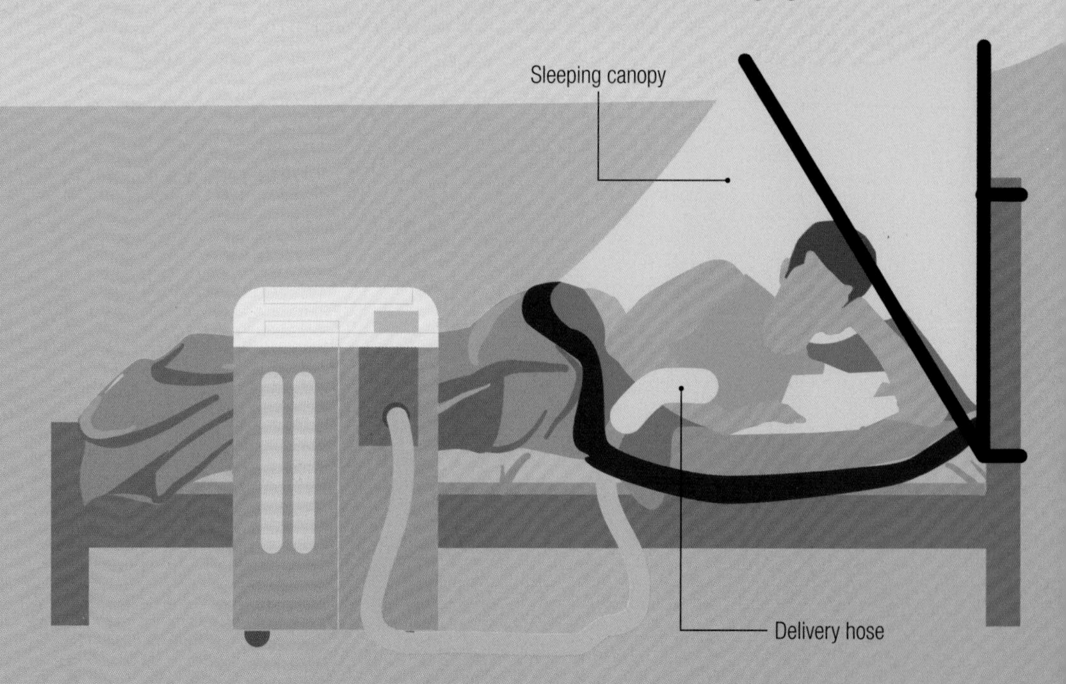

▶ ***Canopies and hoods*** *Cost and ease of use come into play with the altitude canopy—using a smaller canopy is more convenient than a cover for a whole bed and not as expensive. The hood can simulate an altitude of up to 20,000 ft (6,100 m), although most won't need higher than 8,000–15,000 ft (2,440–4,570 m). Canopies achieve the height quicker but are slightly less stable than a full cover.*

cells to carry more oxygen. This tried-and-tested method means that when you return to low altitude for a race or training, you'll benefit from this new oxygenated blood and enjoy a temporary boost in performance—resulting in the oft-heard phrase "live high, train low," which, translated, means sleep and rest in the mountains, then drive down to low altitude to train. Indeed, 95% of medalists at the major championships since 1968 have done exactly that.

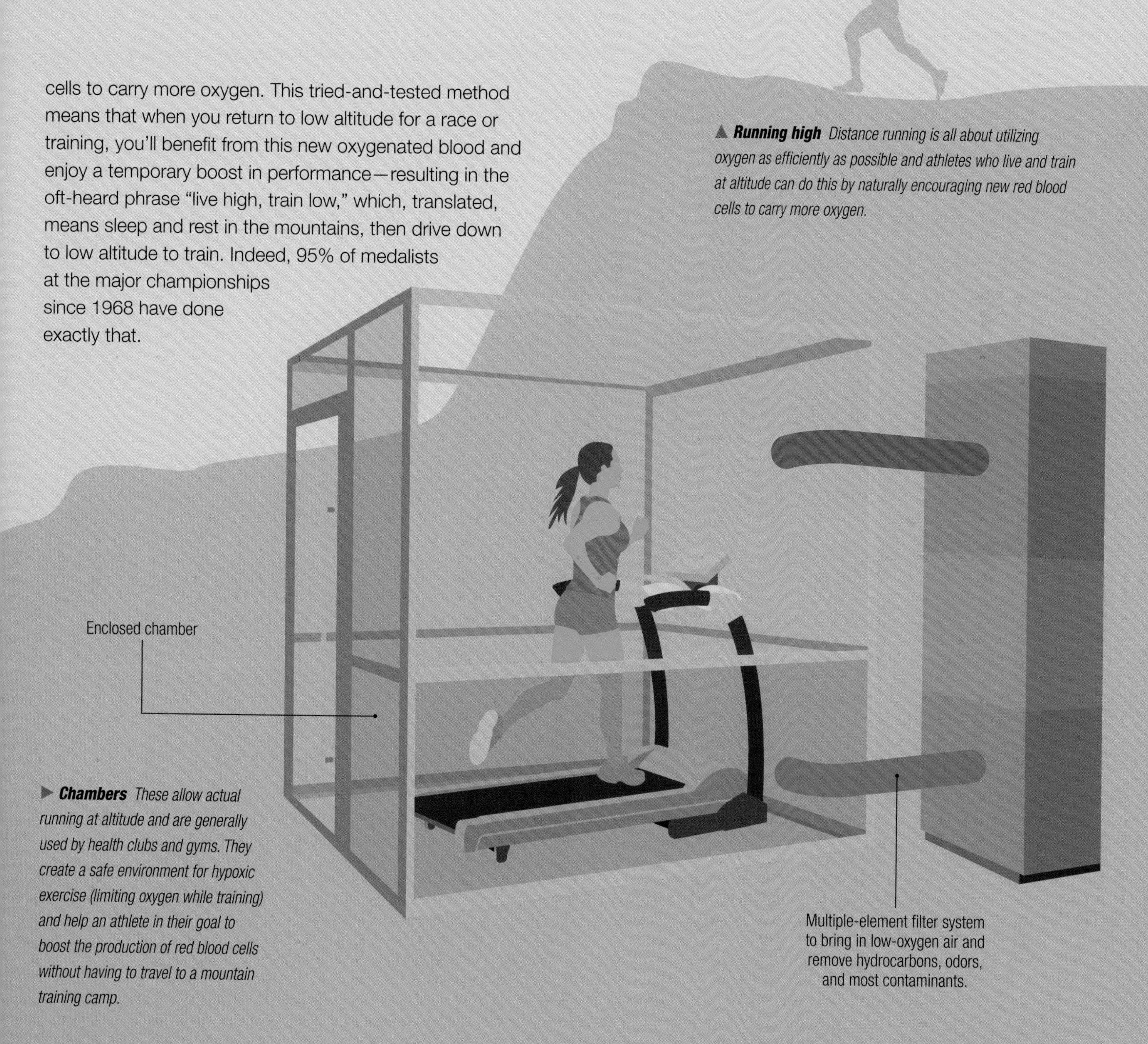

▲ ***Running high*** *Distance running is all about utilizing oxygen as efficiently as possible and athletes who live and train at altitude can do this by naturally encouraging new red blood cells to carry more oxygen.*

► ***Chambers*** *These allow actual running at altitude and are generally used by health clubs and gyms. They create a safe environment for hypoxic exercise (limiting oxygen while training) and help an athlete in their goal to boost the production of red blood cells without having to travel to a mountain training camp.*

Article by **Paul Larkins**

Running is something we all learn to do naturally, surely? First we learned to walk and then went on to run, all by around the age of two. Despite this common starting point, only some people go on to achieve outstanding demonstrations of human performance in running, combining both speed and endurance. Using strategies to manage physical sensations of discomfort and pain and maintain a positive sense of self-belief, they work toward carefully targeted goals, and are often able to achieve far more than their expectations. Such coping strategies to manage discomfort, increase motivation, and raise confidence can not only help athletes run further and faster, but might also make them smarter. This chapter explores the science behind the mind–body connection and how it relates to all kinds of runners, whether preparing for competition, running for health reasons, or training to compete in another sport.

chapter four

running psychology

Andy Lane

Is fatigue driven by physical or psychological factors?

Can I run through pain?

Running and experiencing intense sensations of fatigue are natural bedfellows and, for many people, the notion that running could be anything but an unpleasant experience is one that is difficult to contemplate. However, as you learn to run, and increase the distance and intensity of sessions, you also learn to cope with physical sensations you might initially have labeled as "pain." As you get fitter, running at a certain speed feels easier. For example, at first you might find a 5-km park run a challenge and finish, feeling exhausted, in 30 minutes. After four weeks of regular training, though, running a 30-minute 5 km feels comfortable.

What is going on here? Two parallel changes are occurring. One is a physical adaptation—muscles getting stronger, coupled with cardiorespiratory improvement. The second change is psychological—you are learning what it feels like to run hard. In the initial stage of development, you might have thought "I can't cope here, I'm getting too hot, out of breath... I'd better slow down, or better still, stop!" Over time, this internal dialogue has changed to "I am warming up, take a few deep breaths, feel in control, and hold this effort. I can manage this effort and so let's just keep going." The self-confidence to cope with working at a given heart rate has increased.

▶ ***Hiding pain*** *The message from the body says "this hurts." Physiological systems are being pushed to the limit, but this is competition and important competition, and showing to your rivals that you find the race hurts sends a very powerful message. Runners in this situation say messages to themselves such as "keep positive, relax." They try to hide any message that could motivate their rivals, because they hope they are also hurting. If they can hide their pain and keep focusing on what they are doing, then maybe their rivals, if they are also suffering, will believe that they will not win. If the pain is not worth it, then they will stop.*

We know that experienced runners are good at managing the inner voice that says "you are tired, slow down." For many runners, this voice conflicts with the self-imposed goal to complete a run, or finish in a certain time.[1] There might be a physiological limit that stops the runner making progress, but before that limit is reached, the brain may step in and say "enough is enough." Sensations of fatigue set in and the runner feels the need to slow down.[2] Experienced runners can instead learn to interpret the inner voice message as "you are tired, a great performance could be on" and continue their effort. As you get more experienced in your running, you might start taking greater risks with the amount of effort you put in.[3]

Should I focus on internal or external factors to improve?

What should I think about when I'm running?

There are many different approaches to the question of mental focus during running, and no single correct answer. You could think about your technique, reminding yourself of teaching points such as "run with high hips," "don't over-stride," and "relax, especially your arms and shoulders." Or you could try to focus your mind on things not relevant to running—by listening to music, looking at your environment, or talking with a running partner.

If your goal is to complete a run, or to keep running for a certain amount of time, and currently you find it difficult to run, then having an external focus can help. Distracting yourself by listening to music or holding a conversation is a very effective strategy for slower or low-intensity running.

If you are focused on an external source, though, it means that you are not working maximally. A wealth of research has shown that, as exercise intensity increases, physiological cues overpower any other thoughts. Running hard and fast is accompanied by physical discomfort and your body and brain make sure you are fully aware of what is happening. Your brain is in control and lets you know if it thinks you can't cope.

So if the goal is to complete a run in a certain time or achieve a certain position, then you should focus your mind on internal factors. Running fast involves quicker, longer steps—so one strategy is to focus on your rhythm and cadence, in order to increase your step rate. Concentrating on what you are doing is key for fast running.

It is possible to mix these approaches up, focusing externally for some parts of a run and internally for others. For example, you might focus internally when running up a hill, concentrating on your form, and focus externally for less intense parts of the route. The best strategy largely depends on your goal for the event and the distance of the run.[1]

▶ ***Staying focused*** *Athletes focus on the here and now, on the complex blend of arousal, activation, and technique needed to deliver success, while conscious on being relaxed and activated. The skills need to work automatically and arousal is needed to drive physiological systems. Focus has to narrow and while it helps to see what the others are doing, head movements—even eye movements—can cause concentration errors.*

国盛集团
浦发银行 SPD BANK
浦发银行 SPD BANK

Does listening to music improve performance?

Will music help me when running up that hill?

Music has a powerful effect on the human psyche. This bold statement is evidenced by knowledge that humans have created, listened to, and moved to music for thousands of years. Technological developments have contributed to a rapid expansion in the number of runners listening to music while running. Numerous studies have sought to investigate what type of music has the greatest motivational effects. For example, since the 1990s, Dr Costas Karageorghis of Brunel University London has conducted extensive research on this topic.[1] Over hundreds of studies, he has developed and refined a theory that has real value when selecting music to raise motivation.

Karageorghis' theory classifies certain factors associated with music into two broad areas: intrinsic factors such as rhythm, harmony, and the meaning of the lyrics; and factors external to the music itself but closely linked with it, such as culture, age, and personal relationship with the song. With these factors in mind, Karageorghis' team developed a scale allowing people to rate the motivational quotient of different tracks—the higher the motivational quotient, the more motivating they found the song. From the team's results, the general rule of thumb emerged that a song with a 4:4 rhythm, a fast tempo, and inspiring lyrics will be more motivational.

Of course, musical preference varies, and a song that motivates one person will not necessarily motivate another.[2] Culture, age, and previous experiences all feed in to an individual's choice, and songs with personal meaning in formative years tend to have a greater motivational effect. So a 50-year-old person might find the Sex Pistols' anthem "Anarchy in the UK" motivating, while a 70-year-old might respond more favorably to "A Hard Day's Night" by The Beatles.

Motivating music tends to increase the intensity of arousal, helping the person feel more excited and energetic. Sedative music, on the other hand, can help reduce arousal, and generate a sense of calm. An important part of the effectiveness of music as a performance-enhancing strategy is to identify the emotion you wish to experience (for example, highly charged and excited, or calm and relaxed) and select appropriate music to match that state. Used in this way, music has been found to be a helpful tool in getting your mind right for performance.

◀ ***Run the world*** *Athletes can listen to music for many reasons and gain many benefits. Music can act as a reminder of good technique and performing with rhythm, help increase arousal, or help you relax.[3] Listening to music can help block out distractions, help you gather your thoughts and, with headphones over your ears, put a "do not disturb" sign out to others.*

Is there a link between running and academic performance?

Will running make me smarter?

Preparing for academic work is typically associated with scholarly activities such as reading and writing. The notion that going for a run could help improve intelligence seems a far-fetched claim. At first glance, perhaps it is. However, there is evidence to show that regular exercise, including running, is associated with high scores on intelligence tests.[1] Researchers asked a cross-section of people to describe their exercise patterns, and then asked them to do an intelligence test. The nature of such a research design does not show that if you do more exercise then your intelligence will improve. Equally, it does not mean that attempting to raise your intelligence through traditional methods will lead to you running more or running faster. The research has found that these two variables are related, but does not show that one causes the other. An alternative explanation, for example, could be that more intelligent people tend to decide to exercise more regularly, perhaps because they see value in doing so.

The link between exercise and academic performance is interesting, and the reasons behind the relationship warrant some interrogation. One argument that has found supportive evidence is that exercise is an effective strategy to regulate mood, and better mood helps improve performance. Evidence does indeed show that mood states tend to be more pleasant after exercise, and positive mood has been found to help with creative thinking and coming up with new ideas. However, perhaps surprisingly, effective problem solving is more closely associated with negative mood states, as people in an unpleasant mood tend to search in greater detail and with stronger intensity than individuals in pleasant moods. Some caution is urged, though, as failure to attain important goals when already in a negative mood can spiral downwards into a clinical condition.[2]

The take-home message is that running and exercise associate with mental and physical health, which in turn associate with successful performance in a range of different areas of application, including the academic arena. Running could also be an effective stress-busting activity in periods of intense academic work as a means to improve mood.

▶ ***Personal development*** *The ability to plan, to concentrate, to control emotions and unwanted thoughts, knowing when to act, keeping positive—are all intelligent qualities and all qualities that are developed by running and training.*[3] *Reflecting on what you did well and how you can improve, and learning lessons from the misery of defeat are components of a rich education, and if you realize you have these qualities, you can apply them to other contexts. Running can be a fabulous way of developing self-regulation.*

OHIO

Do emotions lead to poor performance?

Should I control my prerace emotions?

Emotions are a natural to response to situations and something we all feel.[1] When things are going well, or we think they will go well in the future, or went well, we tend to experience positive emotions and these can make us feel energized. When things go poorly, or are about to go badly, or were bad, we tend to experience negative emotions. We can feel sad and dejected when we think we will have a bad run, and these feelings can be intense when running well is important; for example, if we want to achieve a certain time at the London or Boston Marathon. We know the event only occurs once per year, it is hard to get into, and all our friends and family will be watching and will ask about our performance later. When an event is important, we tend to get intense emotions.

We know that emotions have an influence on what we do, and we know that people are aware of how they feel and will engage in strategies to make them feel better.[2] Runners will try to feel in a way that helps them run faster.[3] How should they feel? Most runners like to feel energetic and if they are sad, will try to change that feeling and the thoughts surrounding it. Runners will tend to engage in positive self-talk to manage these emotions.

However, some runners like feeling anxious. They feel that nerves are a source of energy, and when they feel nervous and energetic they equate this with feeling ready.[4] A runner attending a low-key race at which the stakes are low, might feel under-prepared for competition and might therefore attempt to reappraise the race. The athlete might set a goal of achieving a personal best and let others know about the raised stakes as ways to elevate anxiety.

Adjusting your goal from something like "an outright race win" to "doing your best" should change your emotions. Achievement of the goal is under your control alone rather than being dependent on the performance of others. However, evidence also suggests that most runners strive to improve constantly and achieve personal best performances, and so are not likely to set themselves easy goals. Or, if they do set an easy goal, they are likely to feel despondent because they perceive that they are not giving their all.

▶ ***Highs and lows*** *Runners feel a range of intense emotions at the start line of a race, including anticipation and active feelings such as excitement and anxiety.[5] If they feel ready and that they have a chance of performing well, then they might also feel a hint of happiness. If they expect their performance might be poor, then anger and a hint of sadness. Runners experience a range of feelings and all of these feelings can change when the race starts if you start to perform better or worse than you expected.*

RUDISHA

Can psychological skills improve running performance?

How can I keep my mind positive?

Peak performance requires planning, and the mental processes involved are skills that respond to practice, like many other skills. As a rule of thumb, appropriate practice generally leads to rapid improvement, and research shows that many elite athletes have successfully developed psychological skills to help prepare them to perform in training and competition. However, there is also evidence that when athletes first start using such psychological skills they can find it difficult.[1]

Many runners use psychological techniques such as imagery, self-talk, and goal-setting, and do so without formal training.[2] Athletes may engage in a version of self-talk where they speak to themselves about how well they are performing, or about how they are about to perform. In fact, the inner dialogue that fills people's minds is influenced by intense emotions such as anxiety. Research shows that people use psychological skills in competition more than they do in training—so athletes new to the techniques may be trying to learn these new skills at the same time as they are experiencing unpleasant emotions associated with anxiety. In addition, research indicates that managing emotions can be tiring. When these factors are put together, it is perhaps not surprising that some people find learning psychological skills difficult at first, involving an additional physically and emotionally draining task on top of the effort of performance.

When starting to learn how to use psychological skills, it is good practice to use them in training, becoming competent in the techniques and then applying them in competition. In running, research indicates that imagery skills can help manage fatigue—runners mentally rehearse experiencing intense sensations of fatigue and see themselves coping successfully. The notion is that imagery creates a blueprint for the pattern of thinking and behavior required to deal successfully with a situation, so that when those conditions actually occur, the athlete is able to instigate the appropriate coping strategy more quickly and effectively.

▶ ***Performing under pressure*** *The ability to perform to your maximum at the critical time does not want to be a chance event.[3] Mental states wax and wane, but you want to be sure that you will be mentally ready for any major races. Frequent use of psychological skills means that the mental game is trainable, and from training comes the confidence that you can rely on it when it matters. Being able to concentrate on what matters, and manage your emotions and body language so that you are relaxed and look confident to others, come with mental training.*

USA
BUSTOS
Rio2016
KIPRO
IGUIDER
ALGERIA
Rio2016
MAKH OUFI
7
7
USA
Rio2016
CENTROWITZ
2

Does anxiety have cognitive and somatic components?

Is anxiety all in the mind?

Anxiety is a complex emotion. There is a mental side of anxiety, called cognitive anxiety, which describes our self-doubts and concerns about whether or not we will perform well.[1] There is also a physical side to anxiety—somatic anxiety—characterized by increased resting heart rate, sweaty palms, and frequent toilet visits. Indeed, one clear indication of the extent to which athletes experience the somatic symptoms of prerace anxiety is the length of the line for the toilet before an event. There might be 30,000 runners waiting to start a marathon, for example, many thousands of whom are nervously concerned about how fatigued they will be at 20 miles, whether they will manage to run the entire event, or what their finish time will be.

Seeing anxiety as having both a physical and a mental side can help to regulate the symptoms.[2] Somatic anxiety is best addressed through techniques that focus on physical symptoms. Examples of strategies that help with somatic anxiety include relaxation strategies in the form of breathing exercises. A second strategy that is useful for somatic anxiety is to listen to relaxing music, chosen by the runner to have a calming effect. Strategies to regulate cognitive anxiety are focused more on thinking and trying to reappraise or reframe attitudes about performance. Imagining success can help to reduce anxiety, as can changing negative self-talk into positive self-talk.[3]

Somatic anxiety has been found to have an inverted-U relationship with performance—very low and very high somatic anxiety both associate with poor performance. Cognitive anxiety is proposed to have a negative effect on performance—however, many studies have shown that athletes perform well when they feel high cognitive anxiety. Research shows that interpreting feelings associated with cognitive anxiety as negative tends to have negative effects.[4, 5] If an athlete has negative thoughts and experiences negative self-talk, but decides to raise effort levels and ignore the inner voice, then it has been shown to be possible to be high in cognitive anxiety and still perform very well.

▶ ***Different types of anxiety*** *Most runners feel nervous or anxious at the start line. Anxiety comes in two forms: cognitive and somatic. Cognitive anxiety is the thoughts that run through your mind: Will I remember to follow my plan? Will it have in me when it gets tough? Are the other runners stronger than me? Cognitive anxiety can serve to undermine your performance. Runners also feel physical anxiety, which helps prepare for "fight or flight" and can be adaptive to running. Sometimes the best thing to do is get the physical activity under control through focusing on your breathing and having positive images of performing well flowing through your mind.*

adidas

events:
road running

The good news is that physiology means that humans are just about perfectly designed for distance running; it was, it appears, what we were built for. For example, every time our body makes contact with the ground, it does so with a force eight times our body weight, but on impact our foot expands like a shock absorber and our ankle joints act like levers to convert all that force into forward motion. Evolution, perhaps from our need to travel long distances to hunt, mean our body's shape, build, and composition make us ready to tackle events such as the 10 km, which today are hugely popular, with hundreds of thousands running in them every weekend.

But some of the greatest runners will have a far more simple reason for loving road running, and that's its ease of access. While swimming, cycling, and numerous other aerobic activities are undoubtedly as beneficial, they normally require travel and specialist bits of equipment. Road running is about opening the front door and heading out. Olympic champions do just that most mornings. It's that simple.

And that simplicity carries over into events, which vary from standard distances like the 10 km to more unusual events such as downhill races, laps of city centers, or running in tunnels.

Running really is about bringing sport to the masses. Many of us love professional sport and will happily sit and watch our favorite teams from afar; running, however, means you can line up next to world-record holders and Olympians, shoulder to shoulder in a road race. It's something that has barely changed over the years, and probably won't ever. Event organizers understand that it's part of the appeal of their race and elite competitors also realize that road running is an important part of becoming an international standard middle- or long-distance runner.

The great popularity of recreational and competitive running has meant that running shoes are big business, and over the years there have been some innovations in shoe design—and advertising—that have had a major impact on the market.

The development of road shoes

▶ ***Nike Waffle Sole*** *(1972)*
The modern running boom started in the mid-1970s just as Nike's waffle sneaker emerged. Unlike running shoes before, it provided cushioning and grip and was perfect for the road. The early patent described it: "the sole has short, multi-sided polygon shaped studs…which provide gripping edges that give greatly improved traction."

▲ ***Nike Air Tailwind*** *(1978)*
Cushioning took another leap forward in 1979 when Nike again came up with an innovation that changed how runners felt about shoes—air. By embedding air cushioning, the new shoes offered unheard of comfort and response.

▼ ***Nike Free 5.0*** *(2004)*
Barefoot running, or at least a cushioned running shoe offering a similar natural footstrike, is what the Nike Free is all about. By creating as natural a running motion as possible, the shoe aims to reduce the amount of injuries. However, given the road surfaces most of us run on, some cushioning and support is also offered.

▲ ***Adidas Energy Boost*** *(2013)*
The Boost range introduced a range of technologies, the most trumpeted being a midsole made up of foam pellets, which was designed to return slightly more energy to runners on impact.

▼ ***Asics Freak and GT II*** *(1986)*
In a fast-changing shoe world, product is rapidly changing, but Asics Gel is still a mainstay of the company's running footwear. In the 1980s, this cushioning system introduced the ability to absorb shock by dissipating vertical impact and dispersing it into a horizontal plane, and shoes like the Kayano have featured this technology ever since.

Article by **Paul Larkins**

Every day, runners are faced with decisions to make. The typical mind of the runner is always questioning: What if I do it this way? What if I do it that way? Should I include altitude training? What if I stop—will I lose my fitness? This chapter aims to cover some of the practical questions that arise for runners on a daily basis, particularly distance runners, drawing on what is known from scientific research. Of course, there is a lot that remains unknown—research often lags behind the development of new training theories, products, and technology—but decades of research into the physiology and biomechanics of running mean that there are many answers out there. What is certainly true is that each runner is an individual. Genetics, environmental and social factors, personal preferences, and injury history all lead to the individual response and one size of training won't fit all. This chapter should be used as a guide to help make informed decisions and, ideally, accelerate training quality.

chapter five

training and racing

Charles Pedlar and James Earle

What physiological benefits do runners gain by warming up?

Do I really need to warm up before I get going?

Runners and coaches generally advocate a warm-up before a race or training session in order to "prime" physiological systems, enhance performance and reduce injury risk. However, there are no universal criteria defining what constitutes a warm-up. Warm-ups may differ in duration, intensity, and modality, depending on the individual, the type of training session they precede, and the environmental conditions. What's more, the evidence to justify warming up is somewhat limited.

For athletes with certain conditions, such as exercise-induced asthma (which must be diagnosed by a physician), a warm-up has been shown to reduce airway constriction and improve exercise capacity due to a "refractory period."[1] Furthermore, it has been clearly demonstrated that the rate of increase in oxygen uptake during a run is accelerated by a warm-up or priming exercise, which may be advantageous for middle-distance running performance because it enhances the oxidative energy contribution, potentially reducing the rate of fatigue and raising the running speed that can be maintained.[2,3]

The higher the intensity of exercise, the more important a warm-up is. For example, performance in a maximal sprint session will be enhanced by increased muscle temperature—indeed, there is an optimum muscle temperature for maximal force development. Conversely, in a marathon run, which is not an explosive event, performance is likely to be limited to some extent by energy availability in most runners. A prior warm-up has an energy cost, reducing energy availability during the marathon. It makes more sense for the runner to use the first few miles of the event to warm up rather than waste excessive energy on a separate warm-up routine.

The notion that warming up can prevent running injuries is, unfortunately, not supported by a strong evidence base, partly because of a lack of studies.[4] Further research in this area is needed, with well-controlled investigations.

Warming up for a middle-distance race

Marathon race

◀ ***Reducing fatigue*** *A warm-up is thought to be crucial to optimizing performance in a middle-distance race. It increases body temperature, particularly muscle temperature, and "primes" your physiology for high performance by mobilizing fuels and opening up blood vessels. The ability to rapidly increase your oxygen uptake is key, and this is greatly increased through warming up.*

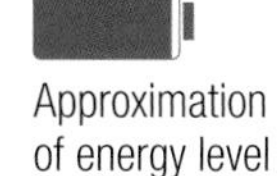

◀ ***Adding fatigue*** *Warming up for a marathon is not as important as it is for shorter races. The sheer length of the race means that there is plenty of time to warm up on the course. Warming up effectively adds more miles to the race with no obvious advantage, and increases the risk of "hitting the wall," which is associated with energy depletion.*

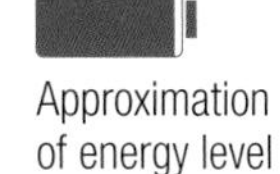

What is progressive overload? → By how much should I increase my training, week on week?

First described in 1950 by Hans Selye, with many variations since, the General Adaptation Theory tells us that adaptation occurs in response to a repeatedly applied stressor.[1] Adaptation refers to the physiological changes your body undergoes through training, as it adjusts to cope with new exercises or loads. The theory remains "general" because there are so many factors that influence the rate of adaptation, including the fitness of the individual at the start of the training program, the quality of recovery following training sessions (particularly in terms of nutrition and rest), the motivations and goals of the individual and, of course, the genetics and epigenetics of the individual.

If recovery can be optimized, then the rate of adaptation to training may be increased, allowing faster progression. A structured training program includes regular markers of progression, and a trial-and-error approach to increasing training volume or intensity according to quantitative measures (for example, blood lactate or heart rate) and qualitative training outcomes (such as ratings of perceived exertion, or RPE). This

The General Adaption Theory applied to training

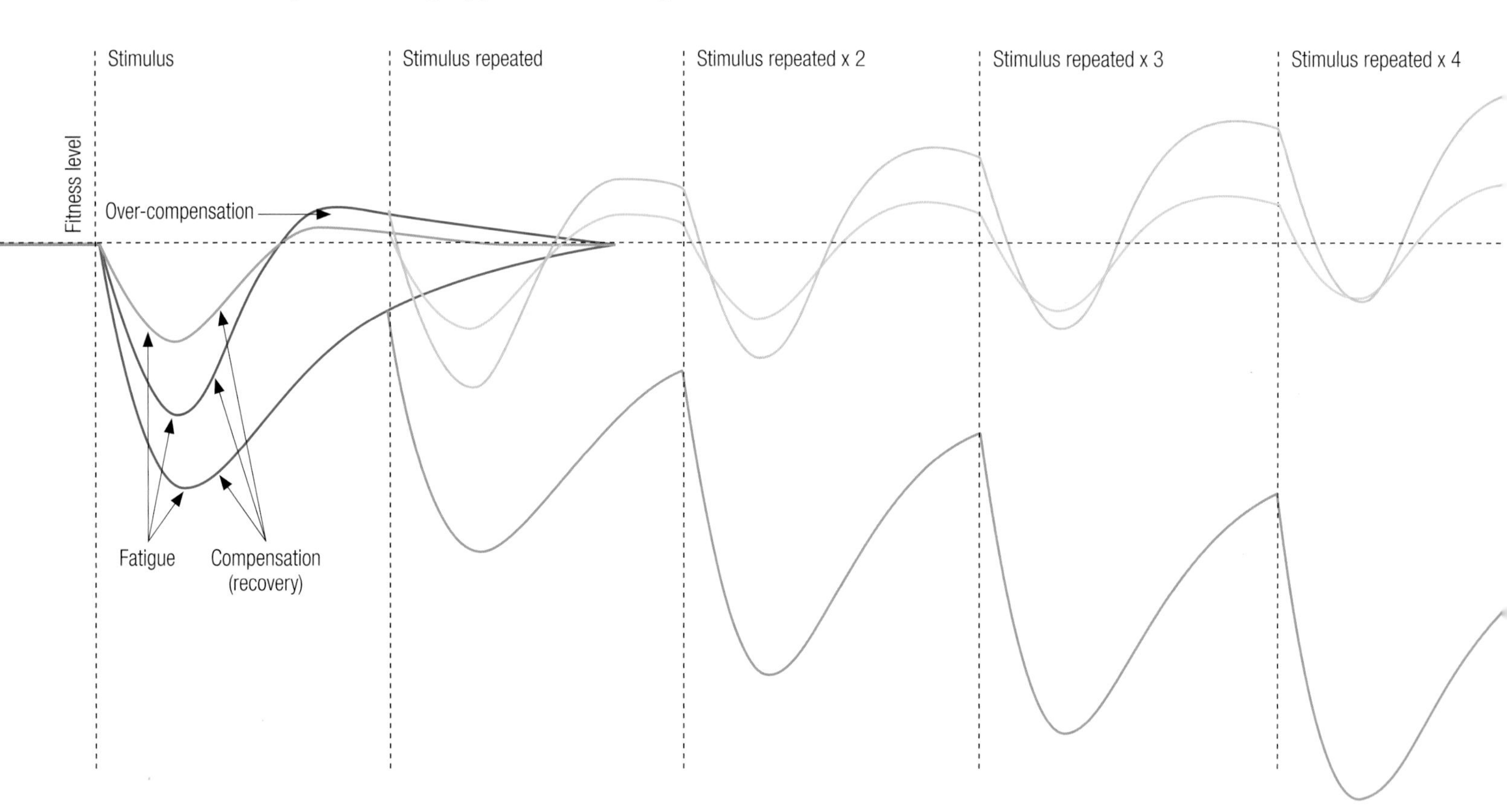

approach is called progressive overload—gradually increasing training volume or intensity over time in order to maintain a desired rate of adaptation. Progressive overload has been found to be more successful at maximizing improvements in lactate threshold in runners than a constant-intensity program.[2]

Like adaptation, optimal recovery also depends on a large number of factors and varies significantly between individuals, but the impacts of a number of common techniques have been investigated, including the use of compression garments,[3] cold-water immersion,[4] and massage. It is universally acknowledged that good recovery practices include appropriate nutrition habits (particularly the intake of macronutrients such as carbohydrate and protein). In addition, although there is a significant lack of research into sleep and recovery from training, it is widely accepted that maximizing sleep quality is essential for optimal recovery.[5]

Undertraining and overtraining tend to associate with increased risk of injury. Interestingly, runners who trained less than 30 km (19 miles) per week for a marathon were found to be at a 102% greater risk of a running-related injury post-event.[6] On the other hand, it has be shown that to minimize injury risk, training volume should not increase by more than 5 to 10% per week.[7] This unit of progression is frequently adopted in training programs for many sports.

▼ ***Achieving a balance*** *Training causes damage to the body and disruption to normal physiology. Broadly, this returns to normal with suitable recovery. The three runners pictured here span the range of training outcomes, from progressive fatigue and underperformance, through to positive gains and performance improvements.*

Optimal
Inadequate stimulus
Insufficient recovery or stimulus too great

Training load level	How you feel	Capacity to train	Action to take
Inadequate stimulus	You should feel fresh and quite rested and the training load will be tolerable. No significant fatigue or soreness.	You can push on and increase training volume and intensity if you want to improve. You are probably doing enough for maintaining health, but not enough to improve your performance.	If you want to improve, you need to gradually increase the training stress by increasing the distance covered or the speed that you run at. Varying the types of runs may be helpful here, for example, including interval training.
Optimal	You're improving all the time, so you'll feel motivated by your improvements. At times you'll feel sore and tired but, because the load and recovery are well matched, this should pass.	You are getting a little more out of your body with each training cycle and your performance is improving. Keeping your recovery practices in place will be essential to continue this progression.	You can continue until you reach your goals, at which point you'll need to rest, review, and set new goals. You might also want to adjust the stimulus if your rate of improvement slows down.
Insufficient recovery or stimulus too great	In the short term you might feel OK, but eventually you'll feel tired and frustrated. You might be seeing this in other parts of your life, too, such as more coughs and colds, a loss of concentration at work, or disrupted sleep.	You're training hard but not seeing the gains. You're noticing that your split times are not improving and you're having to cut some runs short. You may need to redress the balance of training and recovery.	Reduce the duration, frequency, and intensity of training; if the situation doesn't get any better, then you may need a break to get you back on track. Review your diet, sleep, and other recovery practices.

How can HIIT improve running performance?

What is HIIT and should I be doing it?

Although there is no single universal definition, High-Intensity Interval Training (HIIT) is a training method that involves alternating between short bouts of intense exercise and active rest periods. The training stimulus of HIIT is greater, possibly because the time spent at high intensity is longer than could be achieved during continuous exercise. This training method has been shown to improve maximum oxygen consumption (VO_2 max) and to modify risk factors for cardiovascular disease.[1] Furthermore, HIIT training is one of a few techniques that may enhance running economy in distance runners.[2]

HIIT programs are often prescribed from VO_2 max and lactate threshold tests. However, it is possible to judge appropriate intensity by exercising at more than 90% of maximum heart rate or at a rating of perceived exertion (RPE) of more than 15 (on the Borg scale of 6–20). Intervals performed close to or above these intensities have shown the greatest improvements in key determinants of running performance—VO_2 max, lactate threshold, and running economy.[3,4] Such improvements have been reported to occur after four to six weeks of HIIT.

The metabolic contribution and "race pace" of the event for which you are training determine the duration of intervals you should use. For example, a long-distance runner would benefit from long aerobic intervals (1–8 minutes) rather than short anaerobic intervals (less than 60 seconds). An active recovery is recommended rather than a pause between intervals, since it helps elicit and maintain VO_2 max during HIIT, as well as stimulating lactate removal.[5]

Since the popularization of interval training in the 1950s, it has become a universally adopted training method for health and performance. Research suggests incorporating HIIT twice a week alongside long-distance runs.

Typical HIIT session

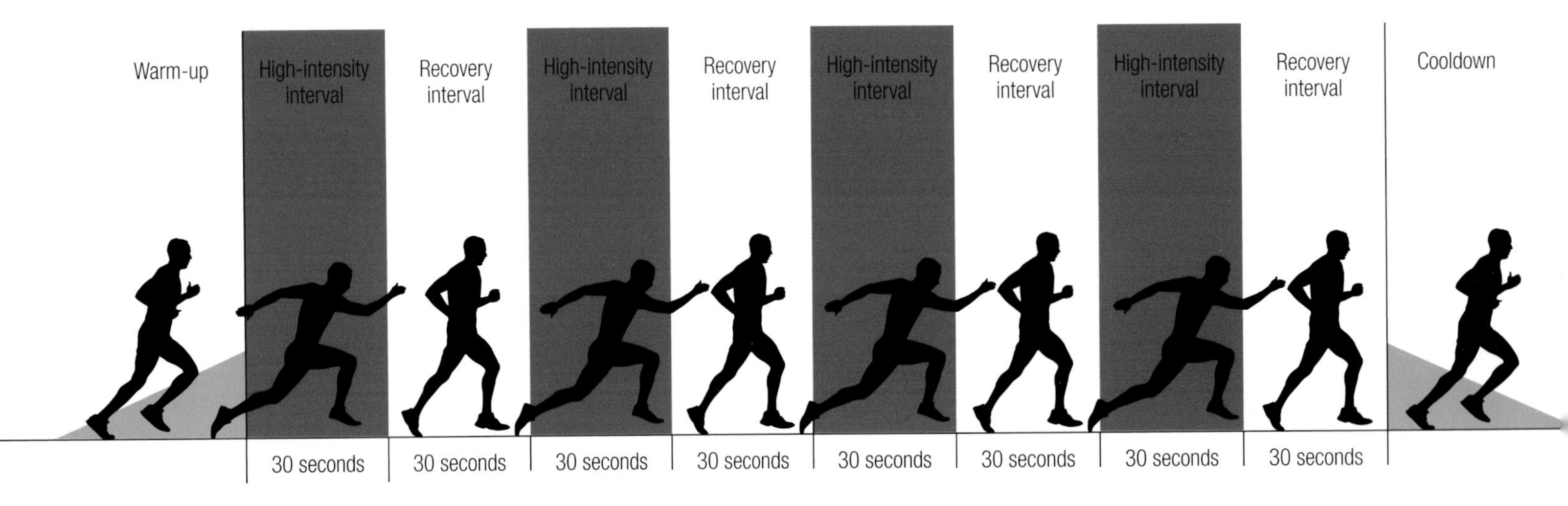

Lactate threshold and turnpoint before and after training

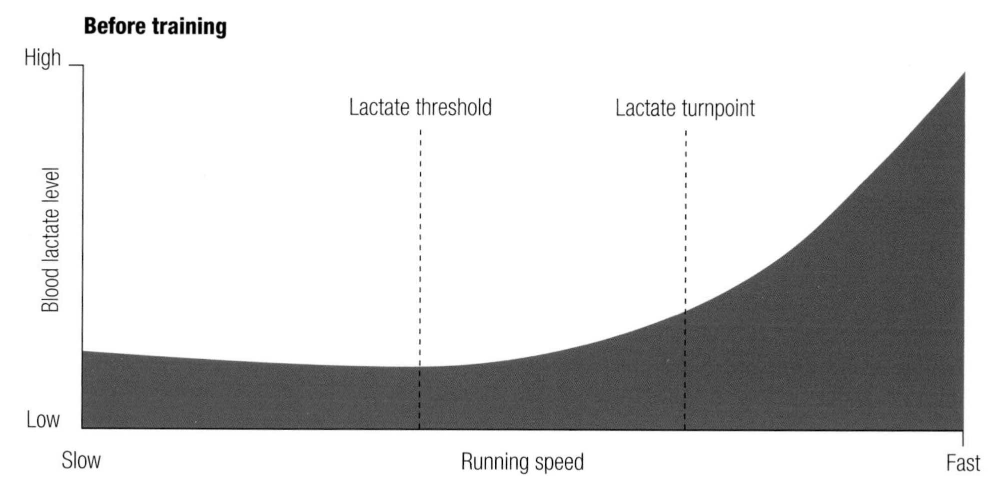

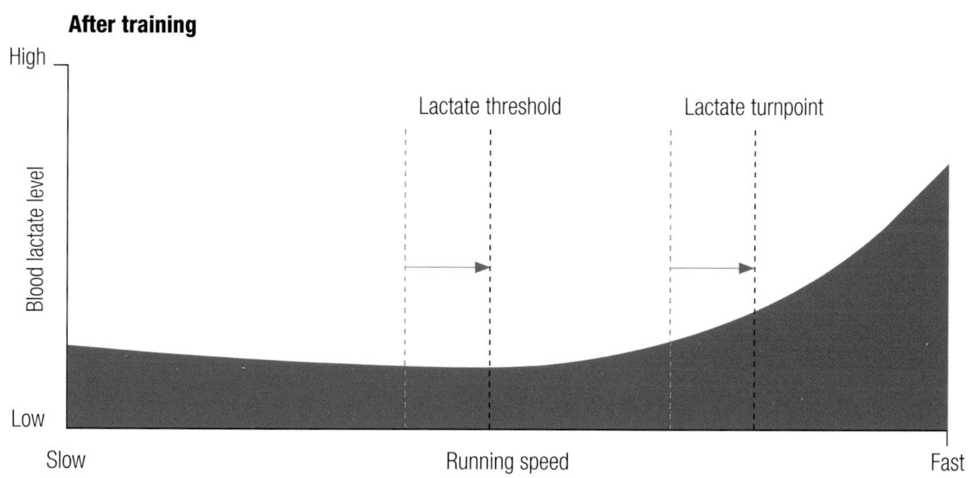

◀ ***Self-improvement*** *The lactate threshold is where lactate production is equal to removal and represents an exercise intensity that can be sustained for long durations. The lactate turnpoint is the point at which lactate production exceeds removal and is unable to be sustained for long durations. Following a period of training weeks with an optimal stimulus and recovery, the lactate thresholds are retested and should have shifted to the right on the curve, representing a lower physical work at a higher exercise intensity.*

◀ ***Repeated behavior*** *High-intensity interval training sessions have periods of work interspersed with periods of active or passive rest. In this example there are 30-second sprints followed by 30 seconds walking back to the start position.*

▶ ***Increasing intensity*** *The Borg scale is a simple tool for monitoring training intensity by using a number to reflect on perceived exertion. Each number represents a heart rate zone.*

Perceived effort when training

Borg scale	6	7	8	9	10	11	12	13	14	15	16	17	18	19	20
Perception of training	Extremely light to light						Becoming hard to hard					Very hard to maximum			
Approximate heart rate (+/- 10 bpm)	60	70	80	90	100	110	120	130	140	150	160	170	180	190	200

Lactic threshold
Approximation based on Borg Scale (orange fading tint indicates margin for error)

Lactic turnpoint
Approximation based on Borg Scale (yellow fading tint indicates margin for error)

What is the optimal marathon pacing strategy? → How should I pace myself when running a marathon?

An optimal pacing strategy can be defined as "the power profile over the course that would deliver the shortest race duration for the athlete on the given day."[1] It is generally thought that an even pacing strategy is optimal for long-distance events, though of course it must also take into account the course profile (that is, whether it is a hilly or a flat course) and weather conditions (temperature, humidity, wind), as well as race tactics.

The Central Governor Model suggests that exercise is controlled by the brain via a series of neural sensory feedback loops from physiological systems.[2] The subconscious brain then modulates exercise intensity based on motor unit recruitment, glycogen stores, cardiorespiratory muscles, and prior knowledge of the end point, to maintain homeostasis and prevent physiological failure. (Homeostasis is the maintenance of a stable internal environment, even as external factors change.) This theory explains why runners report increased ratings of perceived exertion throughout a race in an "anticipatory fashion," even though glycogen stores are never fully depleted.[3] A few studies have identified that runners' pace oscillates naturally, at both a micro and a macro level, as a result of multiple complex signaling processes originating in the brain, aimed at achieving an overall pacing strategy, whether consciously or subconsciously.[4,5] In fact, in these research groups, elite and experienced male runners didn't actually employ the "optimal" even pacing strategy.

Marathon pacing strategies

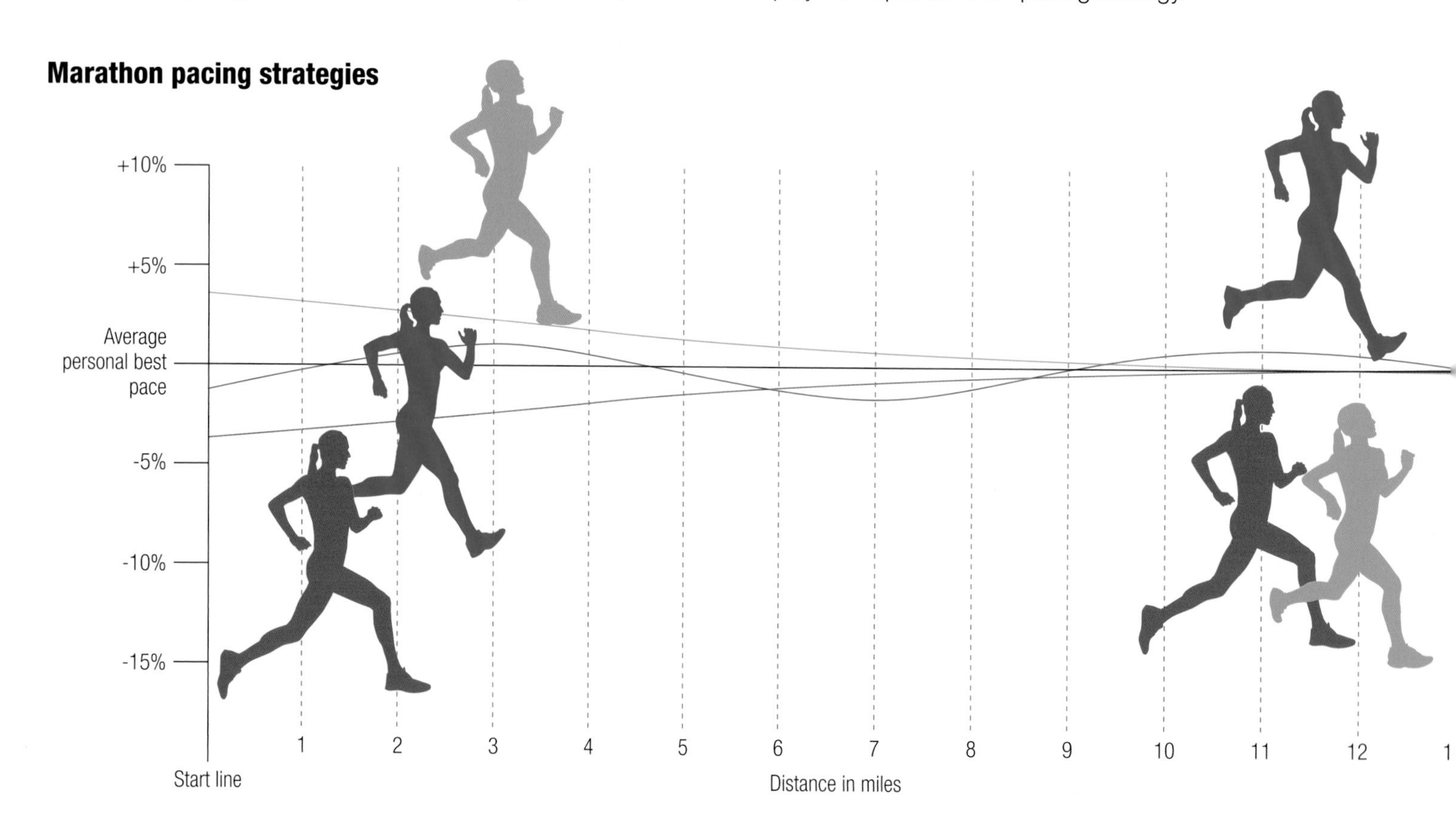

Data from the 2009 IAAF Women's Marathon Championship in Berlin showed the first quartile of athletes had the least variation in mean speed, highlighting the potential benefit of an even pacing strategy.[6] More recently, it has been shown that gender differences exist, with men more likely to employ a "risky" strategy and slow or vary their pace, and women generally adopting a more conservative strategy.[7] This could be caused by physiological and psychological differences between genders.

Clearly, the intensity selected is crucial to the success of an even pacing strategy, and must be predominantly aerobic (oxidative) in order to be sustainable over the 26.2 miles (42.2 km) of a marathon run. Lactate threshold (defined as the first rise in blood lactate and not to be confused with the maximal lactate steady state) is a useful reference point as this generally corresponds to marathon pace and is strongly correlated with marathon finishing time.

Temperature also plays a key role in pacing over a marathon distance. Uncontrolled rises in core body temperature can cause increases in catecholamines, rate of glycogen turnover, and risks of cramp and dehydration, ultimately leading to a reduced pace.

▼ ***Pacing yourself*** *Various pacing strategies are evident when running a marathon, including negative (running the first half of the race more slowly than the second half), even, and positive (running the first half of the race more quickly than the second half). The graph shows how, throughout the course, runners change their pace relative to their personal best time splits. For a marathon, an even pacing strategy is usually the fastest. The lines represent the variation in pace at any particular moment and not the overall time or position—a runner with a negative may have a comparatively good reserve of energy and finish strongly, but they are likely to be behind their potential best time split.*

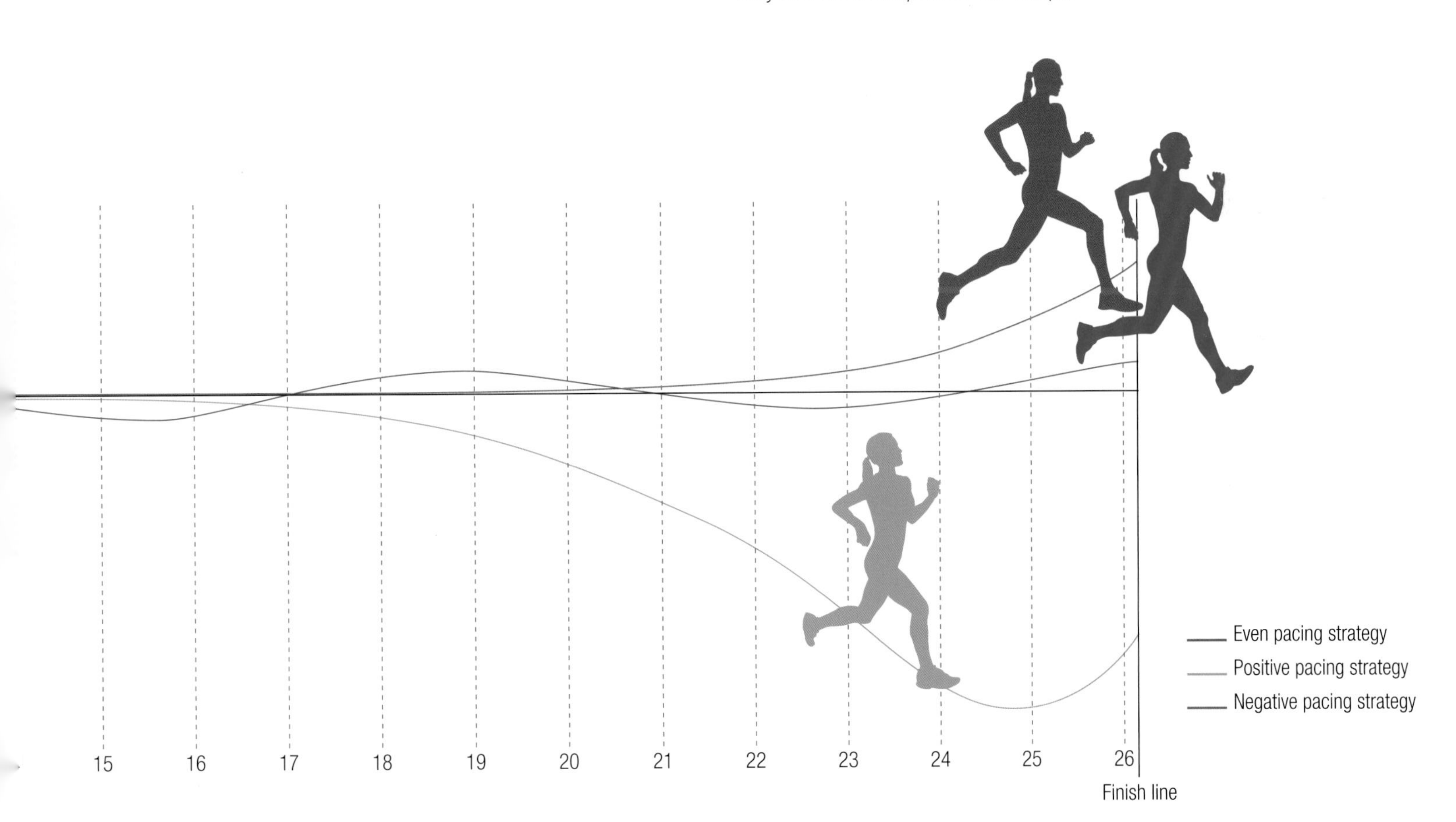

What is the effect of air resistance on running speed?

How much drag is there when I run?

The faster you go, the greater the effect of air resistance on performance, which is why aerodynamics is so much more important in cycling than in running. Nevertheless, it has been calculated that on a calm day the energy cost of overcoming air resistance is 7.8% for sprinting, 4% for middle-distance running and 2% for marathon running. Clearly the greater the headwind, the greater the effect. Runners therefore stand to benefit from sheltering behind others if the opportunity is there. Ultra-distance runners will have the smallest percentage saving in energy, although this will accumulate to be significant over an extended period of time and should not therefore be ignored. For example, it has been estimated that running in a group could save between three and four minutes over a 2 hour 10 minute marathon, so it may be a very advantageous tactic.[1]

Optimum group running positions

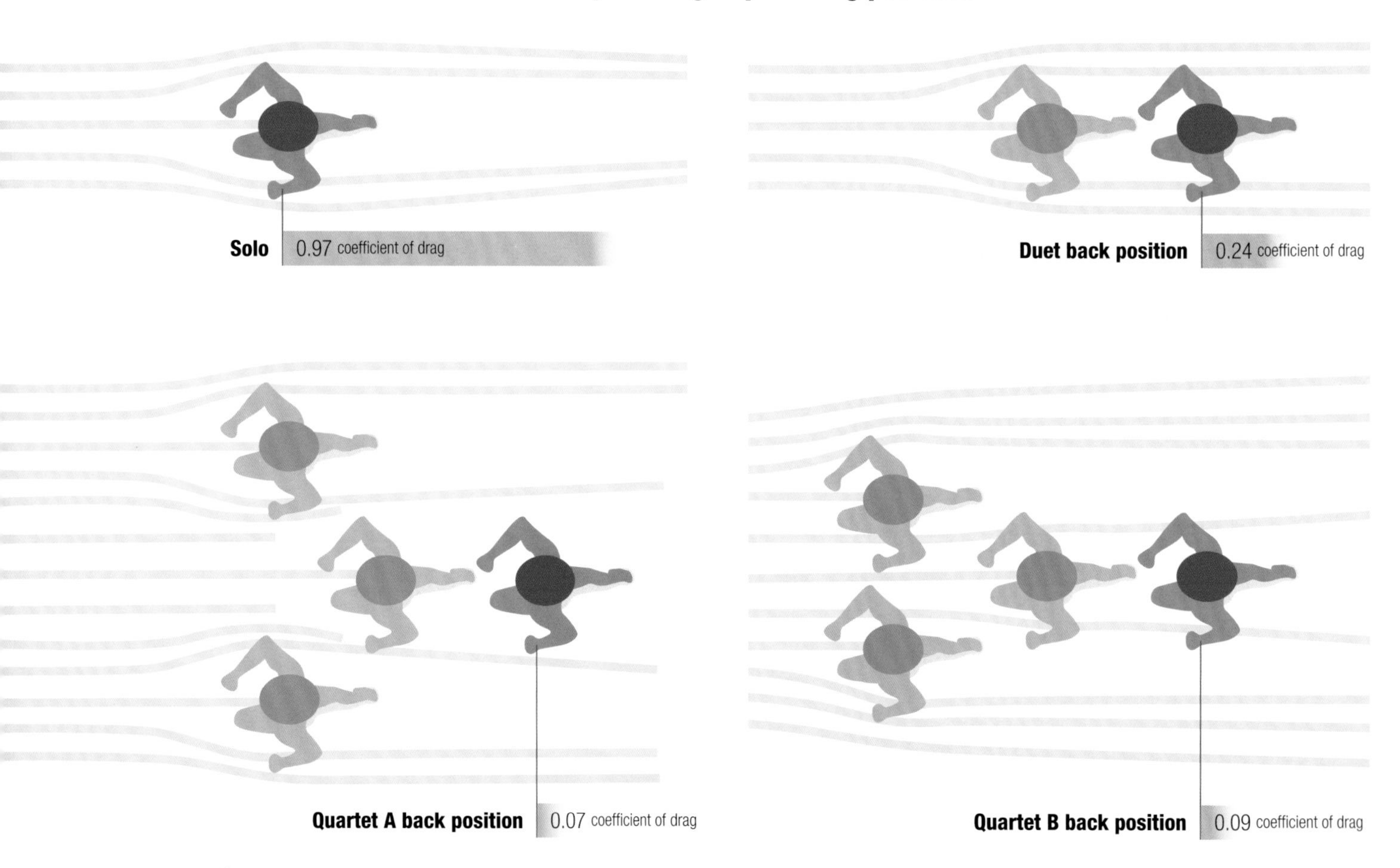

The effect of air resistance has been clearly demonstrated in a study that used an elevated gradient on a treadmill, where the runner is effectively stationary, to replicate the oxygen cost (a measurement of running efficiency) of running outdoors along a flat road. Across a range of running speeds, an average gradient of 1% was required to compensate for the lack of air resistance.[2]

Another way to look at this is to consider the effect of altitude, where there is reduced air resistance due to a lower barometric pressure—in effect, there is less air. Endurance athletes are compromised by the reduced partial pressure of oxygen in the atmosphere, which means there is less oxygen available for uptake. However, sprinting speed is increased because of the reduced aerodynamic drag. At moderately high altitude (5,980 ft or 1,823 m) the lower air resistance is calculated to be worth 0.15 seconds over 100 m.[3]

It is interesting to note that a tailwind of more than 2.0 m/s will nullify a world record due to the advantage it offers—however, this is an arbitrary figure and has not been selected based on scientific evidence.

▼ ***Keeping pace*** *Positioning yourself behind another runner offers an advantage by reducing the drag factor that results from overcoming air resistance. Running solo causes the highest drag factor, whereas running at the back of quartet A would result in an estimated saving of almost four minutes over a 22-mile (35-km) event.*[4]

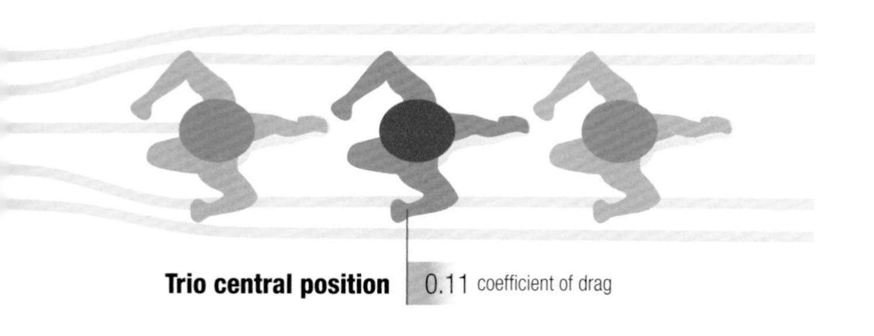

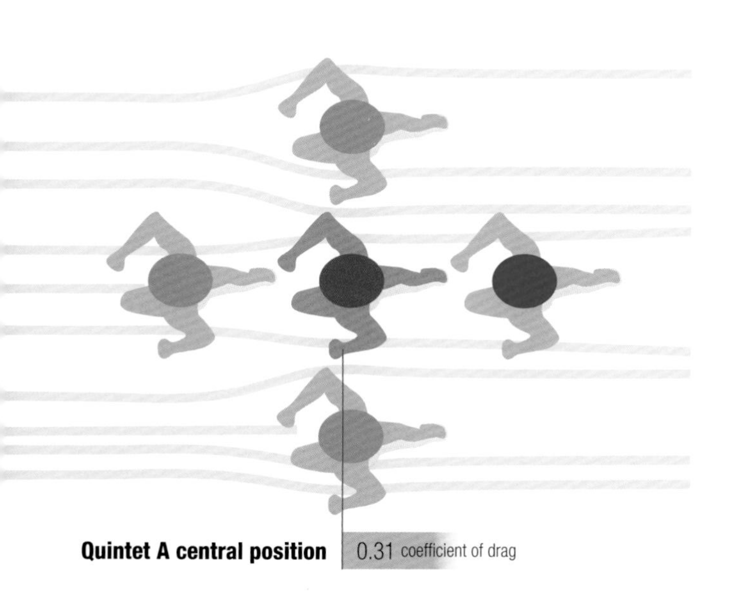

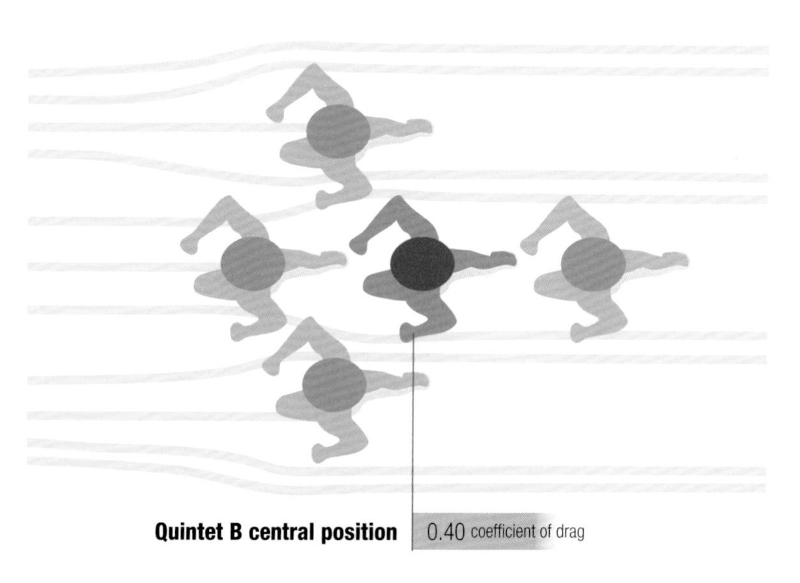

What is the optimum technique for running on a gradient?

How should I run up or down a hill?

Running uphill requires a greater force production per step than running on a level surface and therefore results in a greater metabolic load, as energy is used to lift the body's center of mass.[1] As the physiological demand increases, runners are more likely to utilize anaerobic energy sources with consequential fatigue. In order to avoid possible metabolic acidosis, running speed should be reduced so that heart rate and ventilation rate remain below the level corresponding to the anaerobic threshold.

There is no defined "best way" to run up a hill. Studies have demonstrated how runners automatically modulate their technique for optimal metabolic efficiency.[2] One strategy observed in runners is to increase step frequency and to decrease step length. Furthermore, foot contact time increases with the gradient and momentum is lost, such that there is no advantage in running over walking beyond a gradient of 15%.[3] Mountain runners will typically walk when they reach a gradient where the same speed as running can be achieved.

Hill running mechanics

▶ ***Reaching your peak*** *A person's running form changes when running uphill or downhill. Care should be taken when traveling uphill to avoid leaning forward excessively, which can disrupt your natural gait.*

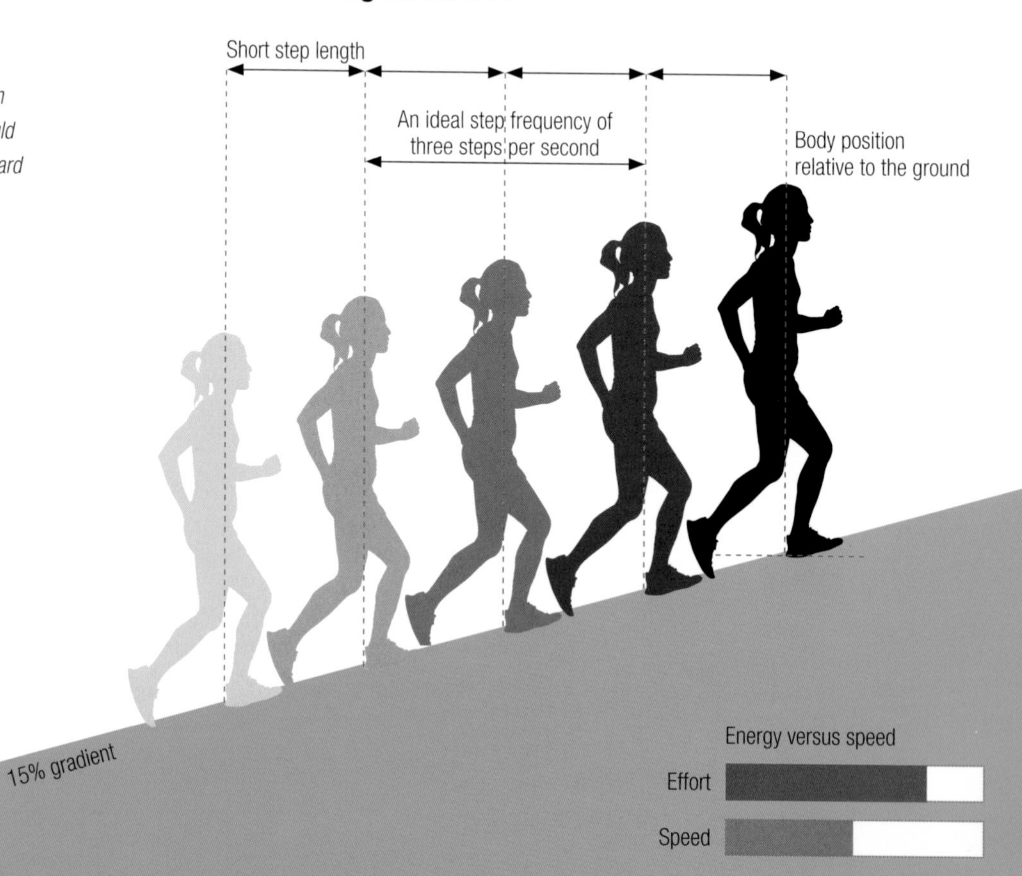

Running downhill causes excessive eccentric muscle contractions (in which muscles become elongated in response to an opposing force), leading to muscle damage, inflammation,[4] and neuromuscular fatigue.[5] Downhill running is trainable, and experienced mountain runners show a better running economy and less muscle damage during downhill running than those who are inexperienced in running downhill.[6] It is plausible that experienced runners may develop superior biomechanics during downhill running, which, in turn, reduces muscle activity and moderates neuromuscular fatigue. Often, downhill running velocity is lower than would be metabolically possible, because the runner slows down deliberately in order to avoid injury.[7] Although further research is required to define the optimal kinematics of downhill running, some have postulated changing the foot-strike pattern to alter the lower limb joint angles and modify neuromuscular fatigue.[8]

Optimal foot strike

▶ ***Coping with the slope*** *When running uphill or downhill the body self-selects an optimal foot strike based upon speed, postural control, gradient, and running experience. The four graphics show: (**A**) a forefoot foot strike when running uphill; (**B**) a downhill midfoot foot strike if on shallow gradient and the runner is trained; (**C**) a trained runner's downhill forefoot foot strike when on a shallower gradient; and (**D**) a heel foot strike when running down a steep gradient, which increases the braking force.*

A

B

What is the optimum weekly training distance for an endurance runner?

Is more mileage in training always better?

It can be helpful to think of training as a kind of medicine. For any medicine, there is an optimum dose—for training, though, this "dose" changes as adaptation occurs, and responses will also be different according to the characteristics (for example, the genetic make-up) and the environment of the individual. Training prescription needs to be delivered in phases, depending on what point in the training–competition cycle the runner is at, and the balance between training and recovery must not be disrupted.[1] Anecdotal data suggest no additional benefit in exceeding 60 to 70 miles (97 to 113 km) of training per week for a recreational runner, while for elite-distance runners the figure lies between 70 and 110 miles (113 to 177 km) per week.

With prolonged training of high mileage and insufficient recovery, glycogen stores are depleted. This results in a negative muscle protein balance and a state of catabolism, which causes weight loss and a reduction in bone mineral density.[2] Subsequently, blood glucose uptake is elevated

The risk of overtraining

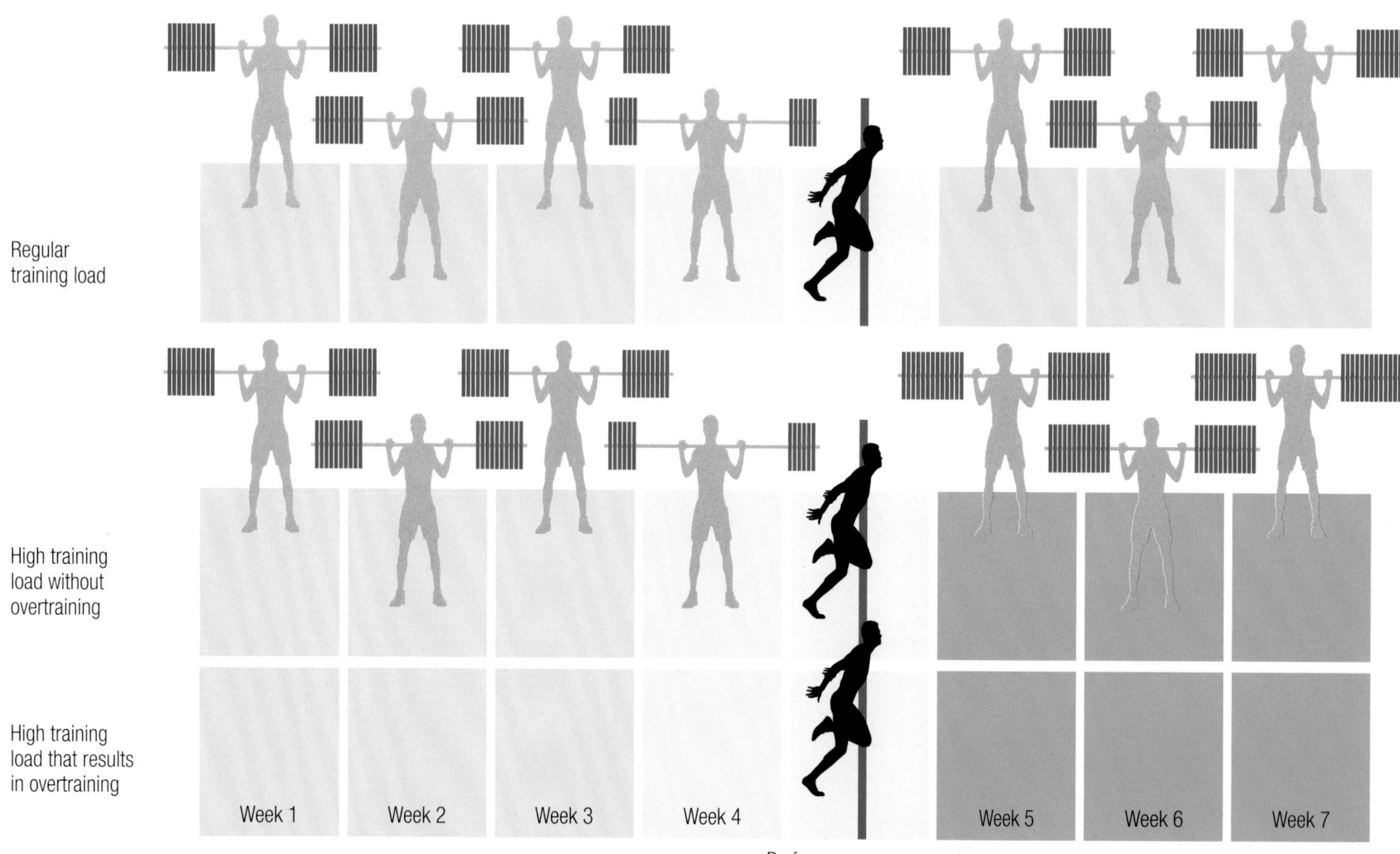

during exercise, which risks premature fatigue. Overtraining has also been linked with suppression of the immune system, and more prevalent upper respiratory tract infections.[3]

Endurance training increases glycogen storage and enhances fat oxidation during exercise. In the last decade, training in a glycogen-depleted state has been periodized into a training program and is of interest to coaches and physiologists. Training when muscle glycogen is low has been shown to activate molecular underpinning that improves the capacity for fat oxidation, enhancing the muscular response to training.[4]

High-Intensity Interval Training (HIIT) presents an effective strategy against constant volume overload. A recent study showed that success in marathon events correlated strongly with average training intensity achieved, rather than with mean weekly training distance, whereas the opposite was evident for ultramarathons.[5] The addition of heavy resistance and explosive (plyometric) training has consistently shown improvements in running economy,[6] with performance gains attributed to increased muscle strength, neuromuscular characteristics, and tendon stiffness.

▼ ***Training within limits*** *It is important that training programs have periods of high training load for physiological adaptation. Without sufficient training variation, the cumulative physiological stimulus results in overtraining and can lead to a short- or long-term decline in performance.*[7]

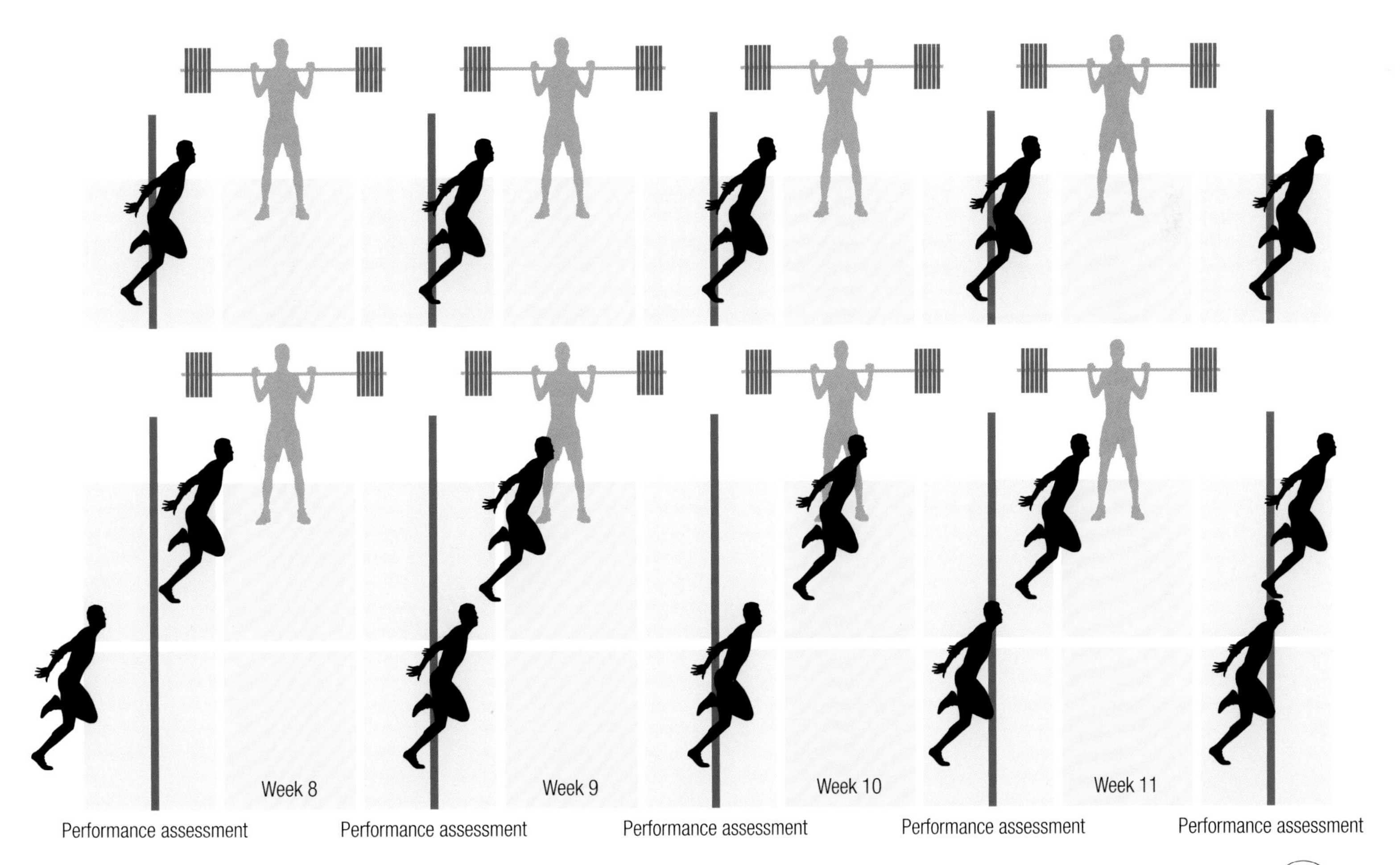

What are the benefits of altitude training?

How high do I train, and for how long, to make a difference?

Continuous exposure to real or simulated altitude can result in many physiological changes that help to combat the low oxygen pressure (hypoxia) that occurs at high elevation. The adaptation that has received most attention is the increased production of new red blood cells (erythropoiesis) and total hemoglobin mass that occurs in a predictable fashion in hypoxia.[1] This is the primary adaptation that enhances the oxygen-carrying capacity of the blood and improves endurance.

As altitude increases, oxygen pressure decreases and aerobic capacity declines by approximately 1% per 325 ft (100 m) gained beyond 3,250 ft (1,000 m) above sea level, with wide variability between individuals. Performance at sprint events improves due to the reduced air pressure but, in long-distance events, performance (and velocity at VO_2 max) declines considerably and in proportion to the altitude gain.

The application of altitude training and the "live high, train low" (LHTL) philosophy enables the athlete to gain the adaptations

Altitude training—live high, train low

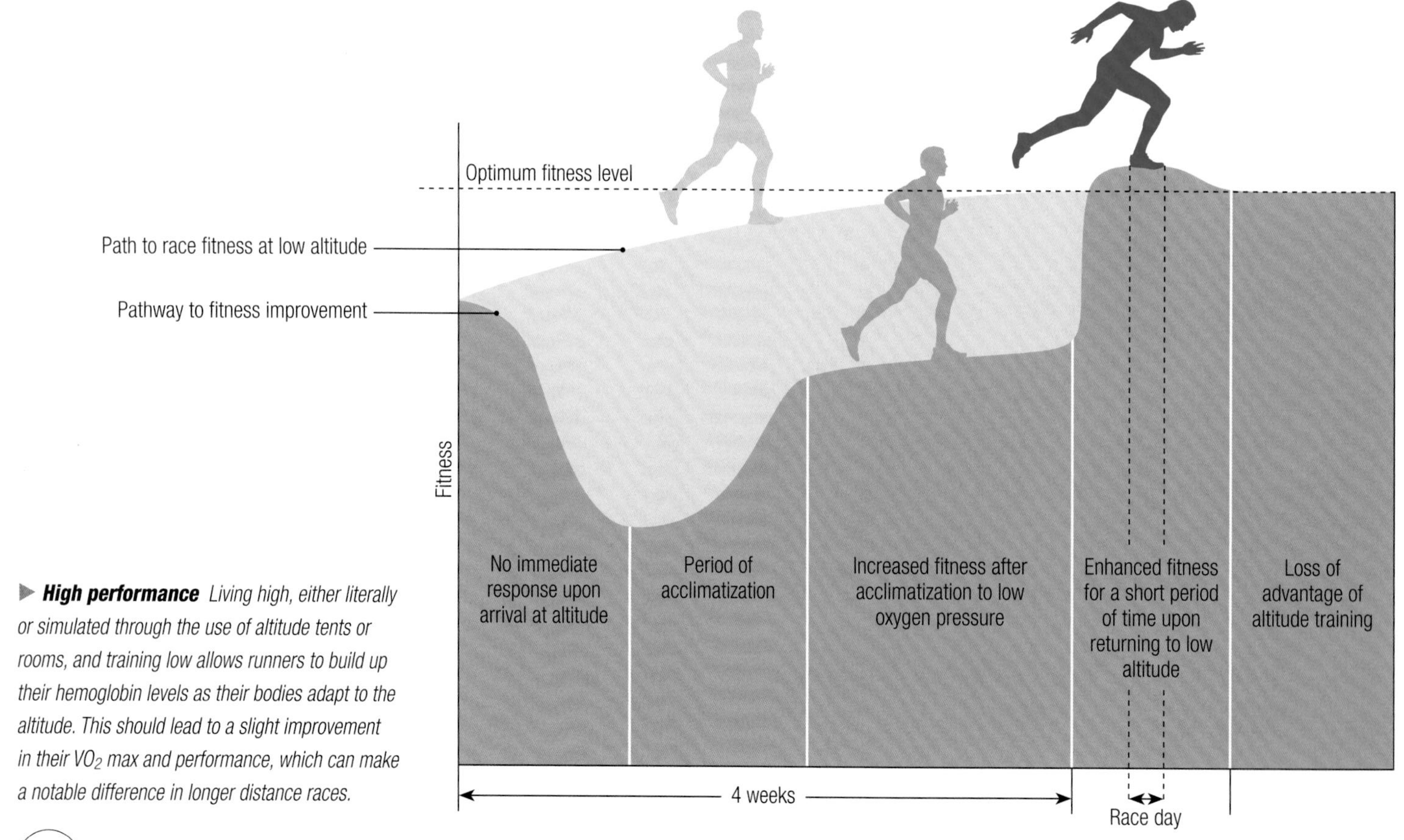

▶ ***High performance*** *Living high, either literally or simulated through the use of altitude tents or rooms, and training low allows runners to build up their hemoglobin levels as their bodies adapt to the altitude. This should lead to a slight improvement in their VO_2 max and performance, which can make a notable difference in longer distance races.*

associated with high-altitude residence, but also maintain high-intensity aerobic training by descending to a lower altitude.[2] "The longer, the better" is suggested for hematological (blood cell) adaptations, with at least twelve hours' exposure per day needed for erythropoiesis to occur. This exposure must be sustained for four weeks at 6,500–9,750 ft (2,000–3,000 m) altitudes for optimal acclimatization.[3] Worthwhile performance improvements of 1–4% have been reported with LHTL programs in elite and sub-elite athletes,[4] though there remain few well-controlled studies in elite athletes. Although small, this effect can meaningfully improve an athlete's chance of winning. For example, in a sub-elite half-marathon, which has a race variability among competitors of around 2.5%, the effect of a training stimulus has only to be greater than the race variability to increase the chance of a medal.

Altitude tents or normobaric chambers, in which the air pressure is kept the same but the percentage of oxygen is reduced, provide financially viable alternatives to LHTL. Many athletes use them for intermittent hypoxic training (IHT) that can be performed at sea level. While evidence of the effectiveness of IHT protocols is limited, and large variability between individuals has been observed, it has been shown that elite runners improved their submaximal exercise economy after eight weeks of IHT.[5] Hence, IHT is often used as pre-acclimatization before further altitude exposure, or in attempts to prolong adaptations after altitude training camps.

It is important to consider some of the possible negative effects of altitude training, and perhaps take advice from a sports physiologist, before making a decision to travel. For example, the running terrain and weather are often highly variable at elevated venues, and altitude can cause iron deficiency and sleep disruption.

▼ ***Marginal gains*** *Researchers followed 22 athletes who lived for four weeks at 8,200 ft (2,500 m) while training at 4,100 feet (1,250 m). Both the men and women in the survey ran 3,000 m 6 seconds (1.1%) faster on average after altitude training despite already being at or close to their fitness peak before the camp. They also saw a 3% improvement in their maximal oxygen uptake.[6]*

3,000 m times after LHTL training

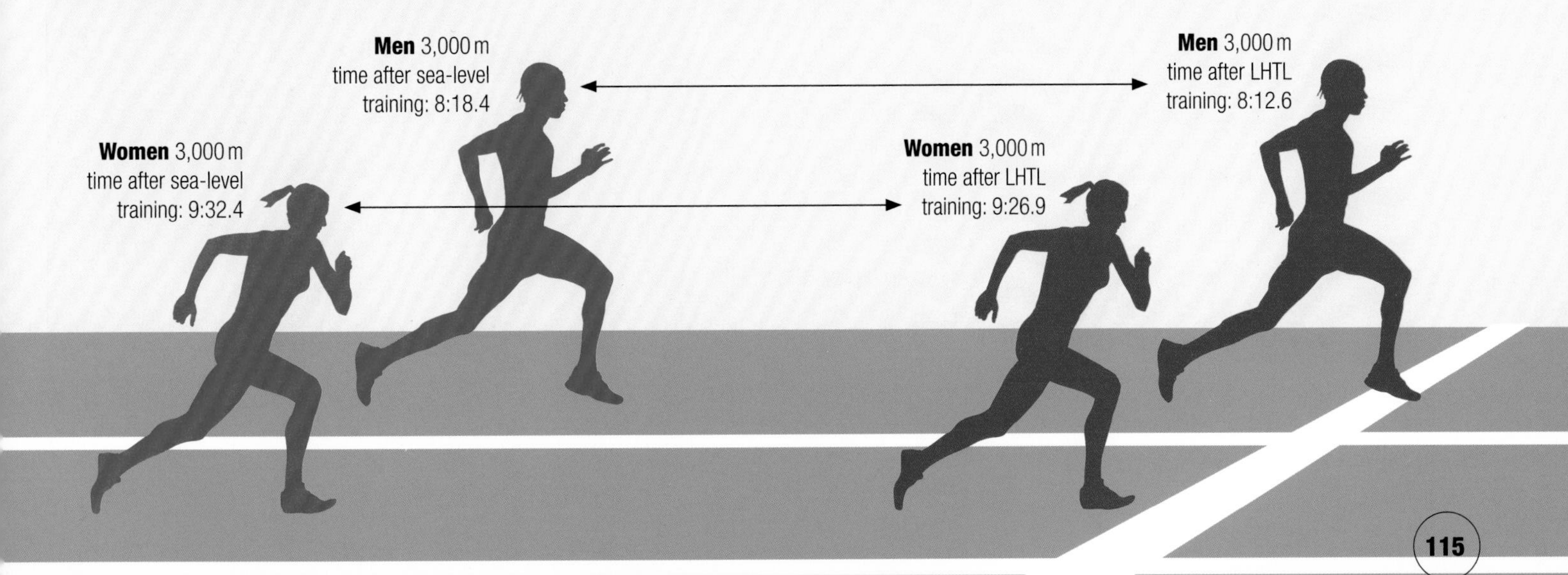

What is the science behind detraining?

How quickly do I lose fitness if I stop running?

Training stimulates a number of positive physiological adaptations, such as increased red blood cell production (erythropoiesis) under systemic low-oxygen conditions (hypoxia), and angiogenesis, the production of new vascular networks at a local level, which also enhances oxygen delivery to respiring muscle cells.[1] The principle of reversibility, or detraining, states that such adaptations will reverse if the training stimulus is reduced or removed, whether intentionally or because of inactivity during illness or injury, for example.

Different systems detrain at different rates. For example, plasma volume (essentially, the water content of the blood) begins to decrease within two days, whereas other structural adaptations, such as heart dimensions, will take longer to reverse. The observed drop in blood volume subsequently affects cardiovascular function, leading to a decline in cardiac output,[2] and maximal oxygen uptake (VO_2 max) decreasing by 4–14% in less than four weeks.[3] With detraining, we also see an increase in resting and submaximal heart rates.

One adaptation to endurance training is the reduced reliance of carbohydrates to fuel exercise. However, within 14 days of detraining the respiratory exchange ratio (RER) increases.[4] This is the ratio between oxygen consumed and carbon dioxide produced and acts as an indicator to energy metabolized—a higher ratio indicates increased carbohydrate oxidation. The glucose-transporter type 4 protein, which transports glucose in the blood and muscle, is rapidly reduced and this is thought to influence RER.[5] Consequently, for submaximal exercise intensities, the energy cost of locomotion is less efficient.

Detraining occurs within four weeks. However, with an appropriate training volume, physiological adaptations can be recaptured relatively quickly with a return to training.

The effects of detraining

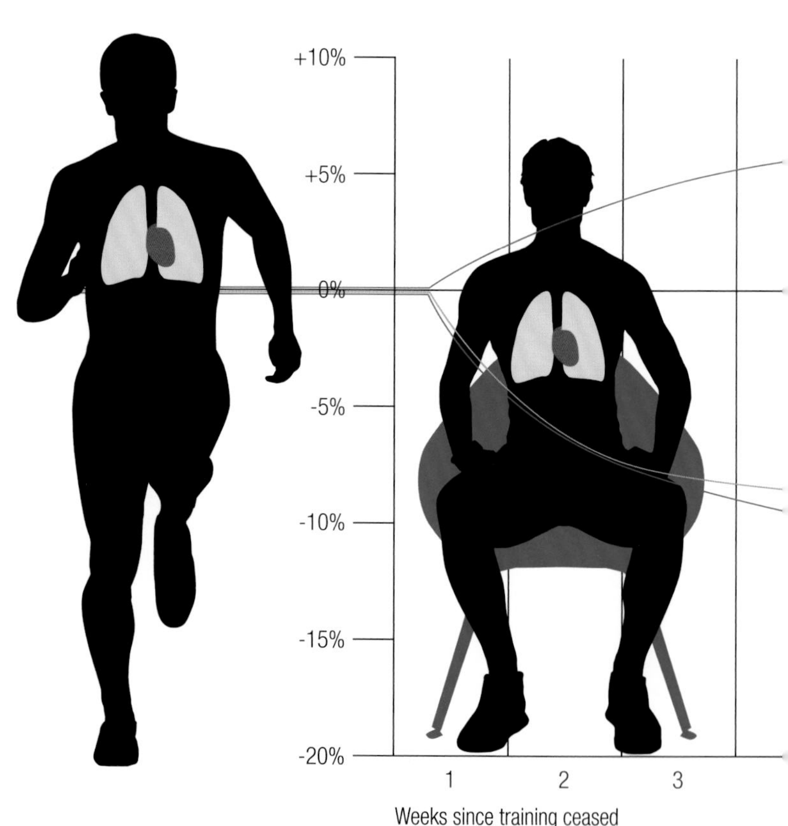

▶ ***Slowing down*** *The effects of detraining on the physiological adaptations caused by training quickly become apparent in runners when they cease training, though the rate at which each measured value changes varies. For example, while a runner's cardiac output may drop by 8% after four weeks, VO_2 may fall by about 10% at this time and continue to fall, dropping by 15% or more after three months.*

Blood changes

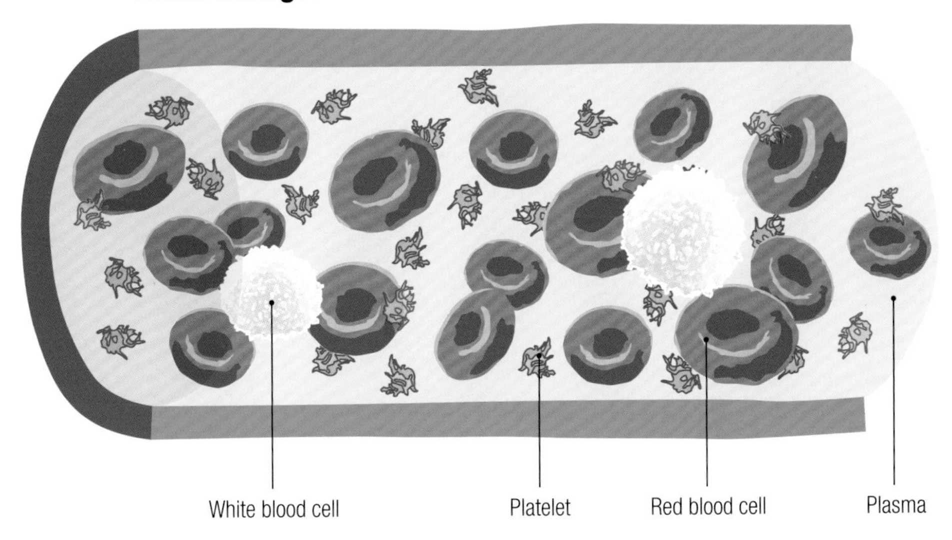

◀ ***Linked effects*** *One of the first changes from detraining is a reduction in plasma volume (caused by a drop in plasma protein content) and blood volume, which in turn leads to a drop in maximum cardiac output and VO_2 max over the following weeks.*

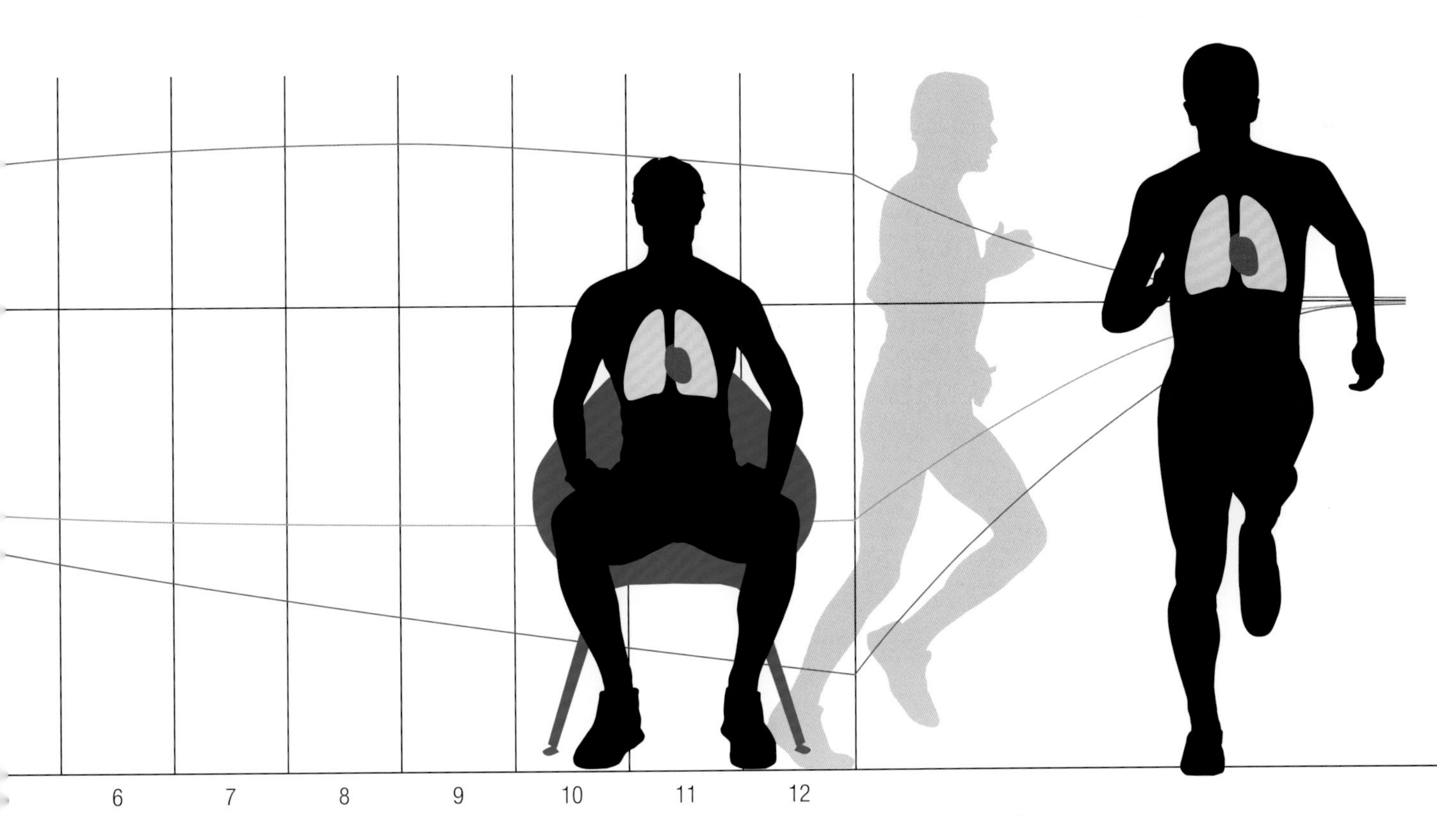

SCIENCE IN ACTION

the importance of sleep for recovery

Sleep is often overlooked as a recovery tool, but is increasingly being recognized as being absolutely key to successful training and adaptation. It is known that sleep disruption increases the risk of injury, illness, and underperformance, whereas sleep extension studies show that more sleep offers protection from those risks. However, it seems that you can get away with disrupted sleep in the short term by taking caffeine or from the increased arousal that is a result of performing, but the cumulative effects of these strategies are all negative.

What happens during sleep? Although the purpose and physiology of sleep still holds a lot of questions, theories of sleep describe its restorative, energy conserving, anabolic, and antioxidant effects. The physiological environment is markedly different from the awake state, with dramatic changes in hormones such as melatonin, cortisol, and testosterone. Learning and memory consolidation are enhanced after sleep, and cognitive performance is best after more than seven hours of sleep. However, it is not just the duration of sleep that is important—poor-quality sleep has the same effects as shortened sleep.

For athletes, sleep can be disrupted by training, particularly training undertaken during the evening. The increased arousal, elevated body temperature, raised heart rate, and muscle damage are not conducive to good sleep. Furthermore, exposure to bright light from floodlights or from phone and computer screens in the evening can delay the onset of sleep, which causes a form of "social jet lag" where the normal sleep–wake rhythm is disrupted and the sleep hormone melatonin remains elevated into the morning. With a similar effect, amateur runners often reduce their sleep by getting up very early to train before work, but this may be counterproductive because some of the deepest, most restorative sleep occurs in the later part of the night. Altitude training is known to disrupt sleep over the first few nights in susceptible individuals. Similarly, there is a newly recognized phenomenon known as "the first night effect," where sleep is disrupted on the first night in unfamiliar surroundings. Napping in the day is a good way for compensating when nighttime sleep is compromised.

▶ ***Any opportunity*** *Napping during the day is a useful recovery tool after competition, as these US athletes at the 1952 Helsinki Olympics demonstrate, or when sleep quality or quantity has been negatively affected.*

events:
trail running

If you're a runner who wants to work on core strength, motivation, balance, speed, and endurance, all on one run, then forget taking out a gym membership and instead hit the trails. Running off-road involves learning a whole set of new skills, while at the same time switching off any concerns about exact distances and precise speeds. Who wants to know their pace when there's fabulous scenery to enjoy, amazing climbs to conquer, or wildlife to spot? Trail running is as much about slowing down and taking in what's around you as it is about working on strength. In a way, improvements in core strength, ankle mobility, and general balance are almost a byproduct of running on unpredictable, soft, rutted, uneven surfaces. Your ankle mobility quickly improves and you learn to read the terrain and change your body position. You also learn to adjust your speed to cope with the slower pace of ascending and the almost frighteningly quick speeds associated with descending. Improvements in balance, coordination, and reaction speed are all byproducts of regular trail running.

▶ ***Finding your position*** *The development of GPS has been beneficial for any runner who uses devices such as watches that use GPS data to track distances traveled and, less accurately, speed. For trail runners these figures lose some of their meaning because the terrain has such an impact on their performance. However, for anyone running away from familiar or well-marked routes, GPS is invaluable as a navigation aid.*

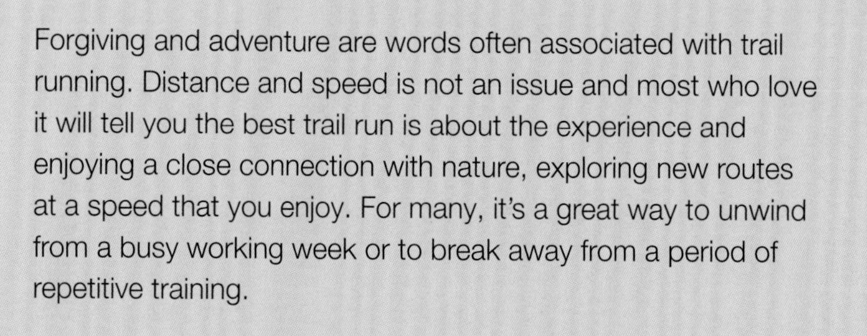

Forgiving and adventure are words often associated with trail running. Distance and speed is not an issue and most who love it will tell you the best trail run is about the experience and enjoying a close connection with nature, exploring new routes at a speed that you enjoy. For many, it's a great way to unwind from a busy working week or to break away from a period of repetitive training.

Of course, should you feel the urge, there are events for the more competitive trail runner, many of which take part in spectacular scenery, be that the Everest Marathon, or around a local lake or national park near you. They give a whole new meaning to racing for many athletes who perhaps otherwise wouldn't want to compete.

Article by **Paul Larkins**

Part of the beauty of running is its simplicity—throw on a pair of shorts and a T-shirt and lace up your shoes and you're ready. However, most runners take into consideration comfort, injury management, and performance when buying shoes, while some won't leave their home without their HRM or wraparound shades. Competitive runners may also look for any marginal gains or recovery aids technological advances may offer.

The past few seasons in particular have seen an incredible leap forward in the technologies used to create performance-enhancing clothing, be it shoes, jackets, or plain old T-shirts. You may find yourself questioning how choosing the right shirt to run in will help you go faster, but such is the innovation behind temperature control and moisture management, it's safe to say shirt choice, for instance, really does make a significant difference.

Shoes, of course, will factor in every runner's equipment choice. There may appear to be an overwhelming range to choose from, but it's actually pretty simple—even the most advanced research will tell you to go with the pair that you feel comfortable with and you'll quickly see the benefit. Be truthful with yourself about what you're going to use them for, how far and how fast you will run, and what kind of runner you are, and you'll quickly find what works for you.

chapter six

equipment

Paul Larkins

Can running shoes help with running form?

Are my shoes right for me?

Think of your body as a cushioning system. As you hit the ground at speed, your heel usually strikes the ground first, then your foot and knees go through a series of movements to lessen that impact. How your foot hits the ground is very personal, and dependent on a huge range of factors from your hip mobility and your age, to the speed at which you're traveling and the way your foot is constructed—do you have high arches, low arches, or flat feet, for example? Then, of course, the surfaces you train on play a role—road, hills, mud, and rock all affect your running style differently, and must be considered in selecting exactly the right footwear.

Most specialist running stores have a facility that allows you to film your stride before you choose a pair of shoes. The film of your running action will show your heel rolling first inwards, then outwards as you drive off for the next step. This inward rolling action as the foot strikes the ground is called pronation.

All of us will display some form of pronation. Some of us, however, tend to overpronate—which means the knees roll further inwards. Excessive movement in this area can cause knee problems, so an overpronating runner should choose motion control shoes to minimize the movement. At the other end of the spectrum is underpronation, where the foot rolls outward on impact. A flexible, cushioned model would be the shoe of choice for a runner, who tends to underpronate. In between, there is the neutral runner whose knees flex to just the right degree and provide just the right impact reduction. Although studies have not clinically proven that motion control shoes prevent injury, anecdotally the evidence suggests that this is the case, particularly with runners who have an unusual anatomy or gait.

There's an additional shoe category that has grown in popularity recently—minimalist or "barefoot" shoes, featuring a very low profile and reduced cushioning. The theory is that by mimicking the "perfection" of barefoot movement, which is the state in which humans evolved to run, you'll minimize injury and improve your speed. Experts are currently debating the precise science of this, but what's not in doubt is that if your body can tolerate the harder, flatter ride these shoes provide, then they present no problem. However, it could be that your body can't get used to these shoes. The bottom line is that you should get yourself to a specialist running store, have them look at your running form, and go from there. The perfect shoe for you does exist—you just need to find it.

Choosing a shoe

Weight (A)	Breathability (B)	Ground contact (C)	A stable midfoot (D)	Flexible midfoot (E)
Lighter is almost always preferable, but support shoes do tend to weigh a little more.	Quite apart from allowing your foot to breathe naturally and keep it cool, a good upper will also let water drain away when you run through puddles.	Clearly if you live in the mountains and run off-road, you'll have different requirements for the grip. You will want a wider pattern or lugs or studs on the sole, because wider means better grip in mud. Equally, the deeper they are, the better they will grip. Look at the surface in between the lugs—it needs to be smooth so the mud falls off.	The shoe needs to hold your foot in place over a variety of different terrains.	At the same time, it needs to allow your foot to move naturally.

► ***Sole wear*** *After a couple of hundred miles in a pair of shoes, put them on the table and see if they have reacted. If they're crushing inward, you're an overpronator, crushing outward means you are a underpronator (or supinator); if they're even, then you are a neutral pronator.*

► ***Footprint*** *The wet foot test will instantly reveal what kind of footstrike you own. When it comes to choosing the right shoes it's vital to know where you should start looking—support, neutral, underpronator. Standing barefoot in the bathroom will tell you as much as a treadmill.*

Arch type and foot alignment

Underpronator

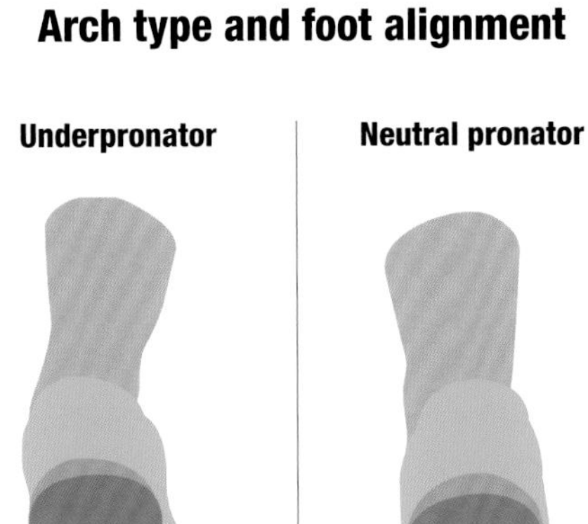

Neutral pronator

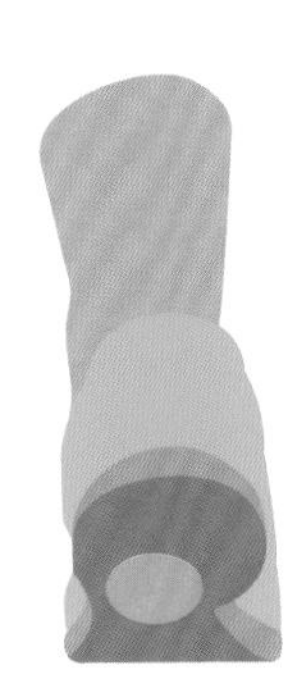

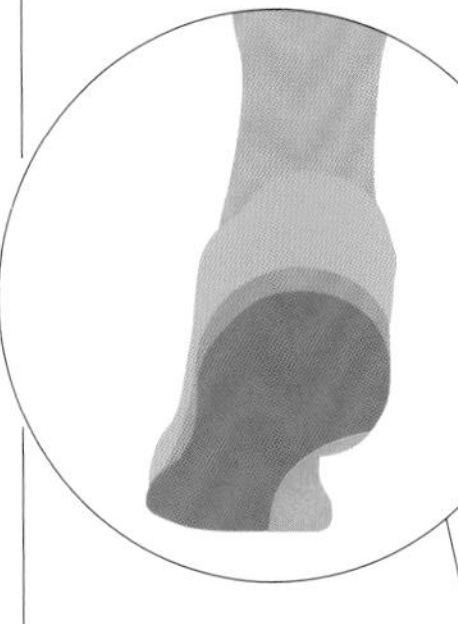

High arch

Normal arch

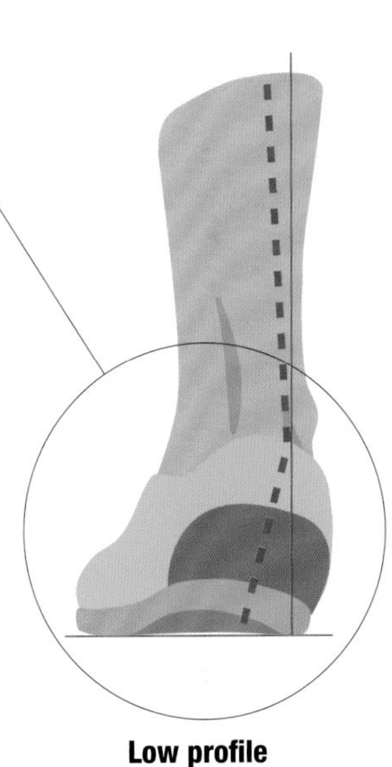

Low profile

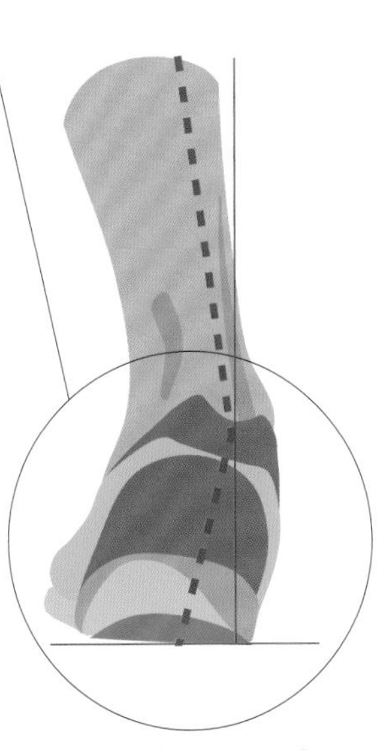

High and narrow sole

▼ ***Spoilt for choice*** *Obviously as a runner, your shoes are the most important piece of equipment. The choices can be bewildering, but by looking at a few different elements you can get a better idea of what you will need. Most good running shoe stores will be able to advise you.*

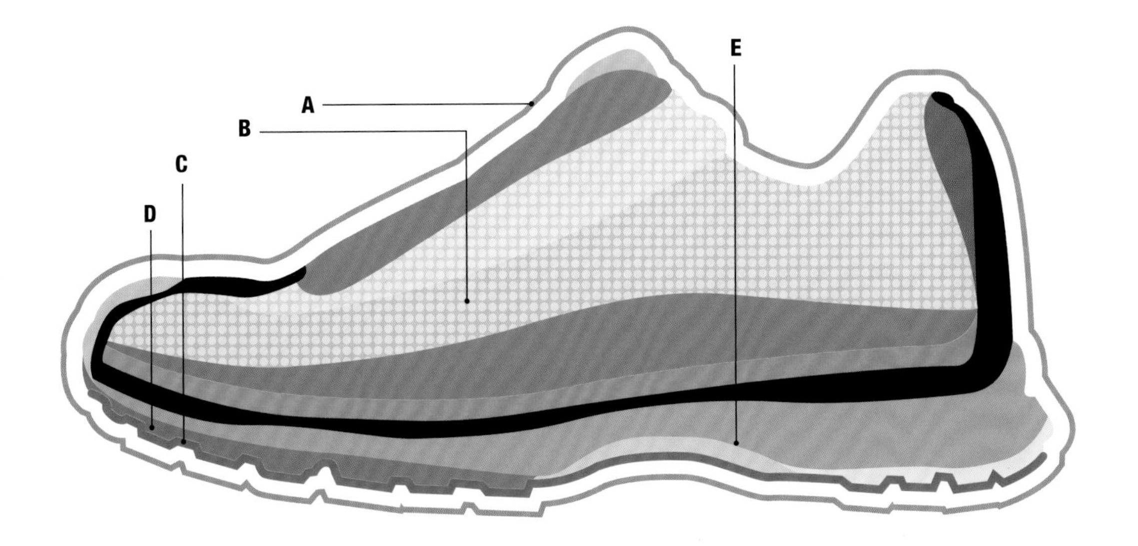

How do lightweight racing shoes improve performance?

Should I buy the lightest shoe possible when racing?

Selecting light, performance footwear for that all-important next race can be a tricky affair, not least because it's an area where science comes head-to-head with less tangible aspects of running, like the feel and look of your gear, as well as plain-old individual comfort needs.

In the cold light of day, there's research aplenty indicating that the lighter your shoes, the more efficiently you will run. For instance, for every additional 3.5 oz (100 g) of shoe mass, there's a 1% increase in oxygen consumption.[1] But equally significantly, research suggests that running off a platform—that is, in a shoe with a cushioned, responsive sole—is more effective in terms of performance than going barefoot.[2] The same research concludes that, thanks to the stiff but responsive sole, such shoes can propel you along the road, and there is actually a 2.1% decrease in oxygen consumption when wearing shoes compared to running barefoot. Of course, double Olympic marathon champion Abebe Bikila, who won one race in shoes and the other just as comfortably when running barefoot, would perhaps have argued this last point.

So how do you choose that perfect race shoe? Most mere mortals need light, flexible shoes that are thin, but not too thin, providing some cushioning from the ground. When making this highly personal choice, you need to think about distance, surface, and your own running history. What works for you one day may not do the job the next. A shoe that feels great over 6 miles (10 km), say, may not provide the cushioning and support you'll need for a marathon, while a marathon shoe may not be the right choice for an off-road adventure over, say, 100 miles (160 km) of rocky, mountain terrain. Before you buy, make sure you run in the shoes on a treadmill, which most specialist running stores provide. Then use them for a long run or two well in advance of your race and make sure they "feel" the same as your regular shoes. Bear in mind that lighter shoes take a while for your body to adapt to, so give it time.

Barefoot and shod foot step length

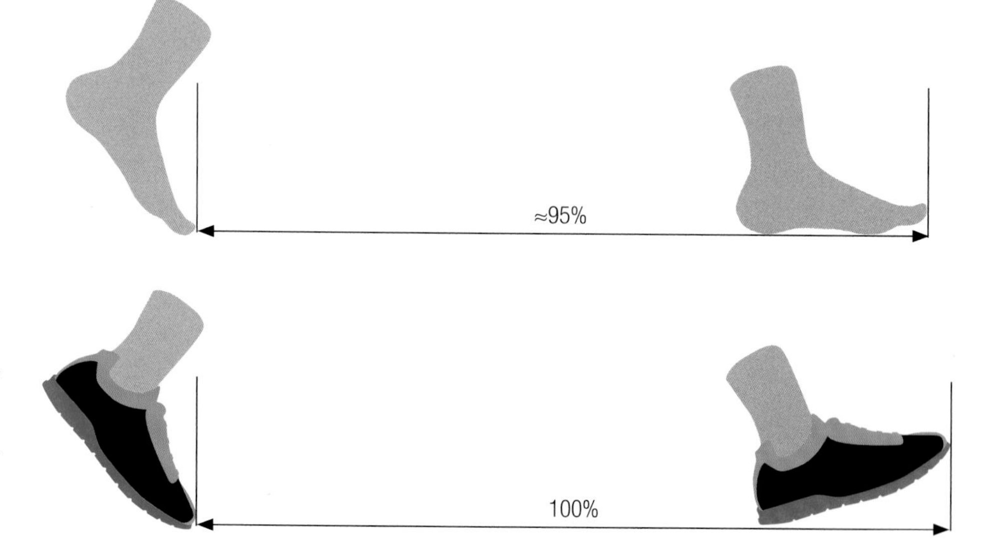

▶ ***Barefoot step length*** *It is widely accepted that runners adopt a slightly shorter step length when running barefoot compared to shod.[3, 4] Interestingly, it has been found that when switching from shod to barefoot running, habitual rearfoot runners adopted a similar strike position to mid- and forefoot runners.[5]*

Submaximal oxygen uptake and metabolic power for shod and barefoot running

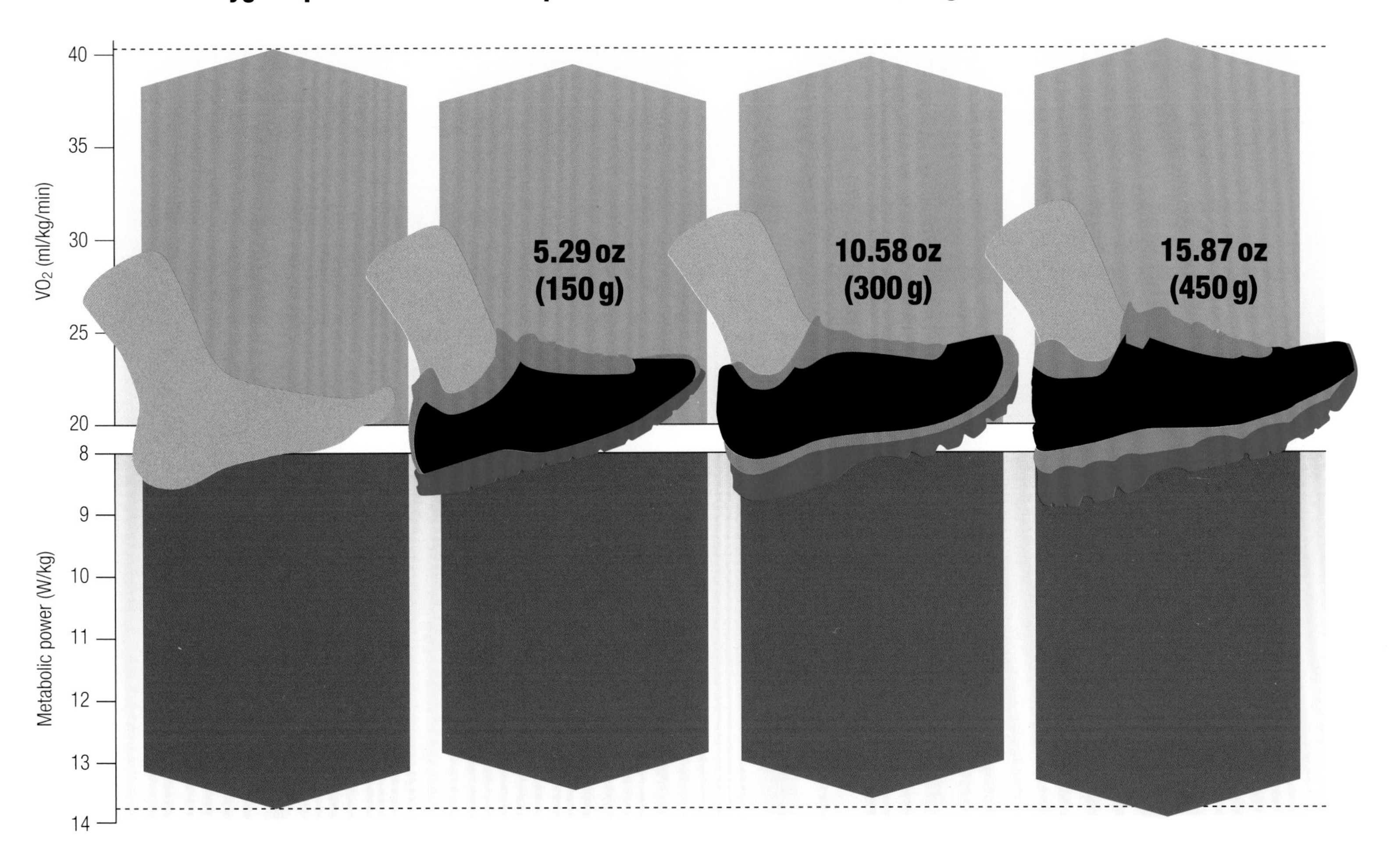

▲ ***Weighty matters*** *A 2012 study compared oxygen consumption and metabolic power when running barefoot and shod. The research supported the view that each 100 g increase in mass raised oxygen consumption by 1%, but also found that, with equal added mass, barefoot running had slightly higher costs in terms of metabolic power and oxygen consumption.*[6]

The pros and cons of switching to minimalist shoes or barefoot running

Pros	Cons
Less weight means less energy expenditure	*Running in light, cushioned shoes can actually offer more energy return on impact than running barefoot or in minimalist shoes*
Encourages forefoot running, which is biomechanically more efficient	*A heel-strike runner may risk injury as his or her body adjusts to a forefoot style*
Minimal cushioning increases the runner's ankle mobility	*Greater risk of suffering an injury from debris such as stones or broken glass*
On the right surfaces in the right conditions, running barefoot can result in faster performances	*Cushioned shoes tend to offer a more comfortable fit*

◀ ***Should I run shod?*** *It's a fascinating argument when it comes to weight of shoes versus performance. Without a shadow of a doubt, research will tell you that simply shedding ounces will indeed help a runner improve, though running barefoot offers no metabolic advantages over running in lightweight shoes. But, you have to counter that with the fact that a shoe's cushioning will also assist when it comes to running fast. So there is a definite middle ground—too light and there's no cushioning and you'll run slower, too heavy and, well, you'll do the same. Add in the fact that cushioned shoes help when it comes to warding off injury and you'll see it's not an open and shut case.*

How do trail shoes improve off-road performance?

Can I get away with using my road shoes for a cross-country race?

Predictably, off-road running shoes should feature one very important element—grip. Good grip means you feel safer as you run over slippery, uneven ground, which in turn means your stride is more relaxed and more natural—and so you'll run quicker.

When it comes to choosing shoes on the basis of grip, there are some key points to consider. Look at the spaces between the lugs or studs on the sole—wider gaps are better for grip in muddy conditions, while a smooth surface between the lugs offers better mud dispersion. Harder compounds are generally used in trail running shoes than in road shoes—in between the lugs, look for a harder density compound, which will act as rock protection. Lug angles differ for uphill and downhill grip, so take this into account when making your selection.

The shoe's fit is hugely important. Running off-road is less predictable, with steep ascents and descents, and as a result your foot will move about much more—good trail-running shoes will address that problem. Waterproofing has to be taken into account too. Shoes will undoubtedly take on water when running off-road, so look to see how they will get rid of it as quickly as possible—a waterproof, breathable upper and perhaps drainage channels are aspects to think about when making your trail-running shoe choice.

As your running terrain increases in difficulty, so you will need a shoe with better adapted elements. More testing terrain requires rugged shoes with harder outsoles for rock protection, and even toe guards. It's details like this that set trail shoes apart—rock guards, waterproof uppers, even tongues designed to keep water at bay.

Of course, if your run is no more adventurous than a few miles on the grass in your local park then, it's fair to say, your normal road shoes will do the job perfectly well.

Effectiveness of trail shoes

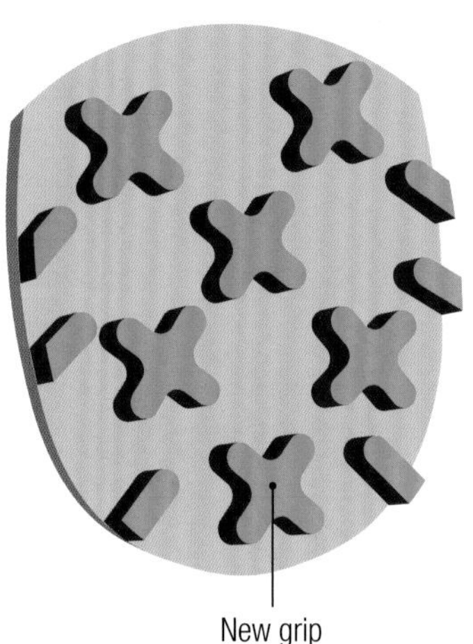

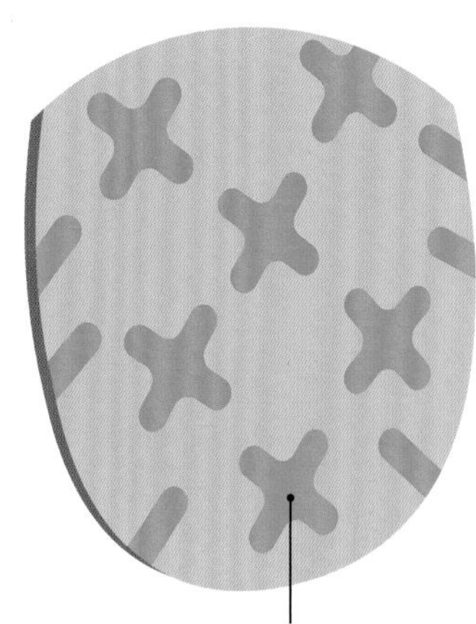

▲ ***Losing grip*** *Aggressive grip tends to be made of softer compounds, so be wary of running too much on the road in shoes with such a grip because they will wear more quickly than if they are worn solely in off-road conditions. Trail shoes also come with harder rock plates to protect from hidden objects underfoot, so it's worth keeping on eye on them for wear.*

The minimalist option

▶ ***Minimal drop*** *Barefoot shoes work off a heel-to-toe drop of around 0.15 inches (4 mm); some are slightly more, some slightly less—but essentially they keep your foot low to the ground and much flatter than normal shoes, which are built on about a 0.3–0.5-inch (8–12-mm) drop. They can take some getting used to because they are so different from everyday shoes, so ease into them slowly with short-distance runs at first.*

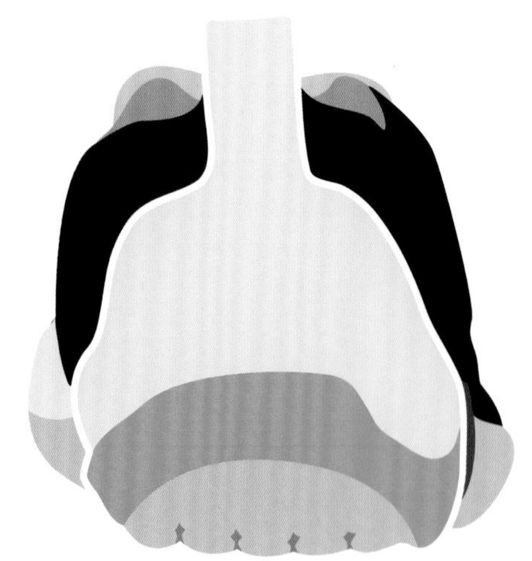

Typical trail shoe undersole features

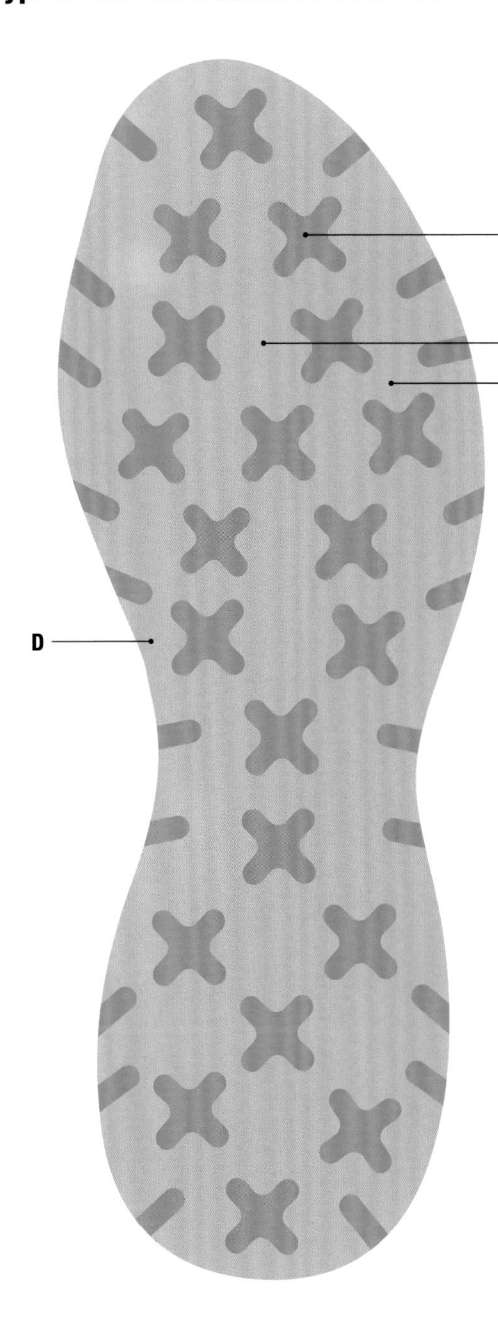

Typical road shoe undersole features

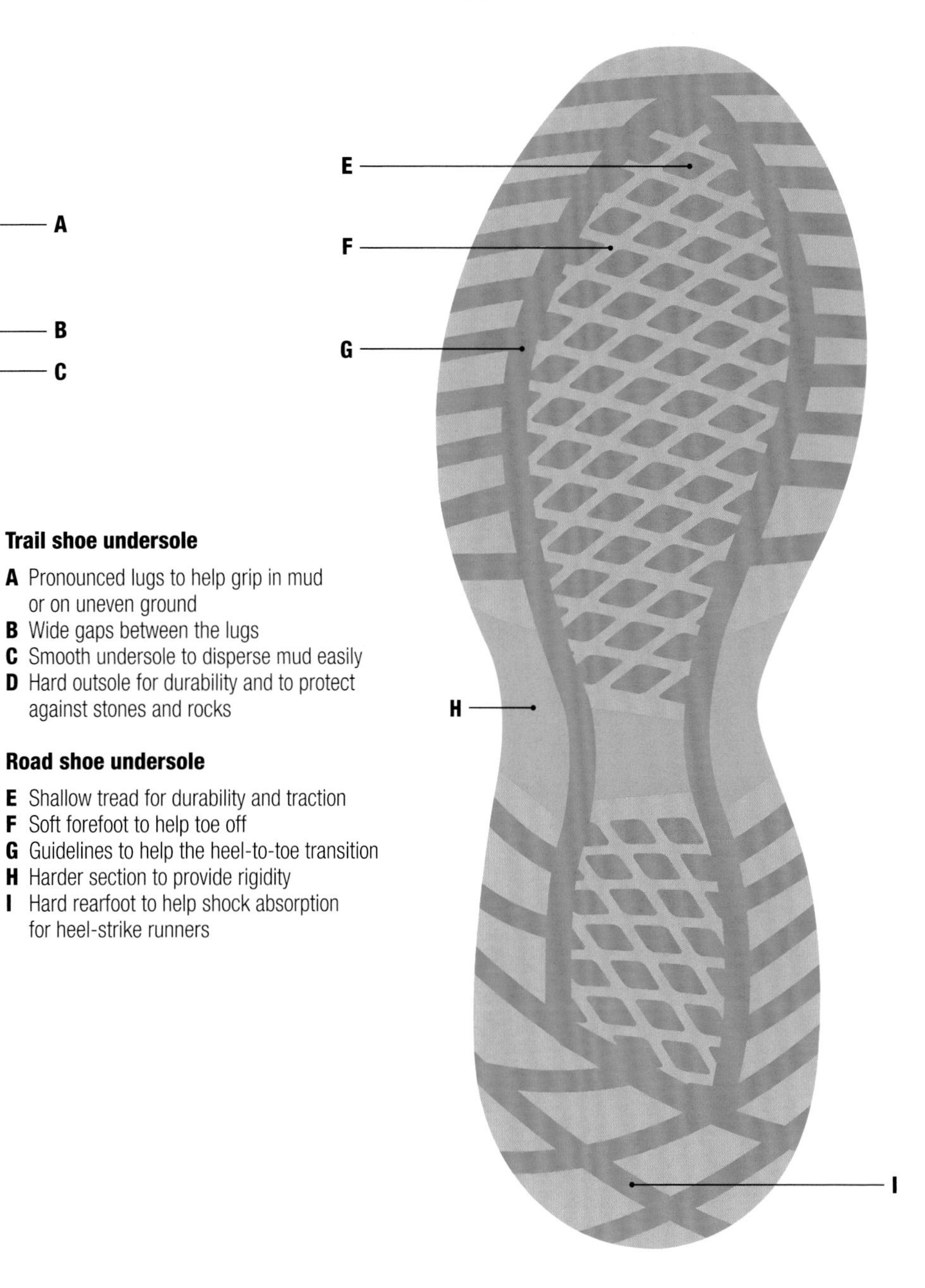

Trail shoe undersole

A Pronounced lugs to help grip in mud or on uneven ground
B Wide gaps between the lugs
C Smooth undersole to disperse mud easily
D Hard outsole for durability and to protect against stones and rocks

Road shoe undersole

E Shallow tread for durability and traction
F Soft forefoot to help toe off
G Guidelines to help the heel-to-toe transition
H Harder section to provide rigidity
I Hard rearfoot to help shock absorption for heel-strike runners

▲ ***Trail shoes*** *An aggressive grip with longer lugs make trails shoes perfect for mud and for wet, off-road conditions where stability and security are as important as weight and response. The best grip is angled to cope with severe ascending and descending.*

▲ ***Road shoes*** *The prominent lugs found on trail shoes would be quickly worn away on harder surfaces. Many outersoles have harder rearfoot sections to provide shock absorption for a heel-strike runner and guidelines to help the transition to toe off from a more cushioned forefoot.*

Can compression clothing improve performance and prevent injury?

Can wearing tight-fitting clothes really help me run quicker?

Improving your running is as much about how quickly your body recovers as it is about how fast your last run was. The faster you recover, the higher the training load you can maintain. This is where compression clothing comes into play. If you were a professional athlete, you'd likely get some light recovery massage after your workout, helping the toxins produced during exercise to be flushed from your muscles. Compression clothing works in a similar manner. Tight-fitting shorts, leggings, or T-shirts increase the blood circulation in your muscles, which in turn increases the rate of blood flow back to your heart and around your body. (Long-distance air travelers will know how compression socks help reduce the chances of deep vein thrombosis by improving circulation in the same way.)

Certainly, research has indicated a link between the use of compression garments and a reduction in muscle soreness after exercise.[1,2,3] This may be not only because the improved blood circulation helps to remove toxins, but also because compression affects the body's inflammatory response to damaged muscle tissue and speeds up cellular repair.[4]

So compression allows you to recover more quickly. The theory is that it also enables your body to operate more efficiently while you're actually running, as lactic acid is flushed away at a quicker rate. And it doesn't stop there. With your blood flowing more rapidly, oxygen is transported to muscle tissues at a faster rate, allowing you to maintain a higher training intensity.

Strength and perceived soreness with and without compression clothing over time

▶ ***Improved recovery***

Compression clothing is particularly useful after performance because it increases blood circulation and allows the body to recover faster and more effectively. More oxygen is circulated through your muscles and the removal of acids and the other byproducts of physical activity speeds up. Researchers have found that athletes using compression garments to recover post-exercise, performed closer to their maximum vertical leap (represented by the position of the toes in the graphic) and also felt less perceived soreness (represented by the size of the ovals) than athletes without compression clothing.[8]

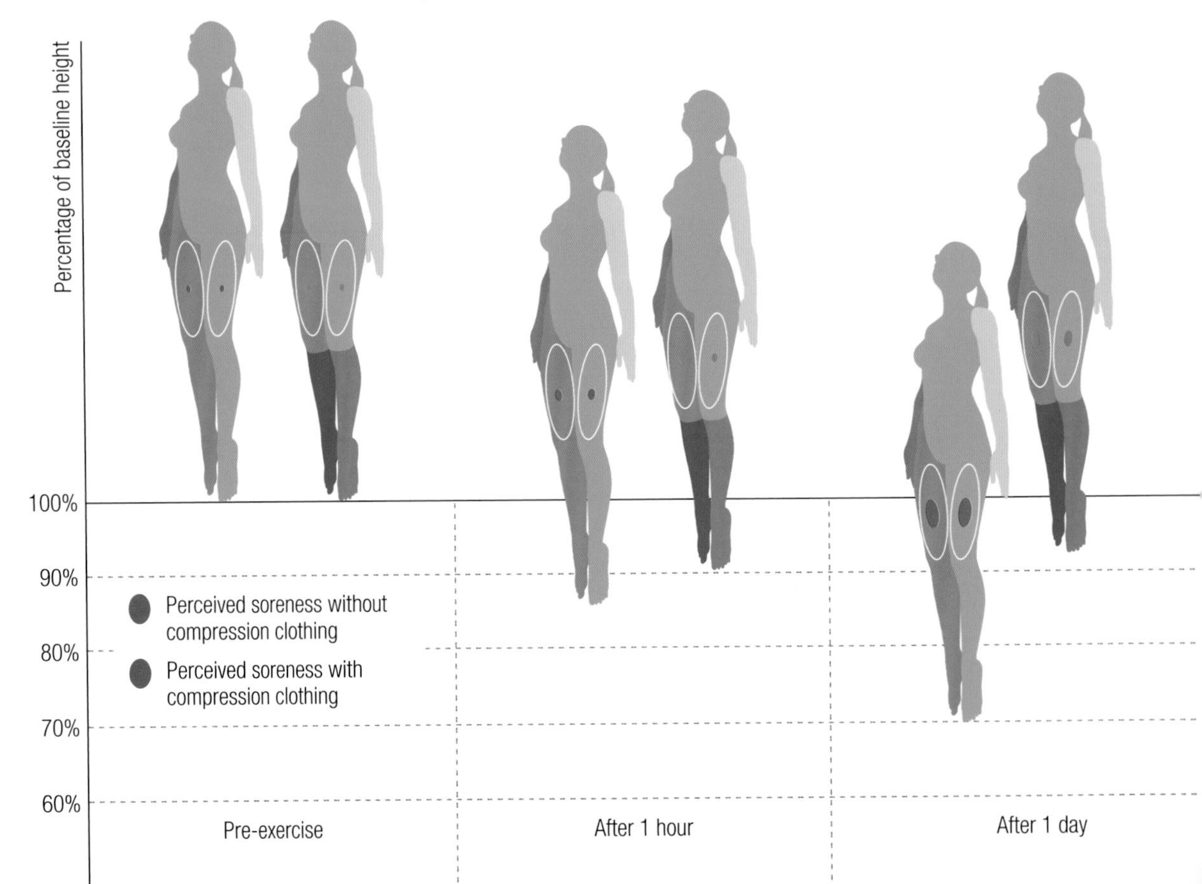

More blood flow also equals more heat, which means greater muscle flexibility and a reduced risk of injury. In addition, it's claimed that compression garments help control the amount of muscle movement during running, thereby reducing potentially damaging impact forces from each pounding stride. However, over-tight garments may restrict the flow of blood and oxygen to muscles. They may also restrict heat loss from the evaporation of sweat, increasing core temperature and the risk of hyperthermia.

It's worth pointing out that you shouldn't expect superhuman powers as a result of your choice of shorts. The gains described here are small—but they could become increasingly significant the faster or further you run. In fact, some studies have found no significant benefits in performance during running at all,[5] though the gains in terms of improved recovery are more widely accepted. Some studies suggest that part of the benefit may be due to a placebo effect—because runners believe compression helps, they experience real advantages from it.[6, 7] Whatever the mechanism, if they work for you, it's worth considering including compression garments in your running gear.

Compression clothing

▶ ***Under pressure*** *Recovery is a key issue when it comes to improvement, and compression clothing plays an important role in helping muscles that have been damaged by exercise return to normal.*

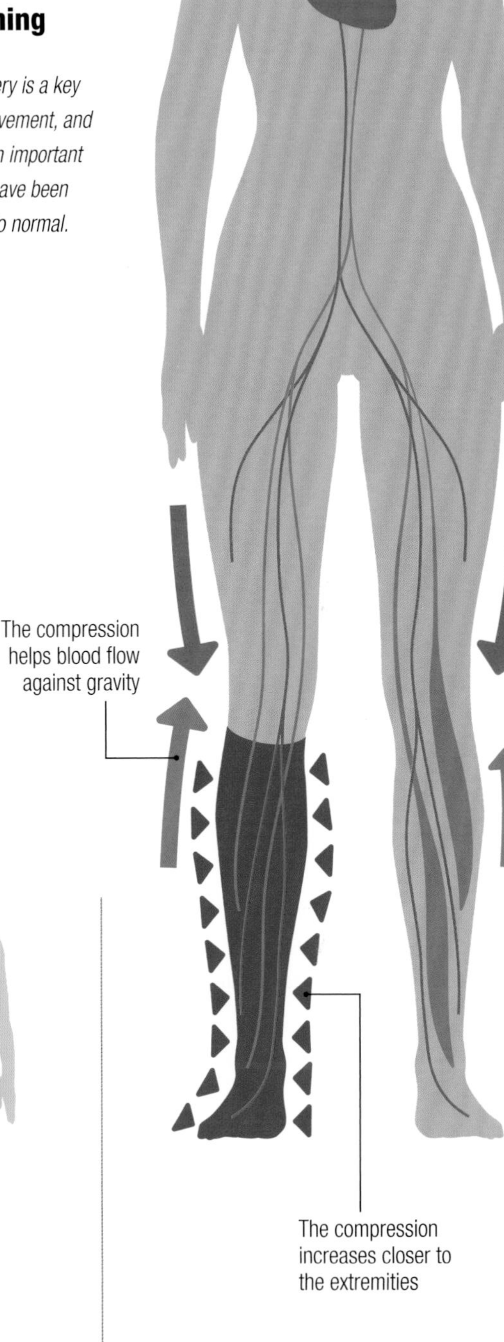

Article by **Jess Hill**

Do different materials make running more or less comfortable?

I can run in any old T-shirt, can't I?

Everybody's wardrobe has the necessary clothing to get you running—in theory, an old T-shirt and a pair of beach shorts will do the job. But while that's true for a week or two, pretty soon you'll work out that heavy, cotton-based clothing will result in you heating up at double-quick speed. How often do you see runners with sweatshirts tied around their waists? That's because your core temperature rises with exercise and that increase quickly becomes uncomfortable if you're wearing inappropriate clothing.

This is where breathable fabrics come into play. Garments made from fabrics that wick moisture from your body and are well-ventilated are essential pieces of gear for every runner. Wicking fabrics allow liquids to pass through to the surface before evaporating. Clothing that does not feature this property will cause sweat to condense on your body rather than evaporating into the air. In warm weather, this leads to overheating, while it can be unpleasantly chilling when your surroundings are cooler.

The market for breathable, wicking fabrics is huge, but essentially you are looking for polyurethane-coated ripstop nylons and polyester. Cotton and wool are not wicking fabrics, but are porous, becoming significantly heavier and more uncomfortable when wet.

Along with a breathable shirt, you might also choose to invest in a breathable waterproof jacket to use during wet-weather runs. These use a membrane with microscopic pores that let sweat wick away from your body, but which are too small to allow larger droplets of rain water to get in. Studies have shown that these membrane materials are around 30% more efficient at moisture control than a jacket that is coated with a waterproof layer.[1, 2]

Cotton

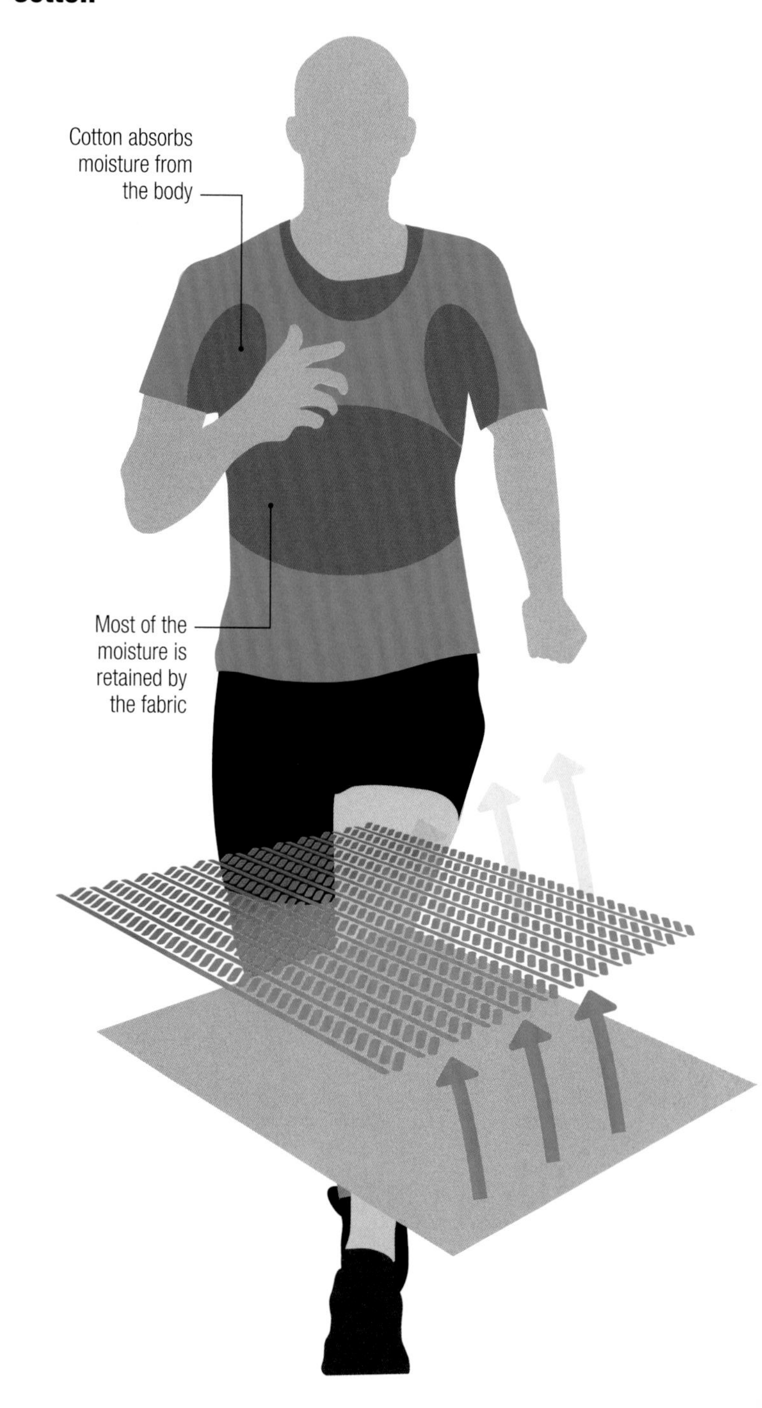

▼ ***No sweat*** *Breathable fabrics allow moisture to move away from the skin and to the surface of the fabric, spreading out the sweat for more effective evaporation and drying. The speed with which the moisture is moved from the skin is influenced by the fabric's yarn weave (tight or loose) and the pressure of the garment: tighter fabrics will force moisture through, while other fabrics rely solely on permeability.*

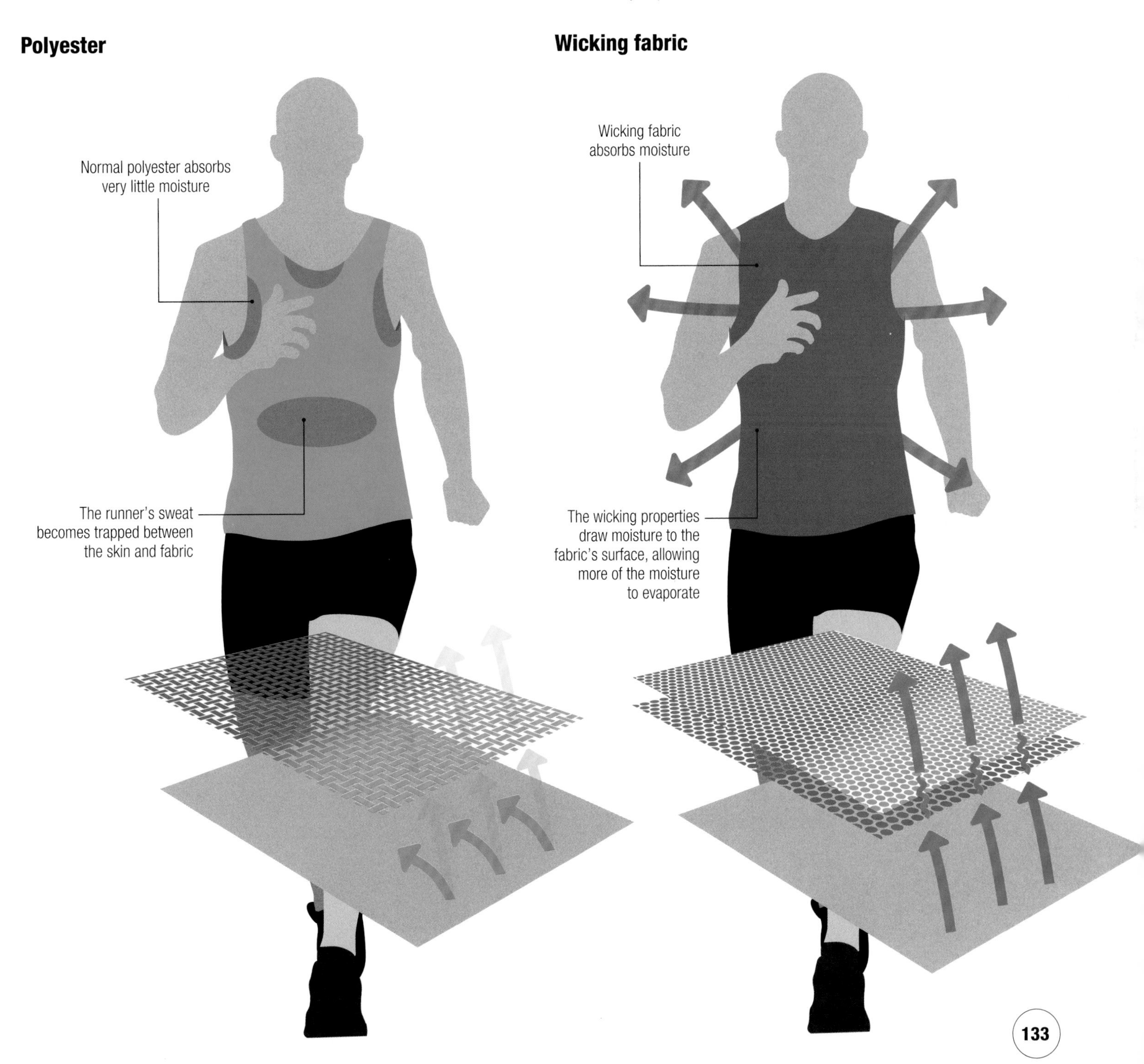

Can a heart rate monitor improve performance?

How do I know if I'm working hard enough?

As you have undoubtedly discovered already in this book, training is all about working to different levels of effort—sometimes that involves easy running at, say, 60% of your maximum heart rate, and at other times you will be going all out in short, sharp bursts at 90% of your maximum. Clearly, accurate measurement of your heart rate, using a heart rate monitor (HRM), is essential in order to manage a training program like this.

Most HRMs tended to use a chest strap containing a sensor that was linked remotely to a watch where the data was displayed, although for more and more monitors now coming onto the market, both the sensor and display are housed in a multifunctional GPS watch. What's really exciting is that many HRMs now have the capability to do far more than merely record your heart rate. Many now analyze the readings and even suggest workouts as a result. Input your height, age, gender, and weight and the monitor will be able to suggest heart-rate zones to work in, and warn you when you're not in those zones—when you're too slow or too fast. You'll then be able to upload the data to dedicated websites, which further record and chart your progress. In some ways, an HRM can act as a personal coach.

Of course, the key is to learn to use an HRM correctly. Every athlete is an individual, and you'll work out quickly what level a "high" or "low training zone" is for you, rather than relying on average values from a book. Applying an even effort is always the best way to run distance events, and a monitor can help you do this even in windy conditions or on undulating terrain. Apply your own data to a race—particularly a marathon where it's essential to start comfortably—and you'll quickly find out just how important and helpful a heart rate monitor can be.

As a side note, be aware that monitors can pick up signals from nearby electrical equipment and from other runners' monitors, so be slightly wary of any unusual recordings.

Recovery rate after vigorous exercise

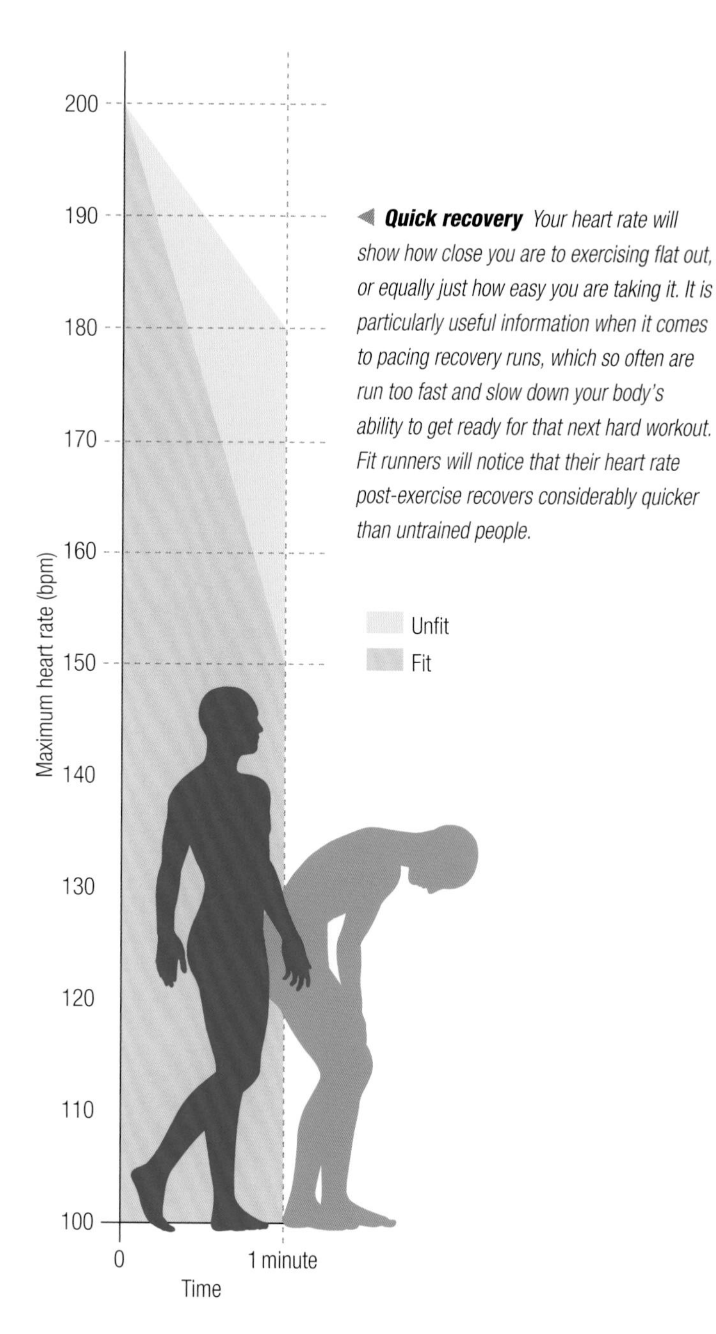

◀ ***Quick recovery*** *Your heart rate will show how close you are to exercising flat out, or equally just how easy you are taking it. It is particularly useful information when it comes to pacing recovery runs, which so often are run too fast and slow down your body's ability to get ready for that next hard workout. Fit runners will notice that their heart rate post-exercise recovers considerably quicker than untrained people.*

Different calculations for estimated maximum heart rate

▼ ***Training intensity*** *Continually monitoring your heart rate will give you a better idea of maximum numbers. The well-known formula of 220 minus your age provides no more than the roughest of guidelines given body size, shape, and background for each and every athlete, and since it was first used there have been many attempts to create more accurate estimates, some of which are shown here. Monitoring a hill session with the last effort flat out will give you a far more realistic upper number to work with. Working to heart rates is a fabulous way to maximize training time and certainly cuts out any waste. Clearly the closer to maximum you are, the harder the workout, and once you have established an upper limit, the lower recovery figures also become more useful.*

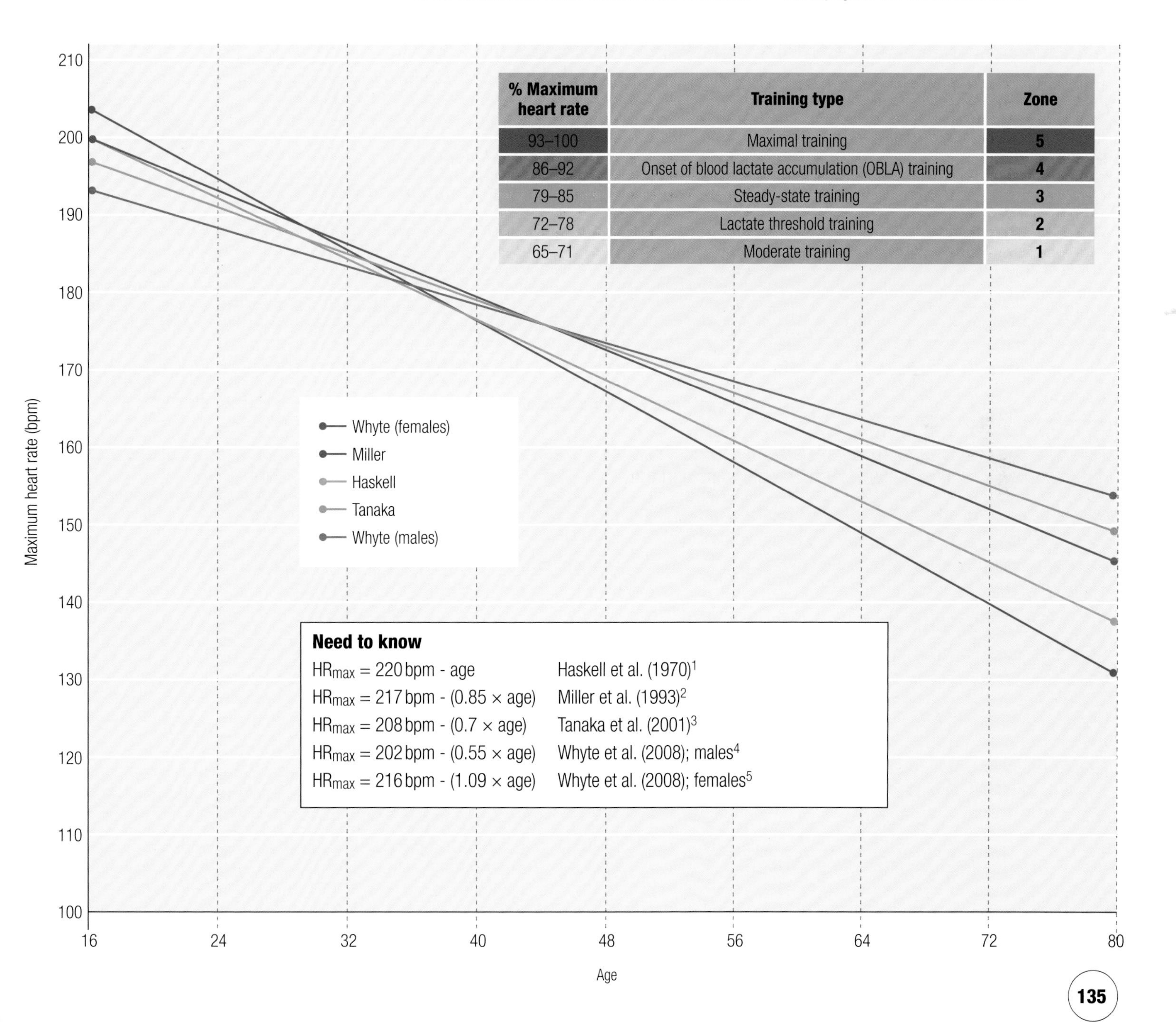

% Maximum heart rate	Training type	Zone
93–100	Maximal training	5
86–92	Onset of blood lactate accumulation (OBLA) training	4
79–85	Steady-state training	3
72–78	Lactate threshold training	2
65–71	Moderate training	1

Need to know

$HR_{max} = 220\,\text{bpm} - \text{age}$	Haskell et al. (1970)[1]
$HR_{max} = 217\,\text{bpm} - (0.85 \times \text{age})$	Miller et al. (1993)[2]
$HR_{max} = 208\,\text{bpm} - (0.7 \times \text{age})$	Tanaka et al. (2001)[3]
$HR_{max} = 202\,\text{bpm} - (0.55 \times \text{age})$	Whyte et al. (2008); males[4]
$HR_{max} = 216\,\text{bpm} - (1.09 \times \text{age})$	Whyte et al. (2008); females[5]

the influence of technology on psychological states

The first decade of the twenty-first century saw a rapid rise in sport-related technology, and this resulted in a number of gadgets that runners now use regularly in training and racing. The most common gadgets include portable music players, GPS watches, and heart rate monitors (HRMs). The rise of HRMs and GPS watches has had tremendous effects on how runners approach training and racing. They can provide details such as running speed, distance traveled, leg turnover (cadence), gradient of hills, and heart rate. Previously, runners relied on distance markers and a stopwatch, running calculations in their heads of whether they were maintaining the required pace.

The popularity of GPS watches suggests that they are helpful. The benefits of having detailed information and feedback can help an athlete learn to judge whether their ongoing efforts will suffice to achieve their personal performance goals. For example, knowing you are running 6-minute miles and that you have sustained that pace for six miles is helpful if your goal is to run 10 miles in 60 minutes. In addition, not having to calculate whether you need to raise your effort means that you can allocate more resources to the physical side of running rather than the mental.

However, the downside of using such devices is that they create a dependency on the technology. Runners can end up checking and rechecking their running time and the watch starts to become their virtual coach. Such a persistent focus on external feedback could cause runners to lose the ability to interpret internal signals such as feelings of fatigue, the intensity of effort from their muscles, and their heart rate and breathing. Focusing on kinesthetic feedback, or running on "feel," is arguably helpful because it allows runners to concentrate on adjusting their technique and correcting errors to improve performance.

▶ ***Running through the numbers*** *There's no doubting the importance that watches for measuring distance, speed, heart rate, and so much more, have on an athlete. Not only do they provide accurate feedback, the downloadable information creates motivation and feedback that a runner can use to their advantage. There's nothing more encouraging than discovering from your statistics the total number of miles you have achieved in a week. Equally, knowing and understanding this information correctly can help ensure any goal setting is of a realistic nature.*

U.S. OLYMPIC TEAM TRIALS
WOMEN'S MARATHON
10
146
U.S. OLYMPIC TEAM TRIALS
WOMEN'S MARATHON
2
52
U.S. OLYMPIC TEAM TRIALS
WOMEN'S MARATHON
17

How can wearing sunglasses improve your running performance?

Does looking good really help you relax while running?

Clench your fists and grit your teeth—you can feel the tension across your shoulders. Now relax; it's a simplified example, but that relaxation is exactly what a pair of sunglasses is contributing to. Running is all about aiming for relaxed, efficient movement and anything that can assist that goal is a definite advantage.

Throw into the equation UV and impact protection as well as vision enhancement through lens tinting and you have a pretty essential piece of gear. Relax your eyes and it is widely accepted that you will perform better.

Start with the obvious and move on from there when choosing your glasses—you'll want a pair that is lightweight, well fitting, and stable. Next, consider technologies that will reduce fogging—an enemy for any distance runner. Breathability is a key factor here as is lens coating. One manufacturer uses a technology that is claimed to improves air circulation, while others opt for lens coating.

Look too for a coating that will protect from 99% or 100% of UVB and UVA rays. UV rays can damage your eyes, just as sunburn can damage your skin. Indeed, Michael Repka, a Professor of Ophthalmology at the Johns Hopkins University School of Medicine says, "we should all be protecting our eyes from both visible light and UV light, throughout our lifetimes, because [sun exposure] can damage structures in the cornea, the lens of the eye and the retina, as well as the skin around the eyes or eyelids."[1] He notes that studies have shown that UV exposure can result in increased risk for a range of vision-related problems, from heightened sensitivity to light and irritation to cataracts and possibly macular degeneration and skin cancers on the eyelid.

Finally, consider lens color. Tinting reduces glare, brightens shadows, increases contrast, and enhances the visual spectrum in natural environments—all of which means you can see clearer as you run through forest trails with dappled sunlight.

Hydrophobic lenses

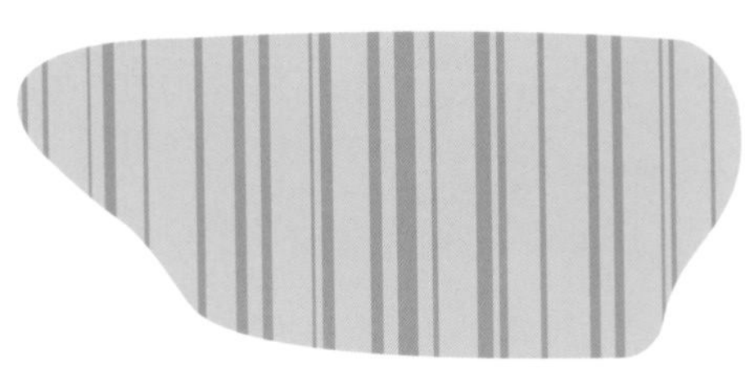

Standard lens

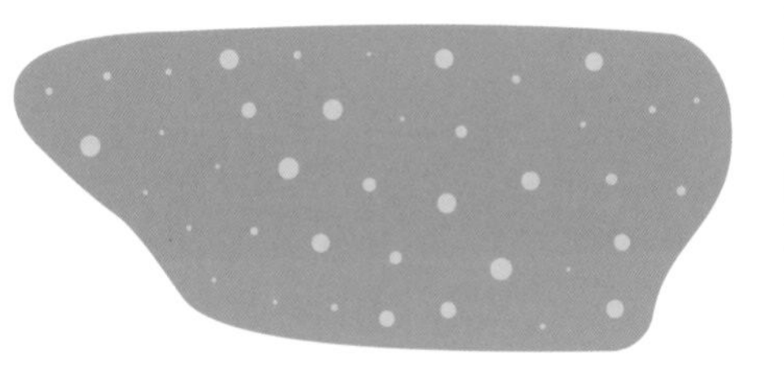

Hydrophobic lens

▲ ***Clear vision*** *Hydrophobic lenses are great for reducing glare and enhancing sharpness; this is especially handy when trail running on unpredictable surfaces dappled by shadow, but is also useful on the road, when staying relaxed is vital for a quicker run.*

Benefits
The molecules in hydrophobic coatings are non-polar while water molecules are polar; as a result, water molecules are attracted to each other rather than the lens, and form droplets instead of spreading out over the surface of the lens.
The low surface tension between the water droplets and the hydrophobic coating means that the droplets are likely to run off from the coating rather than remain on the lens.
Hydrophobic-coated surfaces are anti-static. This means that dust and other dirt particles are less likely to be attracted to them, and if these particles do end up on the lens they are more likely to drop off.

Interchangeable lenses

▶ ***What to look for*** *It is important to consider performance as well as style when choosing sunglasses. You should consider the conditions you are likely to run in and the most suitable lens color for them. If the conditions are likely to change, then interchangeable lenses may be your best choice. It's also important to look at how the glasses will fit, particularly if you'll be running at speed or on uneven ground, and their durability. Sport-friendly sunglasses should be shatterproof and scratch resistant.*

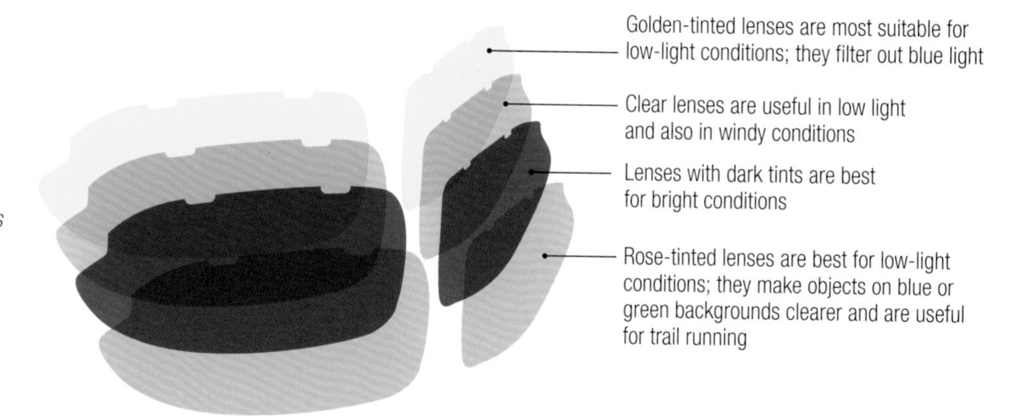

Fit considerations

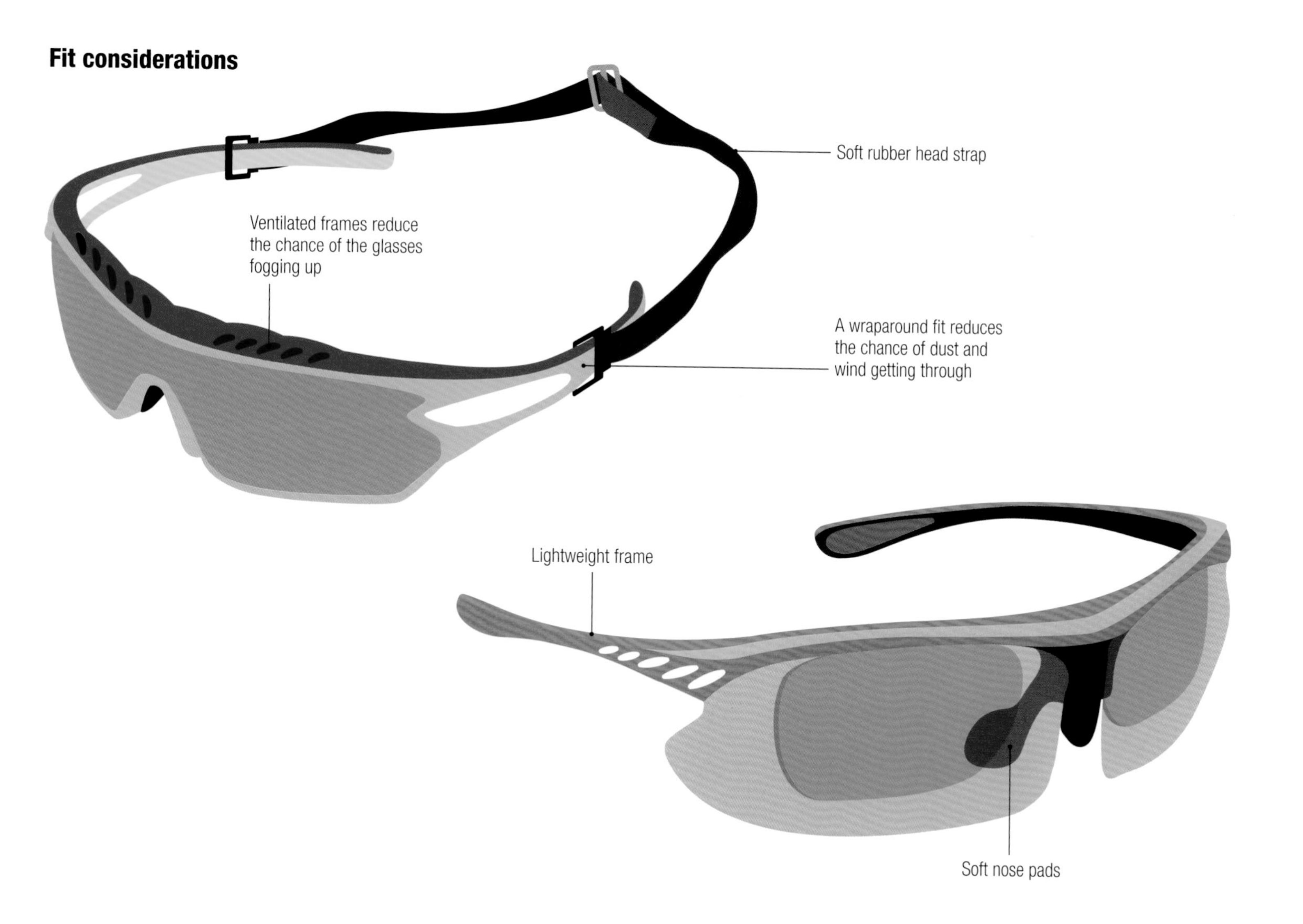

events:
marathons

The organizers of the first modern Olympics, in 1896, invented the marathon to link the modern event to ancient Greece. The race was inspired by the story of the messenger Pheidippides, who ran from Marathon to Athens to tell of victory over the Persians in 490 BCE before collapsing and dying.

The length of the race was later standardized to 26 miles 385 yards (42.195 km). If the modern event was any shorter, the dreaded "wall" would not be such an issue; if it was any longer it would lose its appeal to many runners—although running a marathon is tough and unrelenting, it is a goal that is achievable with the right amount of work.

Obviously, marathon running is about endurance and the simple concept of putting in some miles, but it is important to not underestimate one element, which is the importance of pacing. Run five seconds a mile too fast early on and you'll end up at a walking pace over the last few miles. The distance of the event also means that any pacing errors are magnified.

Establishing that pace before the race is a mixture of art and science. Long runs, a good weekly training routine, and setting and achieving predetermined goals will tell you a lot about your ideal pace, but your final time won't be predetermined. Equally important influences, as Pheidippides would attest, are how your body deals with fuel and access to carbohydrate and fat supplies.

It's not unusual for the best marathon runners to log more than 140 miles (225 km) over twelve to fourteen weeks when training for a race. While logging such high mileage is not for everyone, it's a good concept for anyone wanting to run a marathon, and the best schedules all work to a similar formula of fourteen weeks, building slowly to a long run, training three to five times a week, and including a race pace effort once every ten days.

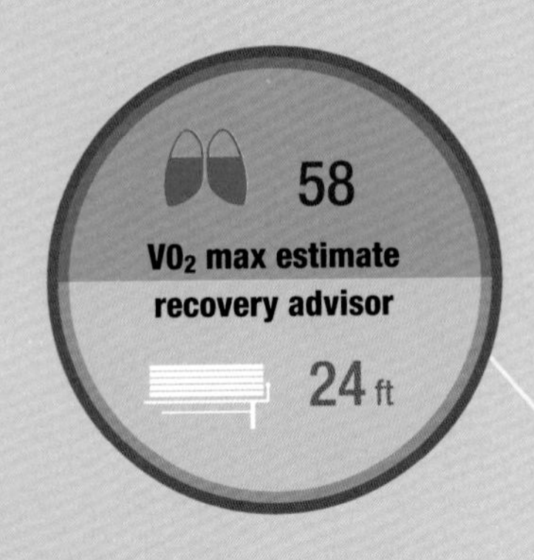

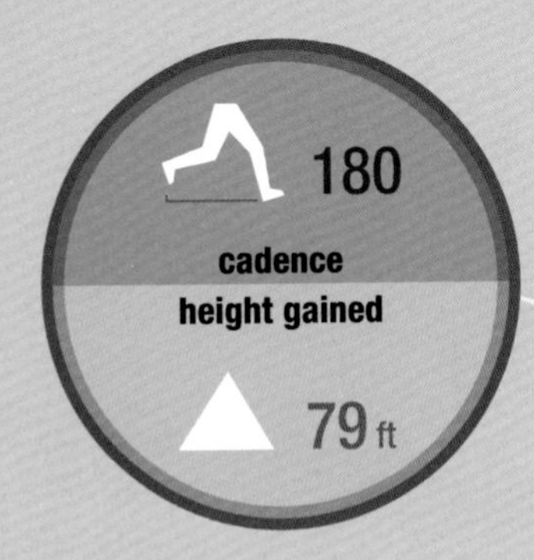

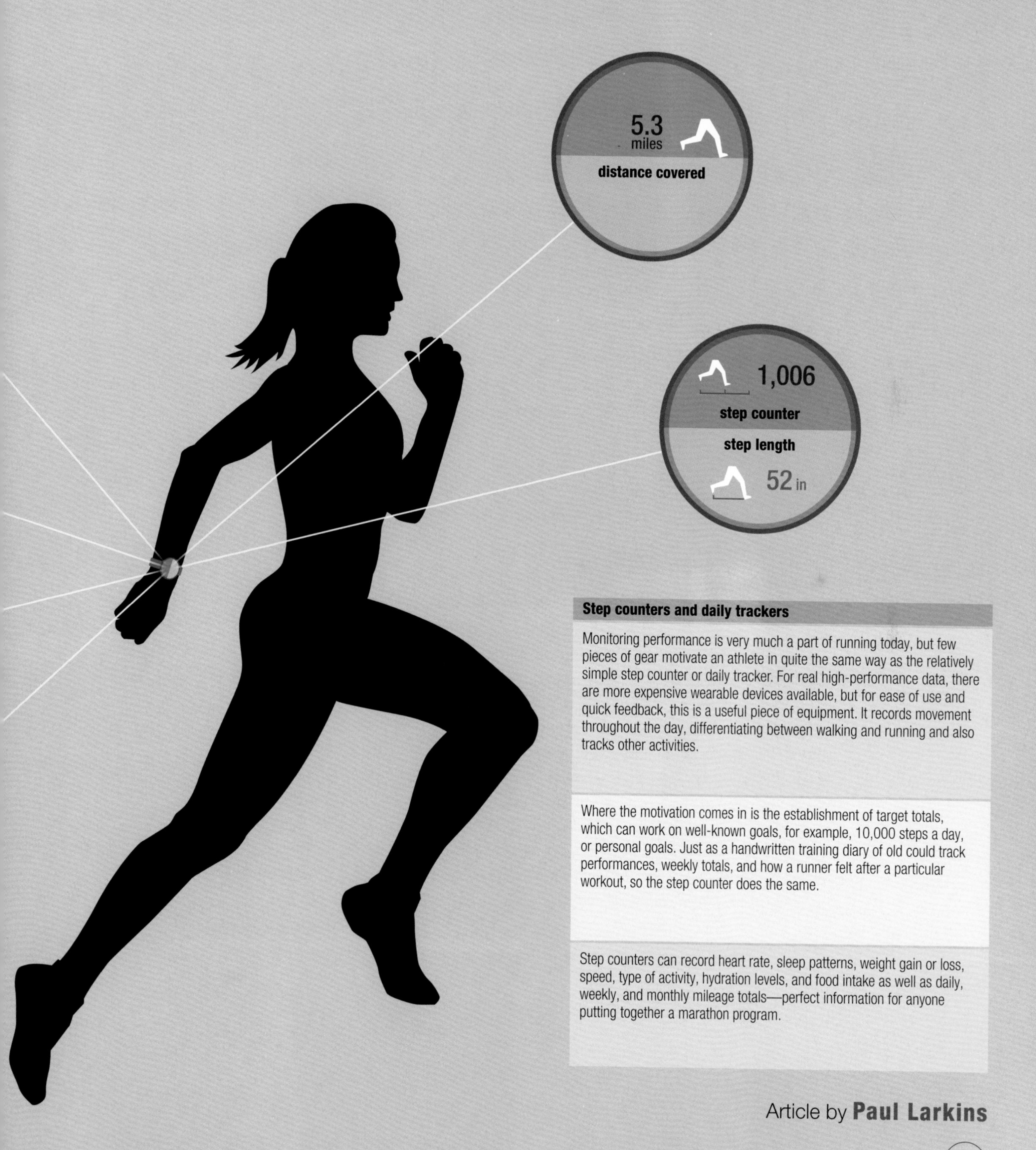

Step counters and daily trackers

Monitoring performance is very much a part of running today, but few pieces of gear motivate an athlete in quite the same way as the relatively simple step counter or daily tracker. For real high-performance data, there are more expensive wearable devices available, but for ease of use and quick feedback, this is a useful piece of equipment. It records movement throughout the day, differentiating between walking and running and also tracks other activities.

Where the motivation comes in is the establishment of target totals, which can work on well-known goals, for example, 10,000 steps a day, or personal goals. Just as a handwritten training diary of old could track performances, weekly totals, and how a runner felt after a particular workout, so the step counter does the same.

Step counters can record heart rate, sleep patterns, weight gain or loss, speed, type of activity, hydration levels, and food intake as well as daily, weekly, and monthly mileage totals—perfect information for anyone putting together a marathon program.

Article by **Paul Larkins**

For a professional runner, staying free from injury is the Holy Grail. Even a day out from scheduled training costs in performance. Every trick, however small the gain, is worth it. This chapter discusses the science behind an athlete's strategy to avoid injury and improve performance, from stretching regimes to punishing ice baths. In it, we discover that proving what works is complex and often the science is vague, since what works in the laboratory does not always translate to the everyday setting. For example, soft tissue release techniques are applied according to how an athlete is on a particular day, not according to what a controlled, repetitive, standardized study dictates. Therefore, although clinical practice is and should be driven by evidence, we cannot discount anecdotal evidence. Until studies better mimic the clinical setting, the experience and intuition of the athlete and the clinician need to be considered. So here's an account of what we know so far, most of it science, some of it anecdotal.

chapter seven

running well

Anna Barnsley

Can controlled movement and muscle balance help in injury prevention?

Will core strength and stability training keep me injury free?

The core muscles, broadly known as stabilizers, are present throughout the trunk from shoulder to pelvis, and not just in the abdominal region. Stabilizers control and sustain forces generated during movement so that both the active (muscle and fascia) and passive (joint) systems can perform with strength and efficiency, to cope with specific load demands under changing conditions. The stabilizers prevent uncontrolled movement (UCM) and sheering forces in the passive system.[1, 2]

Most running-related injuries are insidious—they develop slowly and are not attributed to a specific occurrence. Something starts to feel "not quite right." A niggle becomes persistent—maybe it changes in nature but it just won't go away.

Research indicates that pain is linked to habitual poor movement patterns[3, 4, 5]—the way you run, your form, and your ability to maintain good form when fatigued. And what about the time between training sessions? Research also tells us that sustained poor posture results in muscle imbalances[6, 7]—a disharmony between the low-load, slow-twitch stabilizer system and the high-load, fast-twitch mobilizer muscles.[8, 9] The theory is that persistent poor postural habits lead to undesirable compensations—some structures become tight and overactive while others detrain, become weak and often lengthen.[10, 11, 12] This imbalance in the muscle system affects the way you perform when running.[13, 14] If you are a "weekend warrior," at a desk all week and then sporty at the weekends, it is possible that you detrain your core while sitting and then reinforce the imbalances when running.

Muscle imbalances cause UCM around the joints, and repetitive UCM leads to overuse or overload pathology.[15, 16] In long-distance runners this is a low-force repetitive strain and for sprinters it manifests as high-force or impact repetitive strain.[17] There is growing evidence to suggest that UCM can be linked to the prediction of injury,[18, 19] so training your stabilizer system or "core" in order to correct or prevent UCM is likely to reduce your chance of injury. It adds credence to the statement, "no runner should just run."

However, it is worth noting here that muscle recruitment and growth (hypertrophy) are not the same processes. Because muscles work in synergy, not in isolation, and each muscle involved needs to tension by just the right amount for the task (recruitment), performing isolated exercises such as consciously pulling in your lower abdominal muscles (transverse abdominis or TA) could be counter-effective. This imposes abnormal muscle activity, as does trying to tense individual muscles when running.[20]

Therefore, to be effective for a runner, core training should be functional and combined with neuromuscular and strength training. It should also include exercises that mimic the movements of running.[21, 22, 23] Because strength training itself reduces the incidence of sports injuries to a third, it seems a combination of approaches works best.[24, 25] Finally, as a bonus, there is plenty of anecdotal empirical evidence suggesting that core training also improves performance.[26, 27]

Winding and unwinding theory

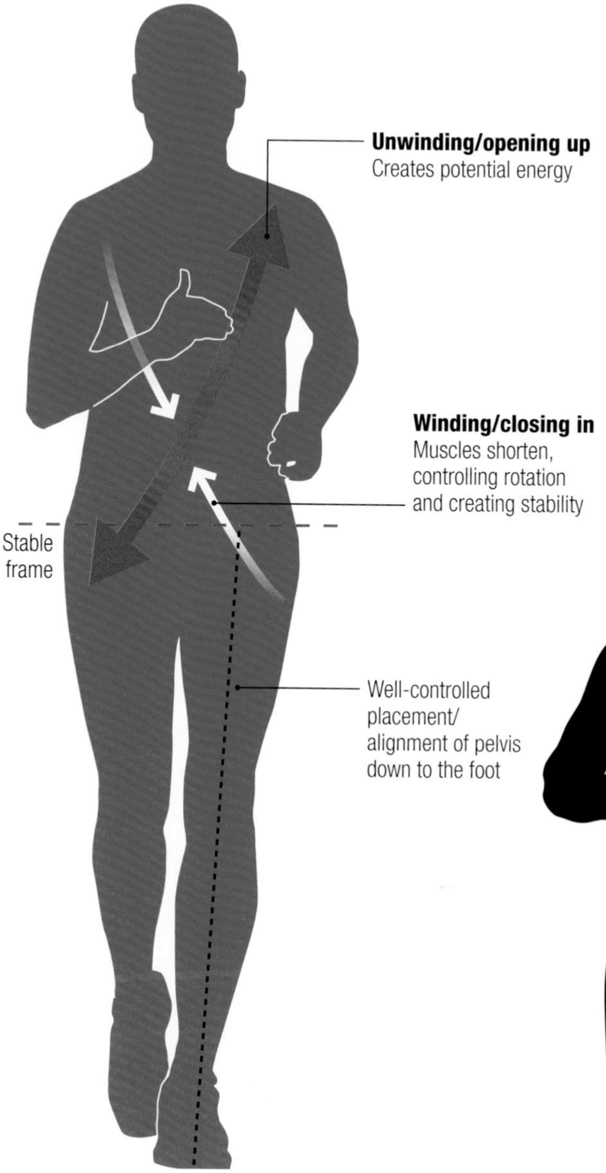

Unwinding/opening up
Creates potential energy

Winding/closing in
Muscles shorten, controlling rotation and creating stability

Stable frame

Well-controlled placement/ alignment of pelvis down to the foot

▲ ***Efficient system*** *Length–tension relationships of muscle/myofascial tissue is important for global body alignment, efficiency, and performance. Muscles with optimal length and tension allow movement to occur around a stable, controlled frame and prevent faulty movement patterns, potentially reducing injury risk.*

Uncontrolled movement

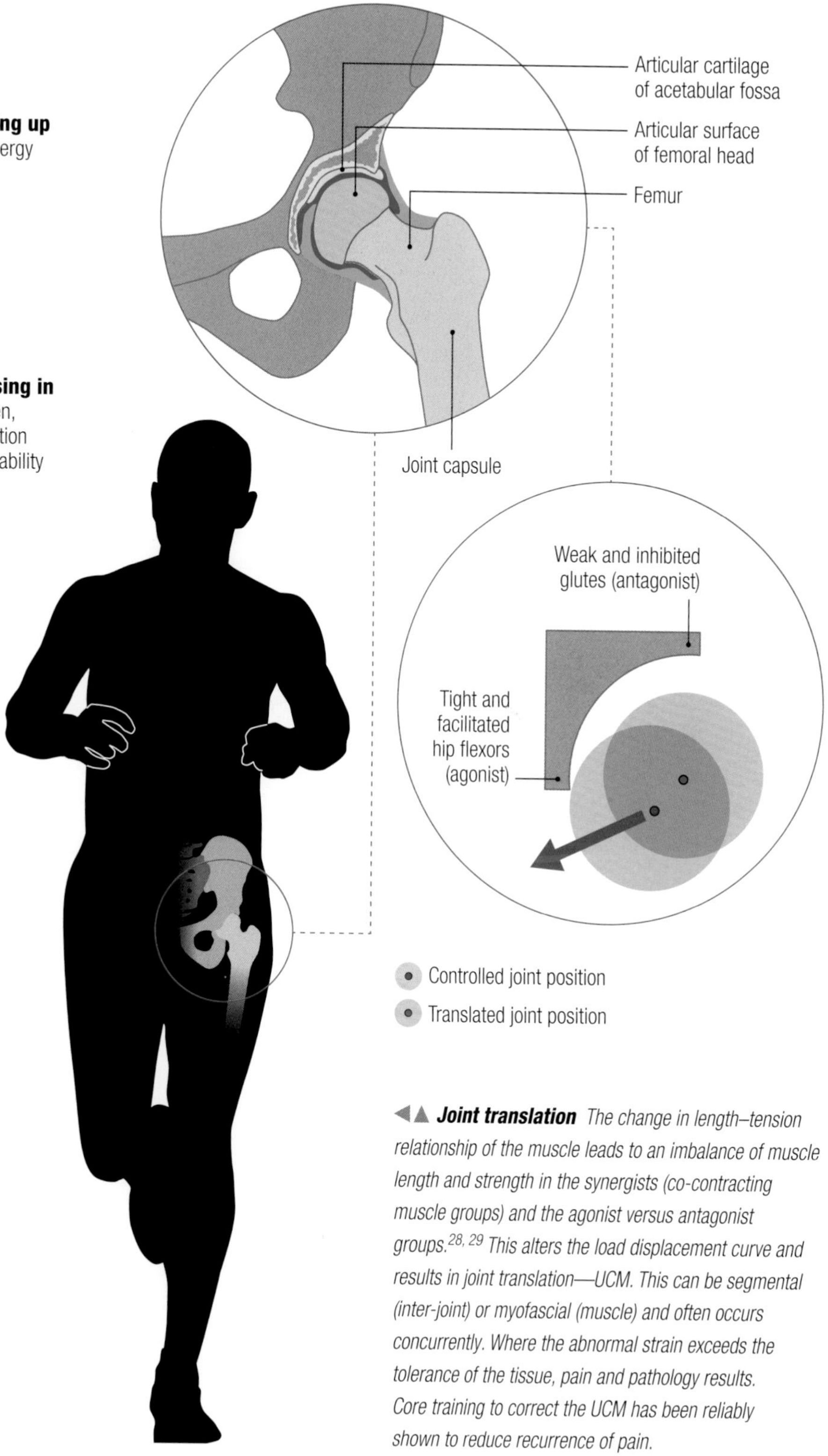

Articular cartilage of acetabular fossa

Articular surface of femoral head

Femur

Joint capsule

Weak and inhibited glutes (antagonist)

Tight and facilitated hip flexors (agonist)

Controlled joint position

Translated joint position

◀▲ ***Joint translation*** *The change in length–tension relationship of the muscle leads to an imbalance of muscle length and strength in the synergists (co-contracting muscle groups) and the agonist versus antagonist groups.*[28, 29] *This alters the load displacement curve and results in joint translation—UCM. This can be segmental (inter-joint) or myofascial (muscle) and often occurs concurrently. Where the abnormal strain exceeds the tolerance of the tissue, pain and pathology results. Core training to correct the UCM has been reliably shown to reduce recurrence of pain.*

Does stretching improve performance and prevent injuries?

Will stretching like a gymnast help my running?

Historically, stretching has been an accepted component of any athlete's program. The mantra has always been to do it or pay for it. Any regime that you put yourself through on top of your training schedule ought to either improve performance or reduce injury, or both—but research throws back conflicting results about the efficacy of stretching in terms of these benefits, and begs the question of whether it is really worth the time investment.

To start with, running does not generally require astounding feats of flexibility. Rather, it is a power- or endurance-based sport, depending on whether you are a sprinter or longer-distance runner. There's a body of evidence suggesting that traditional static stretching (where muscles are held in a sustained, lengthened position for 45 seconds or more), used immediately prior to training, reduces strength and muscle power.[1, 2, 3] This may therefore have a detrimental effect on performance and potentially increase injury risk.[4, 5, 6, 7, 8] However, the duration, sets, and number of each exercise in these studies typically exceeded the accepted norm,[9] so further research to reflect standard practice is required.

Dynamic stretching pre-exercise fares better and has been adopted as the preferred pre-sport stretch method. This is stretching that mimics the motions you go through during your running cycle—calf bounces and lunges, for example. This type of stretching does not seem to adversely affect performance or pose an injury risk, but it is not clear whether it is advantageous in terms of injury prevention or performance enhancement either,[10, 11, 12] and a training program should be based on what it can achieve, not what it doesn't achieve. There is sufficient evidence to show that a thorough, sport-specific warm-up with dynamic stretching reduces injury risk enough to be worthwhile. However, it is not known if these effects are due to the warm-up or the inclusion of the dynamic routine.

There is also still a potential role for static stretching. It can have a positive outcome on performance, particularly in terms of strength and running speed, when used independently of training sessions and on a long-term basis.[13] Also, for an elite athlete—for whom small percentages make the difference between winning or losing, injury, or health—the evidence points toward static stretching achieving a tiny reduction in the likelihood of sustaining a soft tissue injury.[14, 15] If you already have a soft-tissue injury related to muscle imbalance, in which a muscle is functionally short and tight, there is good evidence that stretching is helpful as it increases joint range of movement,[16, 17, 18, 19] releasing muscle tightness to redress movement control.

▶ ***Theory 1*** *When muscles are short, sarcomeres (the functional units of the muscle responsible for the mechanics of contraction) are lost.*[20, 21] *The muscle has a reduced ability to create peak tension at optimal range due to the lower number of actin bridges available to link to the myosin. The main theory of muscle hypertrophy in response to stretch is that it adds sarcomeres in series, thus generating a higher peak force at optimal length.*[22] *However, this research was originally done with animal muscle tissue, which it now seems responds differently to human tissue.*

Muscle fiber before stretching

Muscle fiber being stretched

Additional sarcomeres added in response to stretching

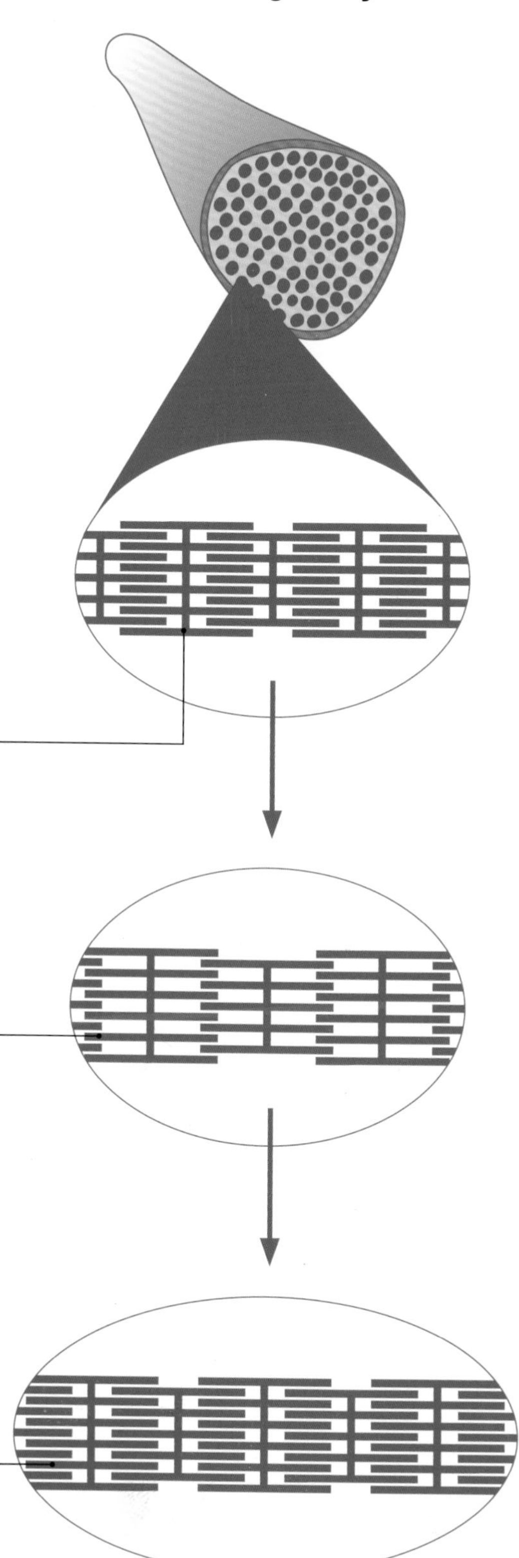

Stretching theory 2

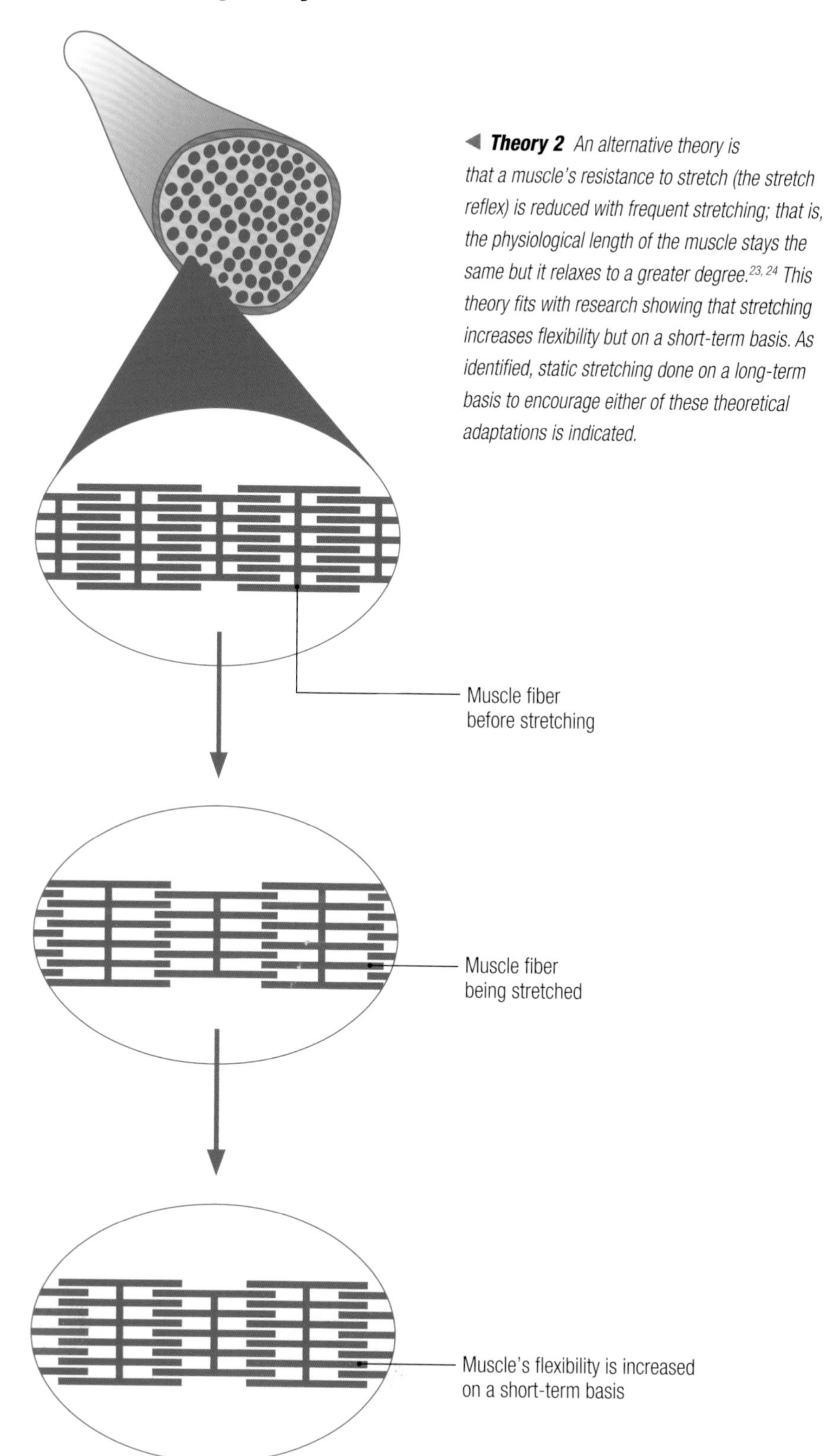

◀ ***Theory 2*** *An alternative theory is that a muscle's resistance to stretch (the stretch reflex) is reduced with frequent stretching; that is, the physiological length of the muscle stays the same but it relaxes to a greater degree.[23, 24] This theory fits with research showing that stretching increases flexibility but on a short-term basis. As identified, static stretching done on a long-term basis to encourage either of these theoretical adaptations is indicated.*

What is DOMS and is it beneficial? → Is the saying "no pain, no gain" true?

Here's a familiar scenario: the elation of completing a tough training session followed by the embarrassing agony of limping down the stairs the next day due to pain and stiffness in your legs, known as delayed-onset muscle soreness, or DOMS. DOMS is caused by an increase in training that involves eccentric muscle activity—that is, when a muscle creates tension to control the rate at which it lengthens, such as when running downhill. It typically occurs the first day after exercise and peaks around twenty-four to seventy-two hours. It resolves over the following five to seven days, after which you can repeat the same training session with far less intense repercussions.

A misconception is that DOMS is due to lactic acid accumulation. Raised lactate concentrations in muscle tissue caused by high-load exercise may be responsible for pain experienced during or immediately after training but, since levels return to normal within twenty to sixty minutes of exercising, this cannot account for delayed-onset pain the next day.[1] Neither is DOMS caused by muscle spasm and associated reduction in blood flow—the theory that led to the adoption of stretching as a means to reduce soreness, which we now know is ineffective.[2]

Although the mechanism is not entirely understood, DOMS is believed to occur due to the mechanical disruption of the units of a muscle, called sarcomeres.[3, 4, 5, 6] Although this sounds worrying, the damage associated with DOMS is physiologically beneficial. It promotes muscle adaptation and growth in response to stress so that an athlete can repeat the same training session with more success and less pain—known as the "repeat bout effect."[7, 8] This effect occurs within five days following a single bout of exercise.

So does this mean that the more pain you feel, the more you stand to gain? An individual's experience of pain is just that—individual. The level of soreness described is not an accurate indicator for resultant muscle growth and varies between professional athletes, and even between those with similar genetics.[9, 10, 11] So how much pain is too much? To date, we don't know for sure, but the best guess is to aim for moderate soreness, since there is some evidence that severe microtearing could lead to a macrotear.[12, 13, 14] And when can you train again? The general consensus seems to be that a recovery period of 3 to 5 days is necessary to allow the adaptive changes to occur.

DOMS timeline

▶ ***Slow burner*** *The characteristics of DOMs and its progression contribute to an understanding of what causes it and how adaptation may occur. For example, the presence of macrophages coincides with the delayed onset of pain, since the products of this activity stimulate nerve endings in muscle.[15] Adaptation theories suggest that the physiological events of DOMs precipitate an increase in the number of sarcomeres in the muscle fiber, improving muscle strength and reducing subsequent damage if the same bout of exercise is repeated. This effect can last for several weeks.[16, 17, 18, 19]*

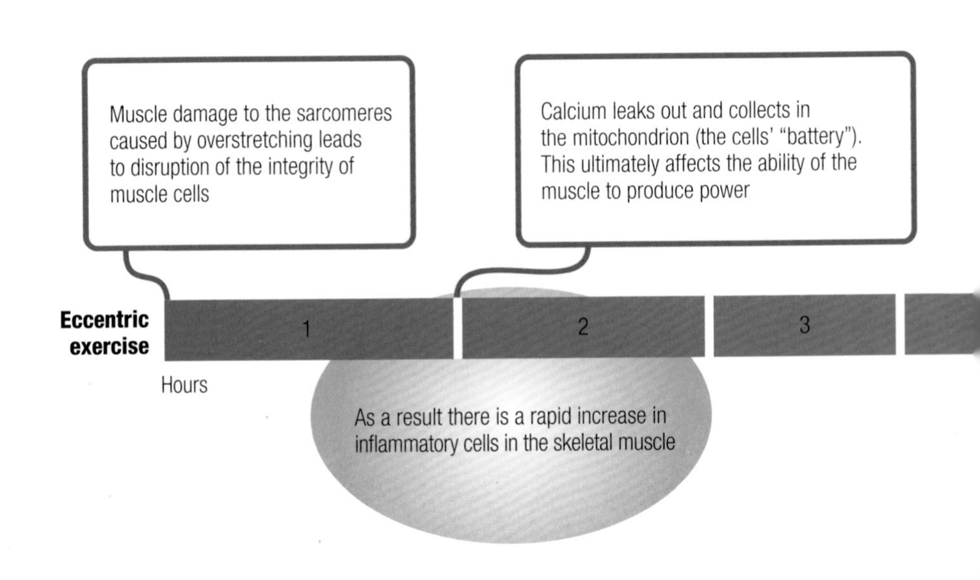

Muscle damage

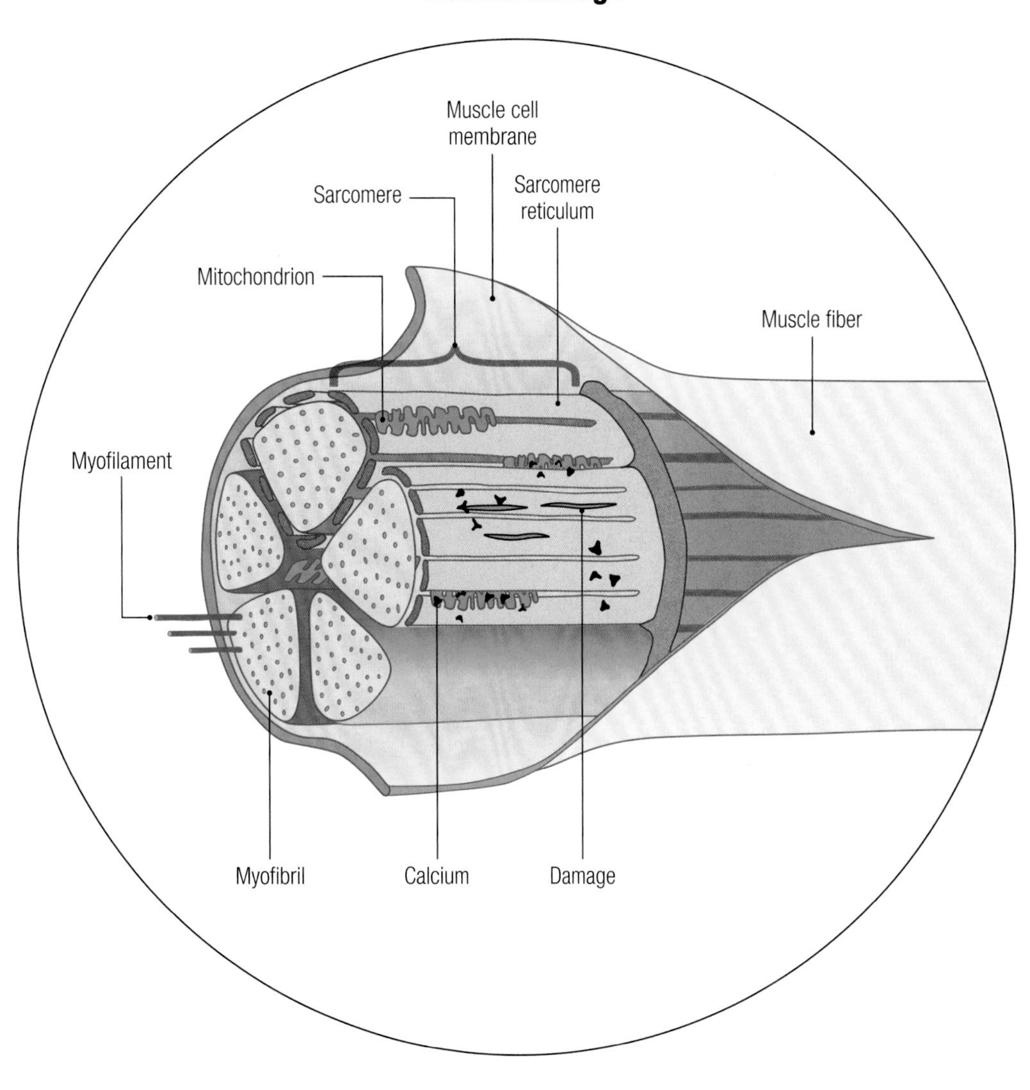

▶ ***What's the damage?*** *With repetitive lengthening under tension, the sarcomere becomes stretched and the integrity of the muscle cell membrane is disrupted.*[20, 21] *Research shows that microtearing occurs in the part of the muscle (the Z-disk) that provides structural support.*[22, 23] *Calcium leakage leads to the breakdown of muscle proteins. This, along with the membrane damage, results in inflammation and swelling.*[24, 25]

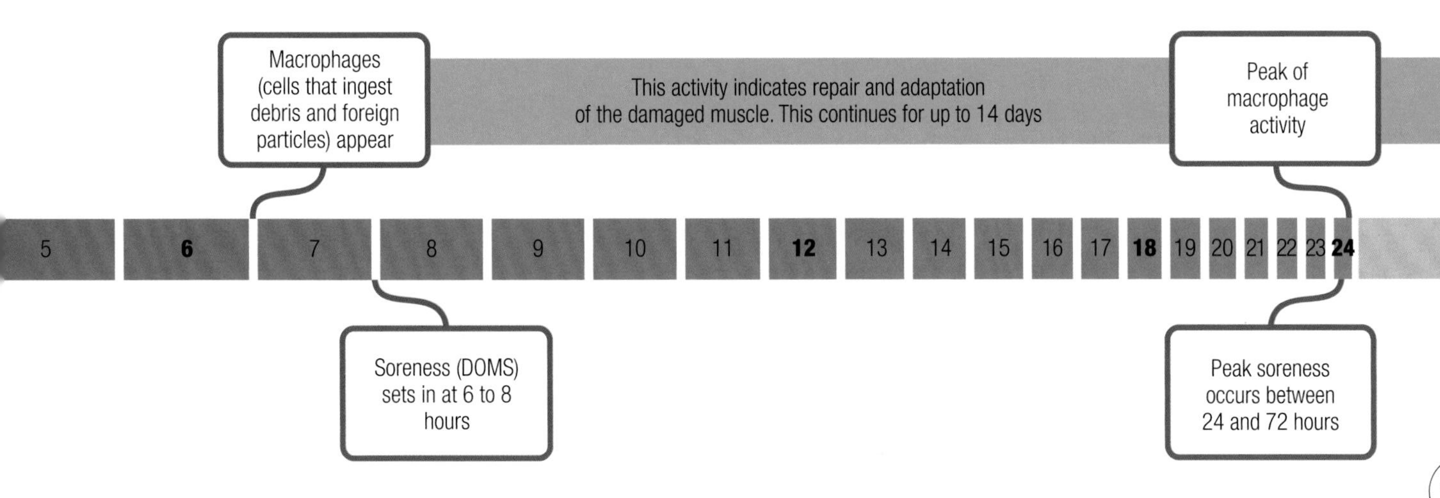

Can sports massage help with injuries or performance?

Can I justify some down-time on the massage couch?

We don't actually know whether or not sports massage really works. This is especially surprising, considering the almost evangelical commitment to this form of therapy within the sporting community. The problem is that studies to date have been equivocal and flawed by bad design, methodological variations, and poor control, so there is currently little evidence to support massage for preventing injury, enhancing recovery, or improving performance.

There goes your excuse for having some "couch time" at least once a week. Or does it? We cannot discount a therapy for which there is so much positive anecdotal evidence on the basis of poor research. Massage is a tricky treatment to evaluate—it tends to be tailored to the individual's presentation at a moment in time and often cannot therefore be subject to a like-for-like comparison. Moreover, there are many different techniques to consider.

So what can be gleaned from the limited quality studies so far? There is greater support for the effectiveness of post-run massage than that of pre-run treatment, especially since there is some evidence that massage used pre-event is more detrimental than beneficial to performance.[1, 2, 3, 4] Some research points towards its use in preventing delayed-onset muscle soreness

Massage techniques to treat injury and dysfunction

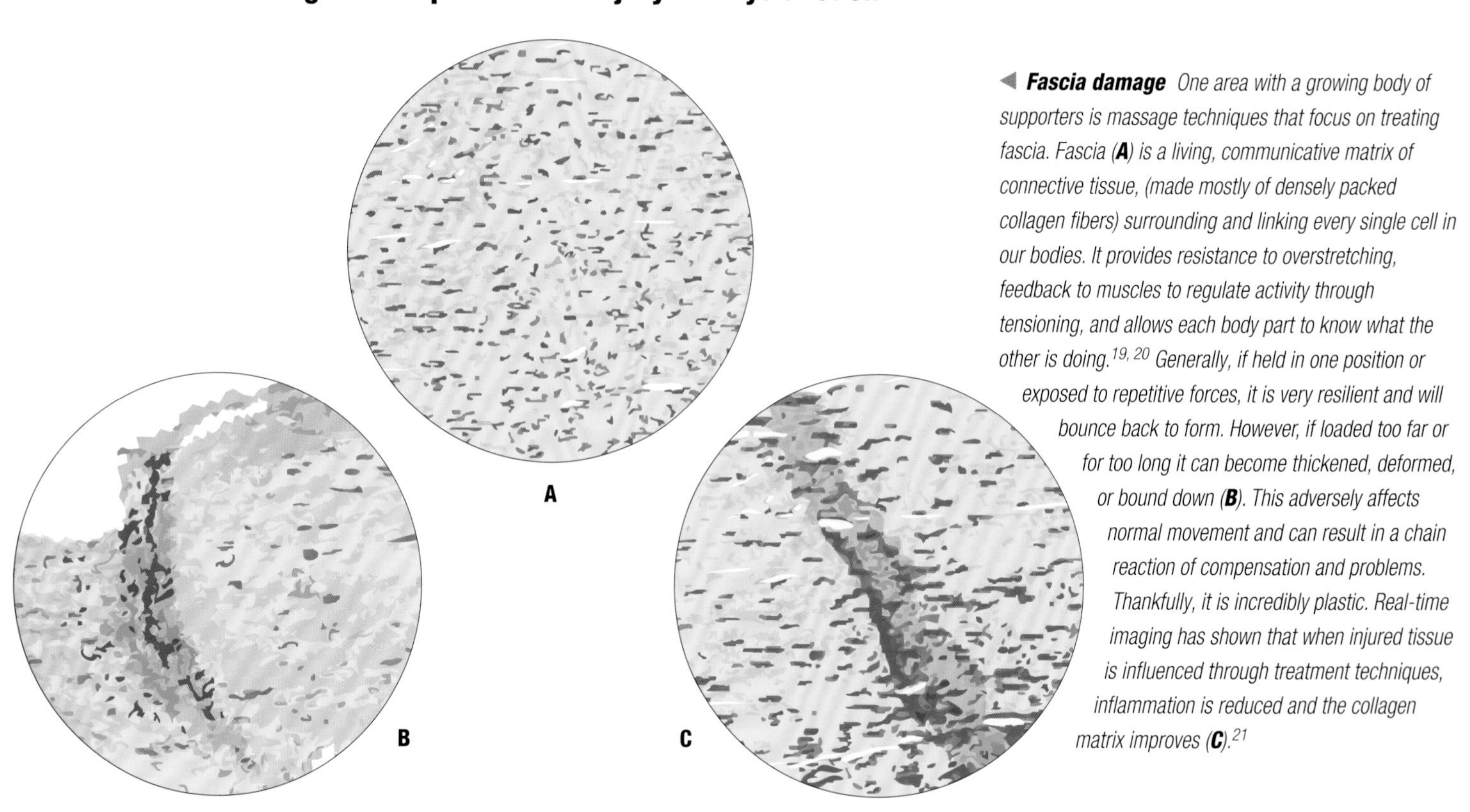

◀ ***Fascia damage*** *One area with a growing body of supporters is massage techniques that focus on treating fascia. Fascia (**A**) is a living, communicative matrix of connective tissue, (made mostly of densely packed collagen fibers) surrounding and linking every single cell in our bodies. It provides resistance to overstretching, feedback to muscles to regulate activity through tensioning, and allows each body part to know what the other is doing.[19, 20] Generally, if held in one position or exposed to repetitive forces, it is very resilient and will bounce back to form. However, if loaded too far or for too long it can become thickened, deformed, or bound down (**B**). This adversely affects normal movement and can result in a chain reaction of compensation and problems. Thankfully, it is incredibly plastic. Real-time imaging has shown that when injured tissue is influenced through treatment techniques, inflammation is reduced and the collagen matrix improves (**C**).[21]*

Soft tissue massage

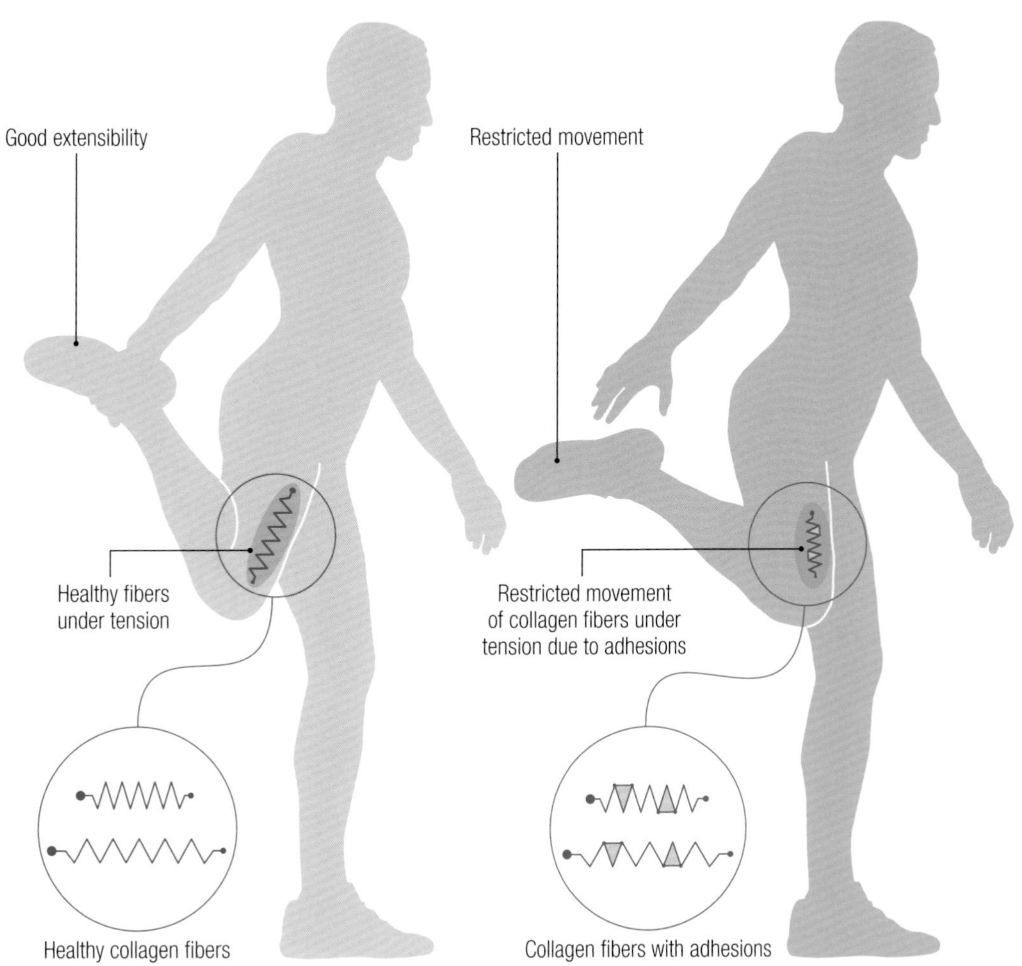

(DOMs),[5, 6, 7] provided it is ideally utilized within two hours after exercise,[8] and that it may also help with reducing fatigue and recovery times.[9, 10] There is also evidence of massage having a positive psychological effect on perceived effort when used as a warm-up technique and the perception of recovery when used after a run,[11, 12, 13, 14] which may be one reason why it is such a popular therapy. Improved mental readiness for sport is certainly a desired outcome.

Massage has also been shown to have favorable effects on immune function. High training loads can lower immunity, making an athlete more susceptible to illness. Some research suggests that massage lowers the stress hormone cortisol and increases natural killer-cell cytotoxicity, which helps to reject any cells that have been virally infected.[15, 16, 17, 18]

▲ ***Preventative treatment*** *High-level training results in sustained increased muscle tone. Elevated tone contributes to muscle imbalance and asymmetry, predisposition to tissue microtrauma, and the development of bulky connective tissue and cross-linkages in fascial tissue, which compromises muscle function. It is also proposed that a prolonged increase in tone impedes delivery of oxygen and nutrients and the removal of metabolites and thus impacts the ability of muscle to recover and repair.*

Soft tissue release techniques used in massage are thought to address these issues through releasing tight and adhered tissue, increasing muscle–joint range of movement,[22] improving circulation, and facilitating lymphatic drainage.[23, 24] Many athletes also feel that regular treatment becomes part of a preventative strategy, both in negating these effects and in giving the therapist the opportunity to identify abnormalities before they become problematic.

In general, most athletes feel that massage is worth their time. The scientific evidence may be unconvincing, but anecdotal evidence is heavily weighted in its favor.

Do ice baths aid recovery after long or intense runs?

Should I take a cold plunge?

Ice bathing or cold water immersion (CWI) has grown in popularity recently, with many elite athletes reporting it is worth the pain. A high-frequency training load, needed to challenge the body to adapt and improve performance, requires a decent recovery strategy and, anecdotally at least, ice seems to help.

However, current research poses more questions than it answers. To start with, CWI emerges favorably as a treatment for reducing delayed-onset muscle soreness (DOMS),[1] but since the inflammatory process associated with DOMS stimulates adaptation in muscle tissue, is lessening this effect a desirable result, or does the ability to undertake a higher training load due to reduced recovery times override this? This is an important question to answer for long-distance runners, for whom conditioning, rather than skill acquisition, is the primary aim. Unfortunately, research into the effects of CWI on adaptation to training is conflicting: it may reduce adaption to resistance training,[2] but it may not when applied following endurance training.[3] Since running has a high inflammatory response, it is suggested that CWI can be used to aid recovery in runners during intense periods of competition,[4] though it is perhaps best avoided during training.

The optimal CWI treatment protocol for runners is based on anecdotal experience more than on empirical data. The best water temperature to use depends on environmental conditions—if the weather is hot, then water as warm as 68°F (20°C) can be effective.[5, 6, 7] In this scenario, CWI theoretically assists performance between multistage events (such as a double road-race event) by precooling, rather than by enhancing recovery. If the water is too cold, CWI can adversely affect performance, as participants feel stiff rather than energized.[8] Most adopt a temperature of 50–59°F (12–15°C) and, in the absence of any compelling data to the contrary, this is currently the accepted practice.

Need to know

Hydrostatic (water) pressure is primarily influenced by depth of immersion. It can be calculated using the following equation:

$$P_{hyd} = P_{atm} + g \times p \times h$$

where:

P_{hyd} = hydrostatic pressure (Pa)
P_{atm} = atmospheric pressure (Pa)
g = acceleration due to gravity (9.81 m/s^2)
p = water density (1,000 kg/m^3)
h = depth of the water (m)

It seems timing is also key—immediate post-event ice-bath treatment, then at least 45 minutes rest before your next run (which should be preceded by an adequate warm-up) appears to work best. Try to run again too soon and performance is compromised and the injury risk is increased.[9, 10] Immersion duration is typically five to fifteen minutes. Theoretically, the ideal immersion period depends on the external temperature, body core temperature, and water temperature but, unfortunately, the research isn't adequate to give guidelines on what works best.

Overall, the use of CWI is superior to passive recovery, however, recent research that compared CWI with an active recovery strategy found that CWI was no more effective in reducing inflammation or cellular stress after resisted exercise.[11, 12] So if you're looking for a good excuse to avoid the icy depths, it may be that active recovery based on light exercise is an alternative.

Cold water immersion

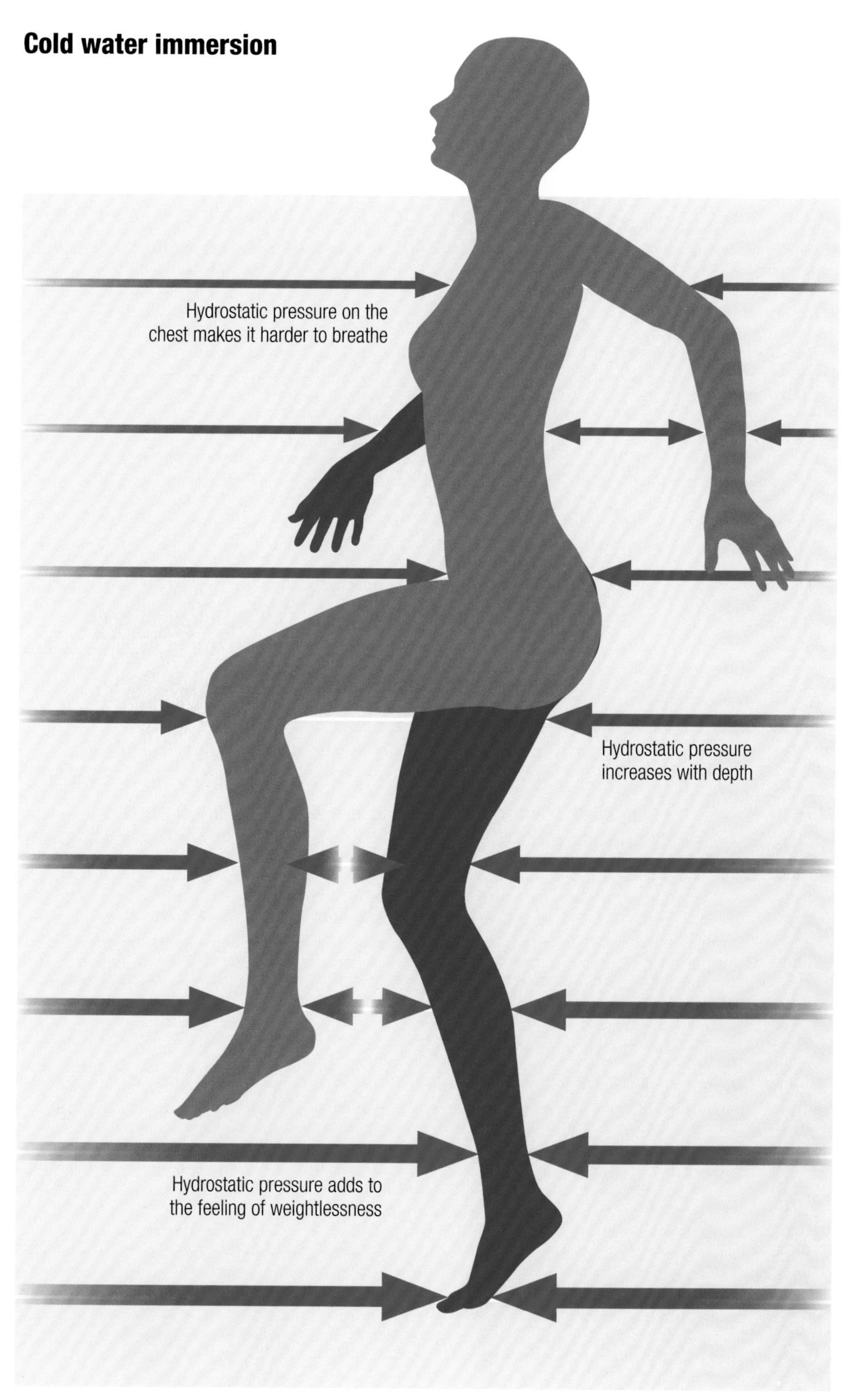

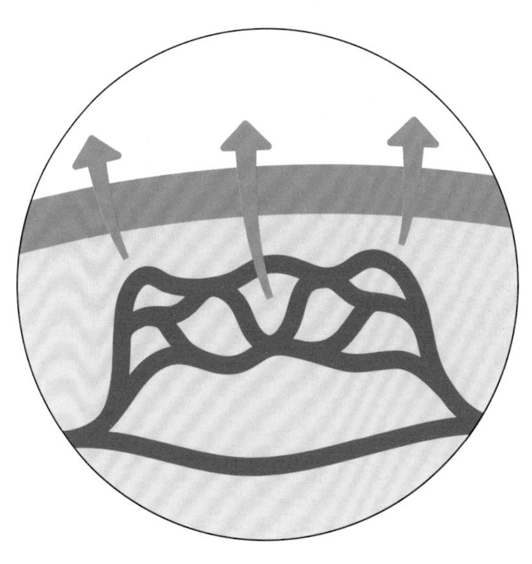

▲ ***How does it work?*** *The theory (and it is just that—the exact rational has yet to be proven) behind CWI, is that the cool temperatures initially cause vasoconstriction (constriction of blood vessels), thus slowing postexercise metabolic activity in the tissues and reducing swelling and inflammation. Following immersion, the tissues warm up, resulting in a "rush" or increase in blood flow, stimulating removal of the by-products of cellular breakdown and improving delivery of nutrients to damaged tissue.[13, 14] Theoretically, CWI also reduces the speed of nerve transmission and increases the nerve tissue receptor threshold (this accounts for the feeling of numbness), which, together, gives a reduced perception of pain.[15, 16, 17] Finally, there is a psychological effect of making the body feel energized and reducing the feeling of post-training fatigue.[18, 19]*

◀ ***Hydrostatic pressure*** *A second potential mechanism contributing to the effects of CWI is the hydrostatic pressure from the immersion itself. The pressure gradient of the water surrounding the body has been postulated to increase cardiac output and enhance blood flow, diffuse metabolites, and increase the delivery of oxygen, hormones, and nutrients to the muscles. It may also limit swelling.[20, 21, 22] It seems that hydrostatic pressure combined with cold water gives better results than cold water alone.[23] However, studies on thermoneutral (body temperature) immersion, where the water does not affect core temperature, have mostly included an element of active recovery, including swimming, stretches, and exercises. Currently it is therefore not known how much hydrostatic pressure contributes to the effects of CWI.*

Is there a causal link between running and osteoarthritis? → Is running bad for my knees?

Long-term damage is a concern of many runners, and the notion that "running damages knees" is a common excuse to not run, but the good news is that the converse appears to be true.

Radiological studies assess damage to runners' knee cartilage as evidence of osteoarthritis in relation to exercise. These studies show bony outgrowths (osteophytes) correlating to physical activity but, crucially, no evidence of narrowing of the joint space in the knee that would indicate degeneration.[1] There also seems to be a strong inverse relationship between cartilage defects and exercise. This is likely to be related to the cyclical on–off pressure on the cartilage when running, which creates a pumping mechanism, promoting blood flow, cell regeneration, and removal of waste products.[2, 3]

It is worth understanding, though, that there are factors that can increase your vulnerability to damaged knees. Runners with a genetic susceptibility may be more likely to develop knee osteoarthritis but it is not running itself that damages them.[4] Weight is the biggest factor in knee joint deterioration because up to eight times your body weight transfers from limb to limb with each step, and fat produces substances that have a deleterious effect on cartilage.[5, 6, 7] But runners are rarely overweight, and if losing weight is the motivation for running, this is great for knees in the long-term. Injury can pose a problem, particularly where the articulating surfaces of the knee joint have been compromised such as occurs with a fracture or meniscal (shock-absorbing tissue) tear. There is a strong correlation between these defects and arthritic changes because they alter the normal mechanics of the knee joint.[8] In short, people who say running damaged their knees probably had injuries or conditions that predisposed them to problems in the first place. "Couch potatoes" have a higher chance (about 45%) of developing osteoarthritis than those that run.[9]

The biomechanical reason for why running doesn't cause arthritis

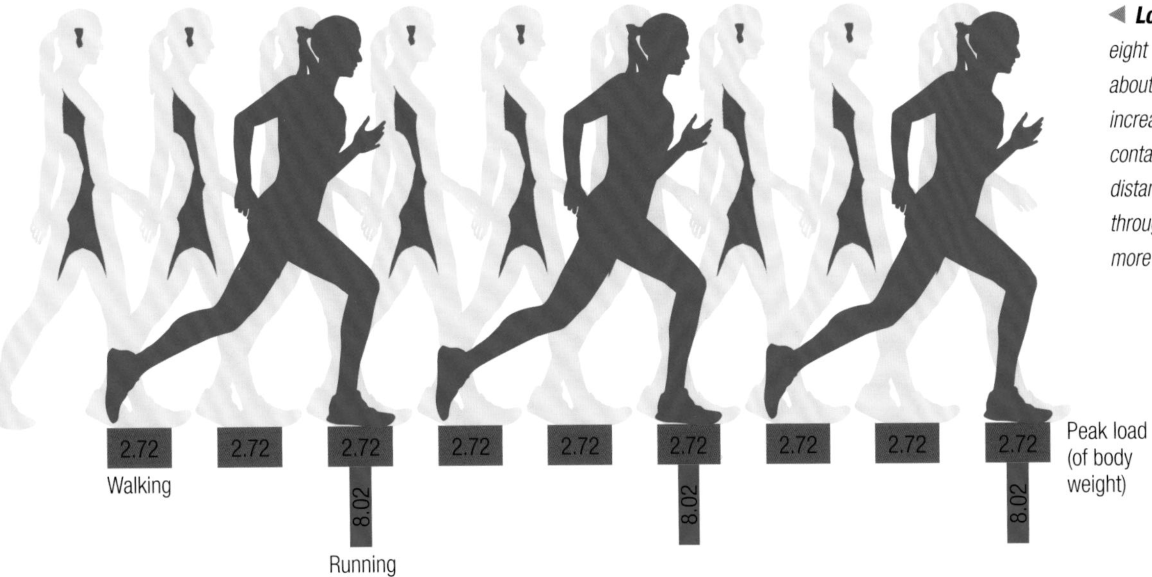

◀ ***Load per unit distance*** *When we run, in general eight times our body weight transfers from limb to limb; about three times as much as walking. However, the increased step length with running, and the reduced contact time with the ground, means that over equal distance, the net difference in load per unit distance through the knees between walking and running is more or less the same.*[10]

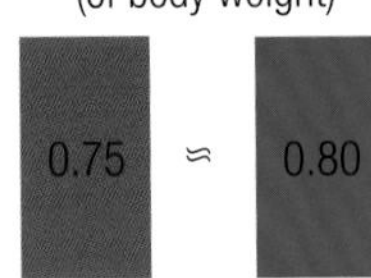

Patella femoral pain syndrome

Gluteus medius

Iliotibial band (ITB)

Gluteus medius

Unstable or asymmetrical pelvis results in poor muscle control

Internal rotation the femur

Lateral pull on the patella

Rotational pull on the tibia

Pronation of the foot

▶ ***Knee Injuries*** *However, before we get too smug, it is estimated that 37–56% of the recreational running population will suffer an injury, and knee injuries are the most common.[11] So, not lasting damage, but enough to cause an issue and possibly prevent you from training. Of these, "runner's knee," or patellofemoral pain syndrome (PFPS), is the biggest risk. PFPS typically originates from the kneecap (patella) dragging laterally or "maltracking" out of its groove on the femur (thigh bone). However, it is worth noting that this popular hypothesis is based mainly on visual assumption and not on solid evidence. Symptoms are grinding and popping under the patella accompanied by pain that we think originates from increased tensile load on the surrounding soft tissues (although there is some emerging research suggesting the bone may be involved).[12] If unchecked, this can lead to wear in the cartilage on the underside of the patella. Maltracking is linked to muscle imbalances of the pelvis and lower limb, leading to over-pull of the iliotibial band at its attachment to the lateral patella, thus dragging it as described. This scenario manifests itself through the kinetic chain in various ways: pelvic asymmetry, weak gluteal and quads muscles, even postural dysfunction as high as the thoracic spine can affect the normal chain of tension and control through the knee.[13, 14, 15] In elite sport, injury prevention is a major part of athletes' stratagem and preempting and addressing faulty biomechanics and poor neuromuscular control is part of an athletes' training strategy.*

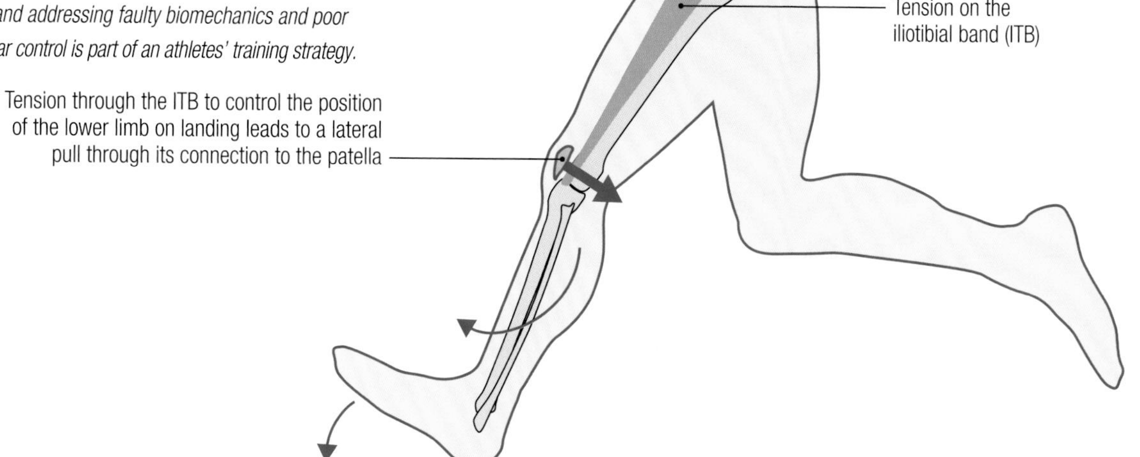

Is it advisable to continue to train when injured? → Can I run through the pain?

Pain is your body's warning system, but is actually a poor measure of injury. The level of pain doesn't always equate to the severity of damage, or even enable you to accurately figure out what's wrong.[1] Having a "stitch," for example, can be excruciating yet does not involve any tissue damage. Scientists are still not totally sure what causes a stitch, but certainly you can run it off with no ill consequences.

Sometimes pain can be a good thing. We know that training that challenges the body sufficiently to generate delayed-onset muscle soreness (DOMS) promotes tissue adaptation and improves performance. Research involving ultramarathoners (covering distances of 2,788.7 miles, or 4,488 km, in sixty-four days) also suggests that you may be able to run through muscle inflammation with no lasting damage.[2] In some circumstances, continuing to run can be part of your rehabilitation and, with the right modifications, may be positive.

Thankfully, there are some rules that are useful to help you gauge whether to grit your teeth or jump into cotton wool. Pain isn't the only indicator of injury. Swelling, a joint that locks or gives way, and tenderness on the bone are all signs that something more serious could be occurring. Some research has also been done to give us an idea of when pain may indicate damage.[3] If your pain is more than 5 on a scale of 0 to 10 (on which 0 represents no pain at all, and 10 is the worst you could possibly imagine), then continuing to train is not a wise choice. If it's less than 5, you could try some strategies to see if you can influence it—reduce or increase your speed, modify your gait, lighten your step, or change your running shoes, for example.

Some injuries don't make themselves apparent until a day or so after damage has occurred, in which case it's best to rest and monitor the response over the next few days. If it persists for several days, you've probably done too much and need to modify your training. Persistent pain will need professional advice.

So it is possible to run through an injury, but it depends on you and your unique brain. Whether you should run through an injury, though, depends on the nature of the injury. If in doubt, get it checked out.

Microtrauma and macrotrauma

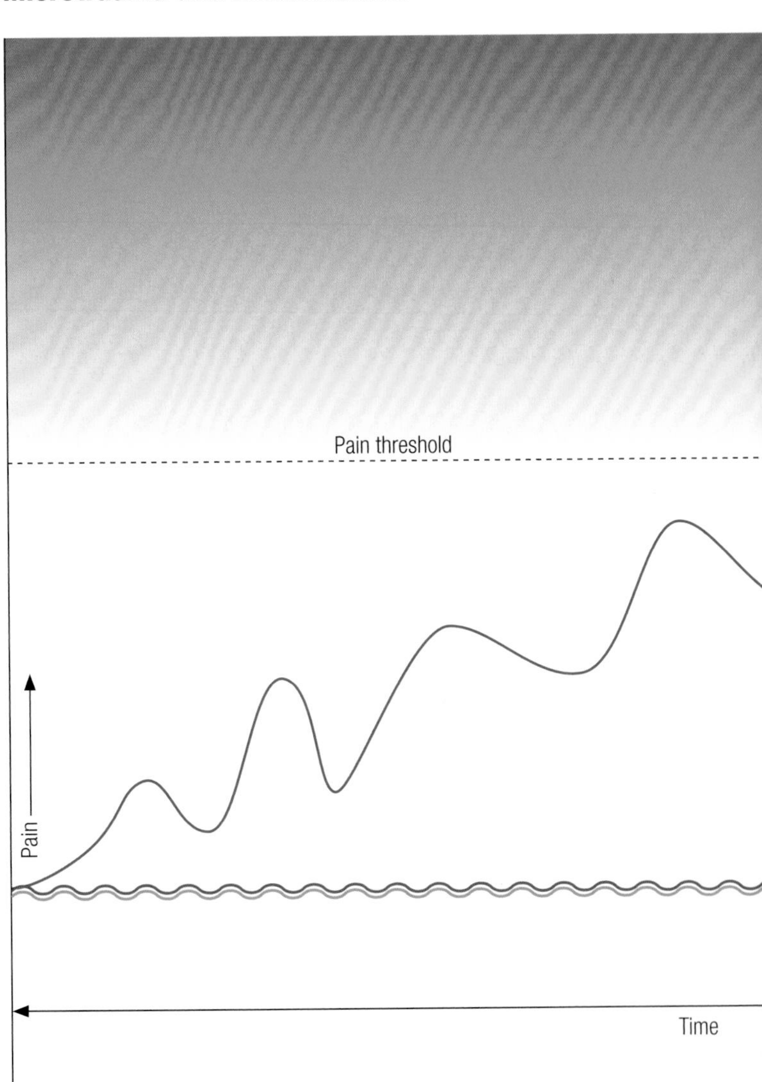

▶ ***Extreme tolerance*** *Cold pain sensitivity is a useful measure used in research because there is a relationship between it and mechanical pain sensitivity. In the experiment illustrated, ultramarathon and other runners were asked to rate the increasing pain in their hand over a period of time when it was immersed in water frozen to 28.4°F (–2°C). Ultramarathoners are masters at coping with extreme physiological stress and provide great examples that pain does not provide a measure of the state of the tissues. Pain can be thought of as an illusion because it is interpreted and mediated by the brain. The thalamus in the brain acts like a switch box, interpreting signals from the nociceptor fibers (nerves that transmit pain information) in the tissues and deciding whether a pain response is necessary and if so, how much pain. This "decision" is based on complex psychological, social, and somatic factors that are unique to the individual.*[4, 5] *The thalamus will only shout if it perceives the injured tissue to be in enough danger.*

Pain—it's all in the mind

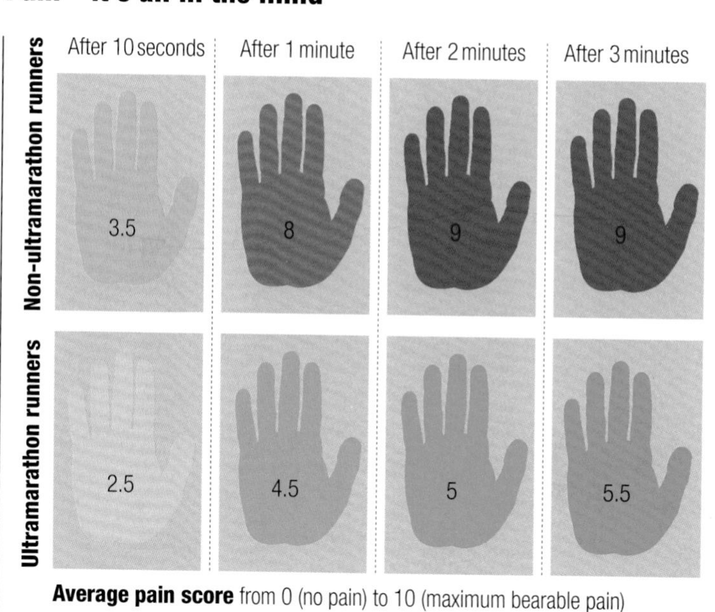

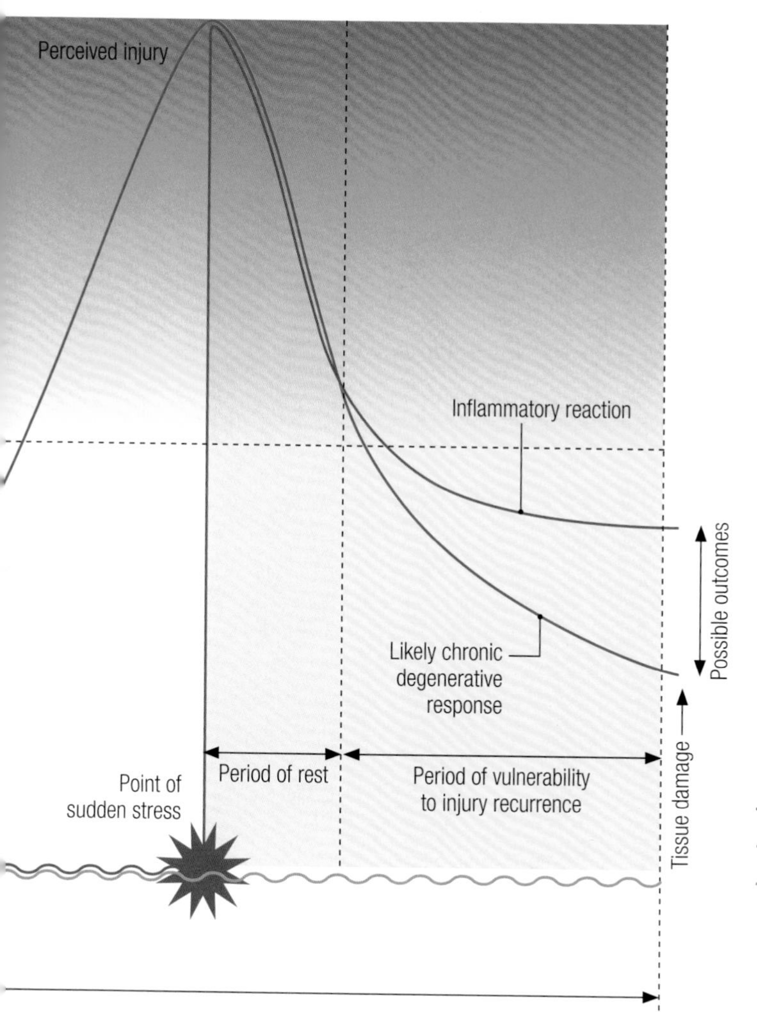

— Path of injury through microtrauma
— Path of injury through macrotrauma
— Path of injury-free training

◀ ***Injury and Inflammation*** *The initial phase in the body's response to injury is inflammation, lasting up to three days. It is important to recognize that inflammation is not a negative event, but a part of healing. Furthermore, inflammation and swelling are not the same. Swelling, or edema, occurs from direct hemorrhaging associated with injury and it reduces blood flow, leading to cell death, if allowed to persist. Therefore, if you were to cautiously jog home having pulled up with a calf strain, potentially you increase the edema and cause further harm. In addition to this, injury inhibits muscle, which can alter your running pattern and can lead to strain elsewhere.*[6]

Where injury has developed over time and involves tissue loading with accumulative microtrauma, the effects of continuing on are trickier to assess. Here, there is a critical limit at which point the load tips the balance from an inflammatory response that promotes positive training effects into overload. Overload interrupts phase two of healing—regeneration and repair—and leads to tissue degeneration.[7, 8] *Therefore, training through pain could result in a chronic condition such as a bony stress fracture or tendinopathy.*[9]

What is the best protocol for dealing with an injury?

I'm injured—should I call the POLICE?

For many, PRICE is a familiar acronym for dealing with the acute, inflammatory stage of an injury, standing for "protection, rest, ice, compression, and elevation." The most commonly used part of this is ice, applied in the form of an ice pack, theoretically to reduce tissue metabolism through cooling, thus slowing tissue damage. However, despite studies showing significant reduction in tissue temperature when ice packs are applied to animals, best evidence suggests that this is not transferable to humans due to complex clinical considerations such as the depth and type of injury.[1, 2]

Neither is complete rest particularly helpful beyond the initial inflammatory period (approximately forty-eight hours). Rest from running may be advisable, but tissues need graded loading within the parameters of pain and function in order to promote the cellular responses required for healing.[3] It's therefore suggested that POLICE be the "acronym du jour," with OL standing for "optimal loading," often achieved using tools such as crutches to enable partial weight bearing. Optimal loading and protection can also be facilitated with tape, used to partially off-load the injured tissue.[4]

It is hypothesized that compression and elevation help to control edema (swelling), thus minimizing cell death and the resulting tissue debris. However, it is recommended that compression and elevation are not used together. Dead cell debris increases tissue protein concentration, which lowers osmotic pressure, pulling water from the blood capillaries and increasing swelling. Swelling causes further cell death and the cycle continues.[5] There is certainly a visible reduction in swelling when compression bandaging is applied. Elevation is believed to ease edema by reducing hydrostatic pressure in the capillaries, although its continuing effect once the limb is lowered again is debatable.[6]

Remember, the aim is to reduce swelling, not inflammation, which is a vital component of healing. The inflammatory process includes the arrival of specialized cells called phagocytes, which clear the area of dead tissue debris and fibroblasts, the cells responsible for generating scar tissue for repair. Therefore, contrary to popular belief, the use of anti-inflammatory drugs (NSAIDs) is not appropriate unless inflammation becomes persistent beyond days three to seven of the acute phase of healing.[7, 8, 9]

▶ ***Mechanotransduction and cryokinetics*** *So—skip the ice? Well, no. Optimal loading is based on the principle of mechanotransduction, the way the cells of the body respond to mechanical stimulus such as exercise. By communicating via an interconnecting network, cells precipitate adaptation or structural change in the soft tissues (muscle, tendon, bone, connective tissue).*[10, 11, 12] *Ice is effective as an analgesic and, 48 hours after injury, can be applied to the injured area to the point of numbness to enable exercises (cryokinetics) to be performed based on these principles. Promoting healthy tissue adaptation through exercise is known as mechanotherapy, a technique supported by high-quality research.*[13, 14, 15, 16] *As a bonus, appropriate exercises performed at the right time, restore range of movement and further assist in reducing swelling.*[17] *"Motion is lotion" as we say.*

Now, here's a thought. Ice baths work on the principle of reducing core temperature and so may be more effective than topical ice packs. They give pain relief through numbing and, if the water is deep enough, they also tick the compression box through the mechanism of increased hydrostatic pressure. The buoyancy of the water promotes optimal loading since exercises can be performed with reduced weight through the injured limb.

POLICE—dealing with the first phase of injury

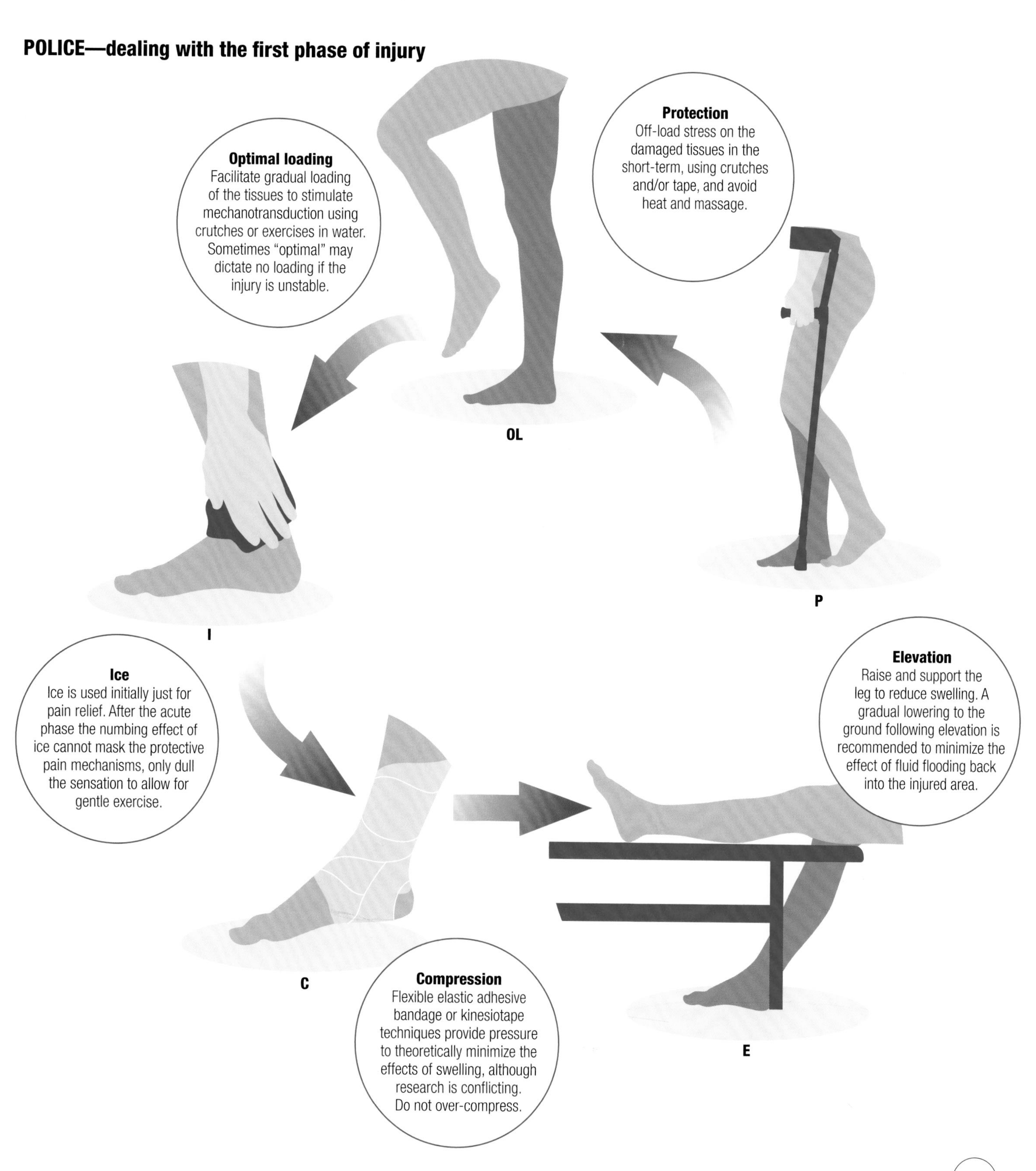

SCIENCE IN ACTION

can running injuries be prevented?

Estimates of incidences where runners have been injured vary, however, it is thought to be between 37% and 56%—high enough to be a significant issue. Screening, a tool designed to assess intrinsic factors such as foot posture, alignment, and muscle control to predict and then correct "faults" that may leave an athlete vulnerable to injury, is popular in the professional sports community. But how useful are these tests?

Many trainers rely on the Functional Movement Screen (FMS), a series of seven tests. This assessment was successfully used for footballers and soccer players[1, 2, 3] and is now widely adopted by the sporting community. However, research has found that when applied to competitive runners, the FMS composite score is a poor predictor of injury. What is useful, however, is that two components of the test—the deep squat and active straight leg raise—were found to accurately assess the likelihood of injury.[4]

Studies into patellofemoral pain syndrome (PFPS), the most common of running injury, have identified some key risk factors such as hip muscle weakness and iliotibial band tension.[5] Research into the assessment of core stability has also been shown to be useful, as has the assessment of foot posture as a small, but significant, predictor of injury.[6]

Another recent study found that preventative methods, such as strength training, reduce the incidence of injury to a third of the predicted rate, and proprioceptive training reduced them to a half,[7] suggesting that identifying key strength and proprioceptive deficits in runners could be useful for screening. Gait analysis has also been shown to be useful when the results are used intelligently to work on improving running technique.[8]

To date, there is no tried and tested injury prevention screen for runners. Injury occurs when the cumulative load on the tissues exceeds what they can cope with. There are different risk factors for each injury and each individual runner—there lies the challenge in developing screening that works for runners as a "one test fits all."

▶ ***Under observation*** *Screening can be a useful tool to help identify some possible faults and weaknesses that may lead to injury in runners, but there are limitations to what can be identified in the lab.*

By what mechanisms could use of a foam roller improve running performance?

Will a foam roller make me a better runner?

Foam rollers first appeared in the late 1990s and now have an extensive following in the athletic community. Their popularity is mainly attributed to their use for self-therapy, to release tight soft tissues and "trigger points"—localized, tender knots within muscle and fascia (connective tissue surrounding muscles and organs). They are also used to relieve delayed-onset muscle soreness (DOMS).[1, 2, 3]

Rollers can be made from molded foam or polystyrene. Foam is harder and tougher, while polystyrene rollers are cheaper but have a shorter life, as they deform with use. They come in different sizes, as a full or half cylinder, and are light and portable, making them useful for the traveling runner. Some more expensive, multilevel, rigid rollers have been shown to have favorable results in comparative research.[4] Foam rollers are also used as a tool to improve proprioception, training the body's sense of position, motion, and equilibrium to develop balance and coordination, which is beneficial for injury prevention and rehabilitation.[5, 6]

The main use of the foam roller is to supplement or replace hands-on manual therapy in releasing tight fascia. Fascia is a three-dimensional matrix of connective tissue, surrounding and integrating every system in the body. In muscle, it can become deformed or bound-down through overuse or injury, forming adhesions. This can theoretically be resolved by applying pressure to the tissue, changing the formation of fibroblasts (cells which make up fascia) such that the integrity of the tissue is improved.[7, 8, 9] Critics say that the pressure needed to alter adhesions is torturous and that rolling over a piece of foam is not sufficient to achieve this effect. However, those that have tried this on a tight iliotibial band will testify that the sensation is, if not torturous, pretty unpleasant. Also, there is an emerging body of evidence showing that fascia is such a responsive system that even light touch can influence it.[10] There is a question mark over whether foam rolling is specific enough, and it does seem to be impermanent—there is a need to repeat the treatment frequently to maintain the effects. In theory, if you continue running with a faulty movement pattern, fibroblasts may just reform in the old, defective configuration.[11, 12, 13] Fascia experts advocate following up release work with exercises that utilize the improved movements achieved through treatment.[14] Theoretically, this will establish better habits and achieve longer-term results.

Destabilizing tool

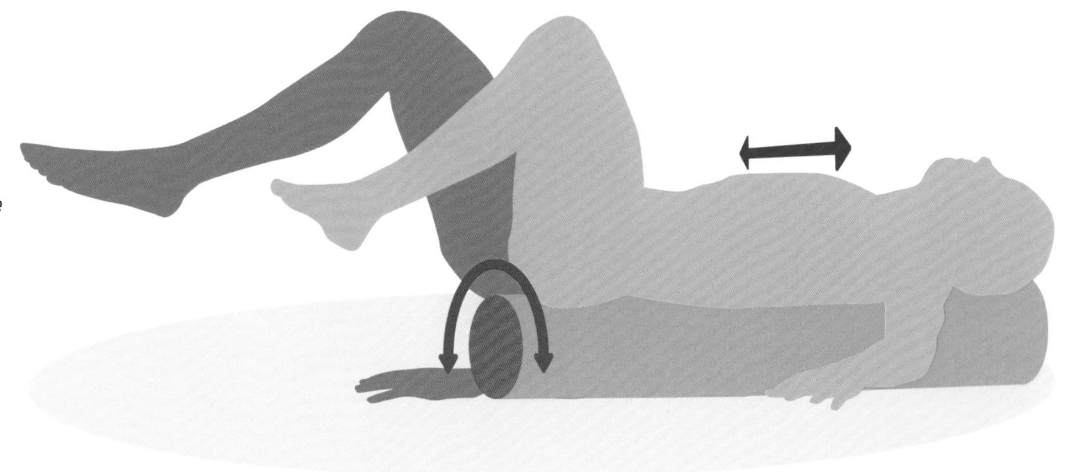

▶ ***Improved reactions*** *The roller can be used as a destabilizing tool to train the body to react more quickly and efficiently to external factors that may upset balance, for example, running on uneven terrain. There is some evidence citation to suggest that a proprioceptive training regime has a role in injury prevention (although this research is not specific to runners, there is reason to assume it has crossover)*[15, 16, 17, 18] *and also in rehabilitation of injuries that result in instability, such as ankle sprains.*[19, 20]

Proprioceptive tool

▶ ***Soft tissue work*** *As a proprioceptive tool, foam rolling is particularly interesting. The majority of receptors responsible for proprioception are found in the subcutaneous fascia. These receptors are stimulated via pressure from the foam roller, directly influencing muscle activity and tone, and changing the relationships between tissues. Soft tissue work therefore not only effects the alignment of the fascia but also the sensory perception of movement, achieving an immediate change in movement quality.*[21, 22]

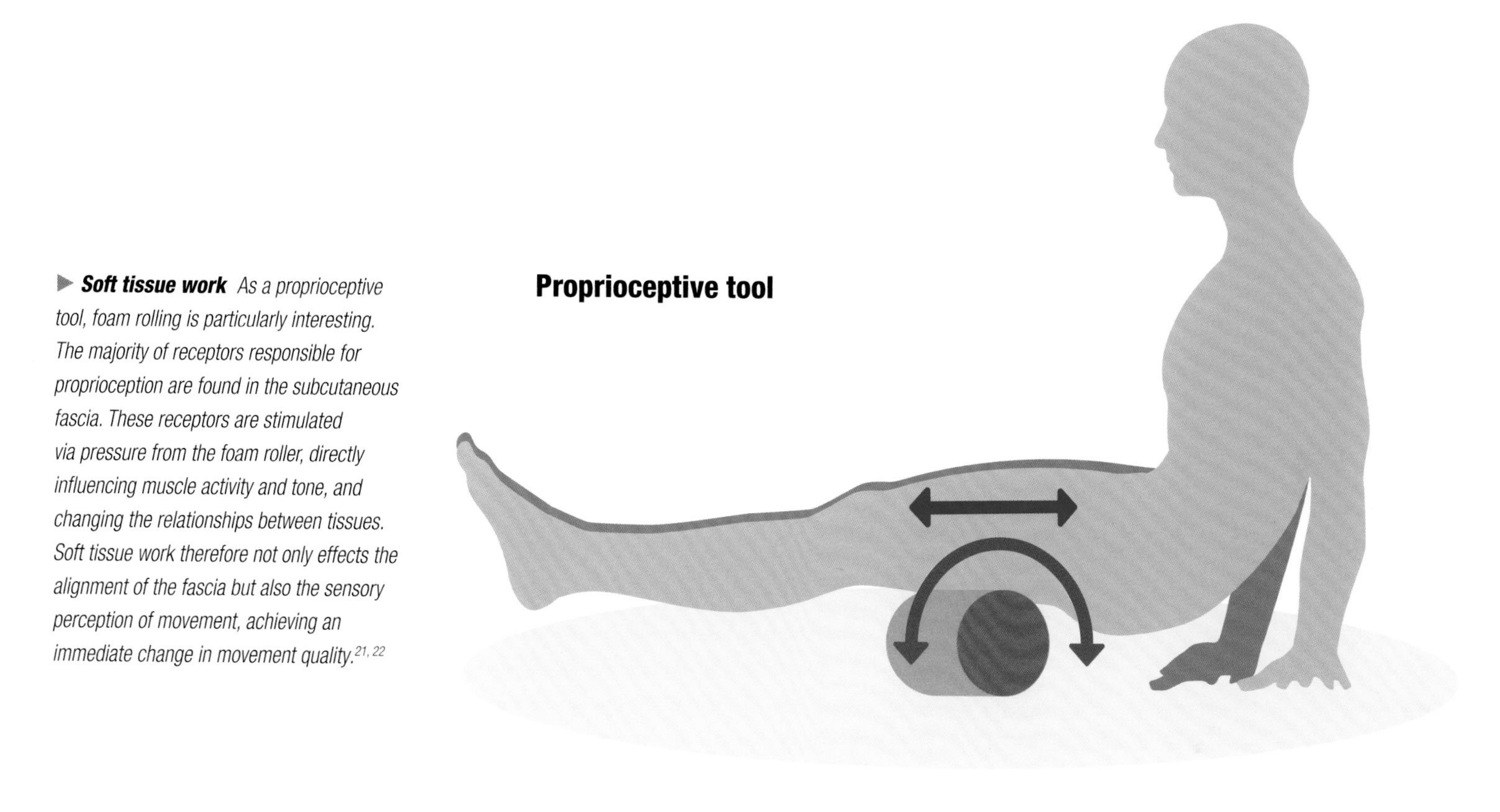

A tool to increase flexibility and reduce DOMS

▼ ***Positive research*** *Research on the hamstrings and quadriceps supports the use of foam rolling over dynamic and static stretches to improve acute flexibility.*[23, 24, 25] *The advantage is that, unlike static stretching, foam rolling does not have a detrimental effect on performance when used prerun.*[26] *It also has a positive effect on reducing the postrun/event symptoms of DOMS, although this does not aid subsequent performance.*[27, 28]

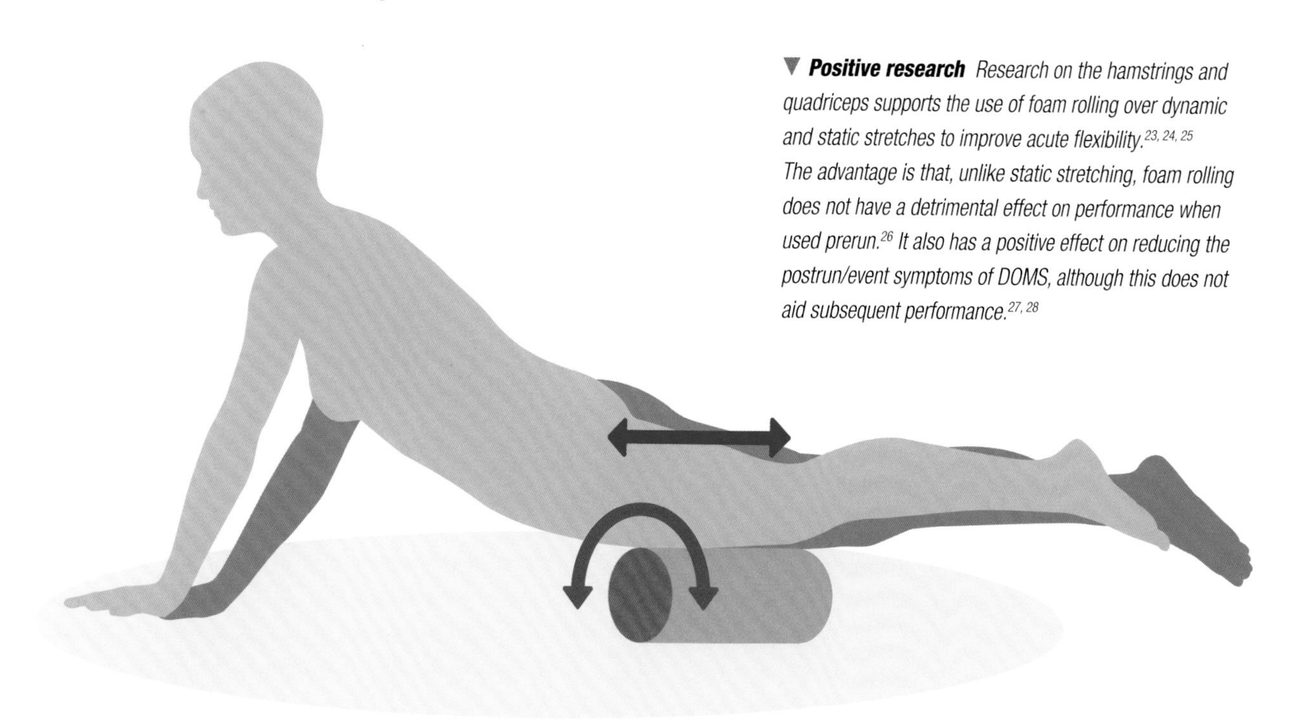

events:
ultras

For runners wanting to explore events beyond the marathon, the world of ultra running awaits. Distances are officially longer than 26.2 miles (42 km), with 50 and 100 miles (80 and 161 km) being popular, but by no means the longest. Runners are drawn to these events partly because of their love of training, which often means they've logged thousands of miles, but also because of the dedication required.

Ultra running does require good planning, great support, superb equipment, and, of course, great hydration and an ability to take on food while moving. It's very much about the

Hydration methods

total package and nothing can be overlooked. In non-ultra running, while top athletes follow a planned training program, amateur runners may be more inclined to ah-hoc training. This isn't the case for ultra runners, who, at every level, must have the discipline to follow a strict training schedule, not least because following a carefully designed program will increase a runner's resistance to injury, Interestingly, most ultra runners love the event because of the variation it creates. Long runs, strength work in the gym to improve posture and core strength, interval training, and, in particular, just pure and simple hours spent on their feet are all important. While so much of running is about numbers, specific distances, and an unrelenting pace, training for ultras is about getting out and spending the day running and walking, teaching your mind and body how to deal with continually moving for eight hours or more.

▼ ***Hydration systems*** *There are many different ways of staying hydrated when running. The choice is up to the individual just as long as their chosen method allows the runner to stay hydrated throughout their run. (**A**) For road races organized under IAAF rules, water will be available at least every 3 miles (5 km), which will save competitors from carrying their own water unless they choose to. However, for other events such as trail running, and when training, runners will have to take their own supplies. (**B**) For shorter events or training sessions, a single bottle may suffice. You can buy grips that make it easier to keep hold of the bottle as well as keeping it cool and offering storage for small items such as keys. (**C**) For longer events, particularly ultras, you'll need more than a single bottle. Some runners use a hydration belt that typically allows them to carry multiple bottles around their waist as well as provides storage space for larger items such as a phone. (**D**) An alternative is to use a backpack or vest system. Some just provide an alternative way of carrying bottles and personal items. Others dispense with separate bottles and have a reservoir or bladder with a connected tube to allow you to hydrate with minimal effect on your step length or frequency.*

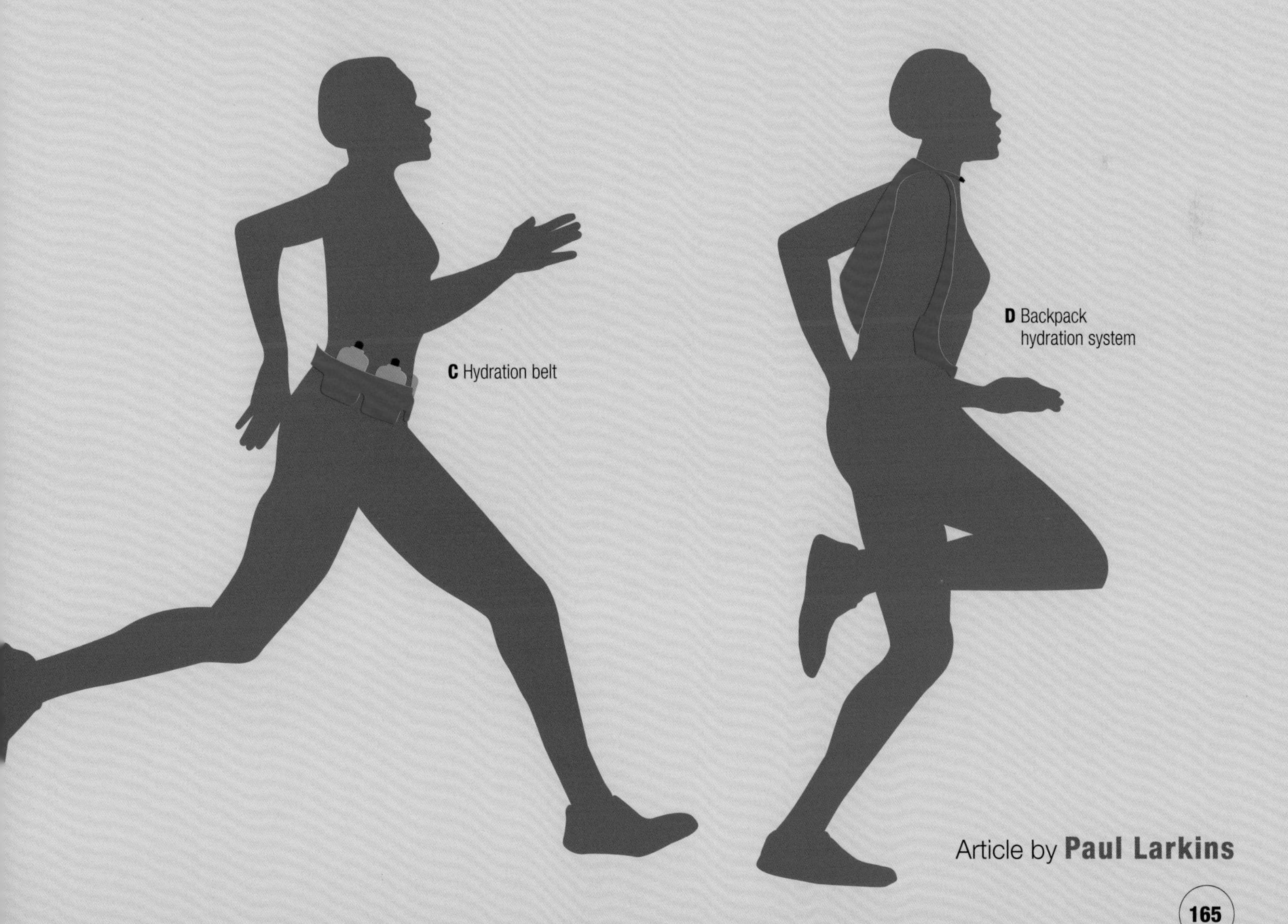

Article by **Paul Larkins**

Every runner has a reason for putting on their running shoes to train or race. For some it is simply to improve fitness, for others it is all about performance. But whatever the goal there is an underlying expectation that running will deliver the expected returns, perhaps through faster performance times, or possibly through increased health and longevity. Some of the undoubted benefits of running have been clearly documented in scientific studies, backed by data and statistical probability. But despite this growth in scientific data, there are still many questions that remain unanswered. Just how fast can the human body run? Does running really extend life expectancy? Can we ever expect to see the very best female runners defeating their male counterparts? For every question that science answers, another is raised. When compared with the amount of time that human beings have walked and run on the Earth, the sport of running is still in its infancy—what times will humans be capable of achieving for distances from 100 m to 26.2 miles in a hundred, thousand or million years' time? These are questions that only time, with some help from science, will be able to answer.

chapter eight

the big questions

John Brewer

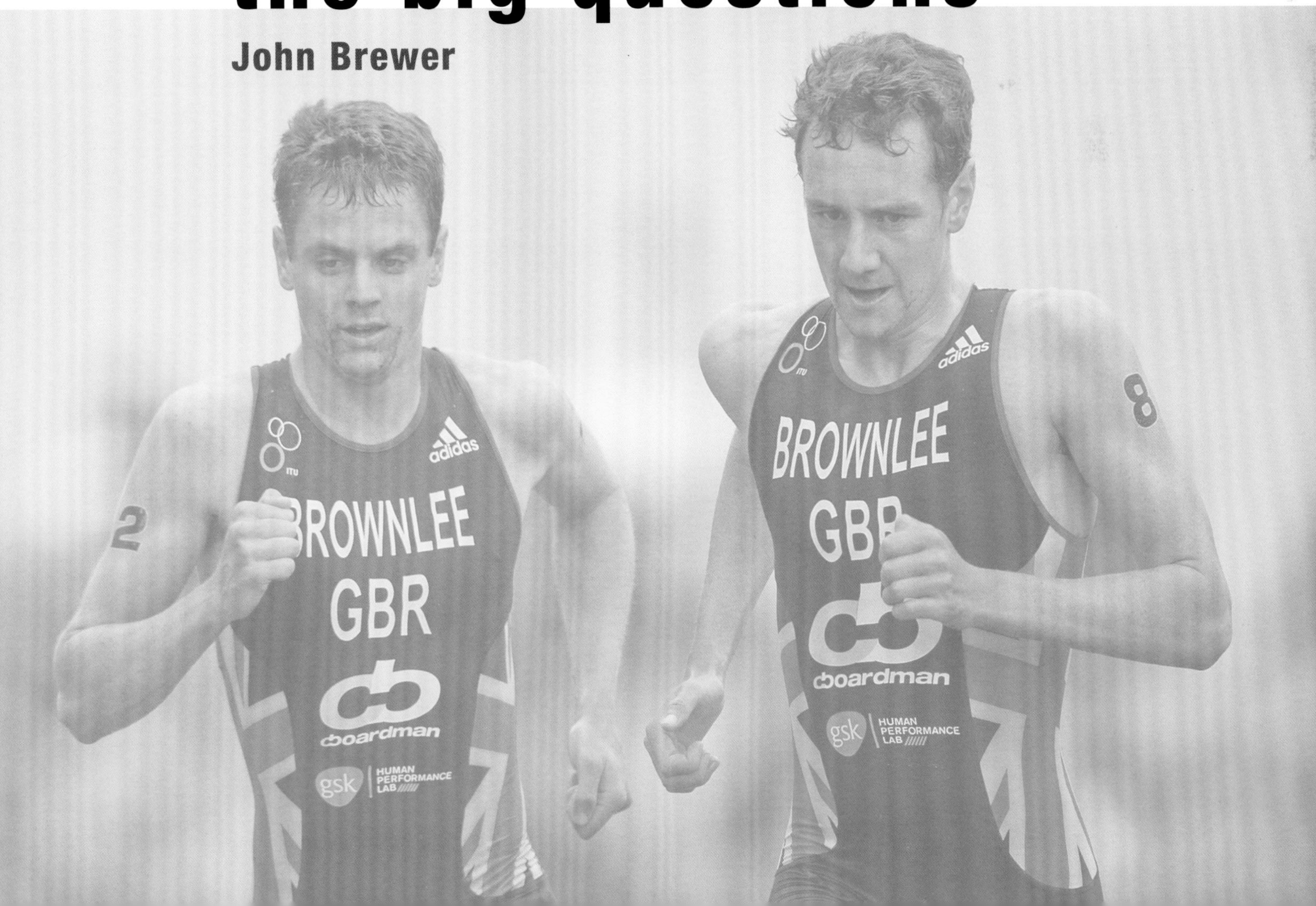

How much of an athlete's running ability is genetically determined?

Does running run in the family?

It is often said that if you want to be a world or Olympic champion, you have to select your parents carefully. This reflects the fact that, at the very highest level, our genetic characteristics are a prerequisite for success. Having the right physique, an appropriate distribution of muscle fiber types (slow-twitch fibers for endurance runners, fast-twitch for sprinters) and the capacity for high rates of oxygen uptake are all essential if you are going to be the very best.

But simply being born with the right genes is not enough—nature has to be underpinned with nurture, through great coaching, dedication, and a lifestyle that is conducive to excellent performances. There also needs to be great support from sport science, medicine, and nutrition—areas that on their own may have a small impact but, when combined, help to make the marginal difference between success and failure at the highest levels.

Although the vast majority of runners do not have the genetic characteristics needed to enable them to become Olympic champions, that should not deter anyone from training to become the best that they can be, and at the same time performing very well. The human body evolved to run—running was an essential part of our "hunter gatherer" lifestyles from many thousands of years ago, and with training and application, we can all develop the capacity to run, and to run successfully. Some of us will always be better than others, but in the domain of "recreational" rather than elite running, our genetics are less of an issue, and training, dedication, and application are more important.

The probability of being the perfect endurance runner

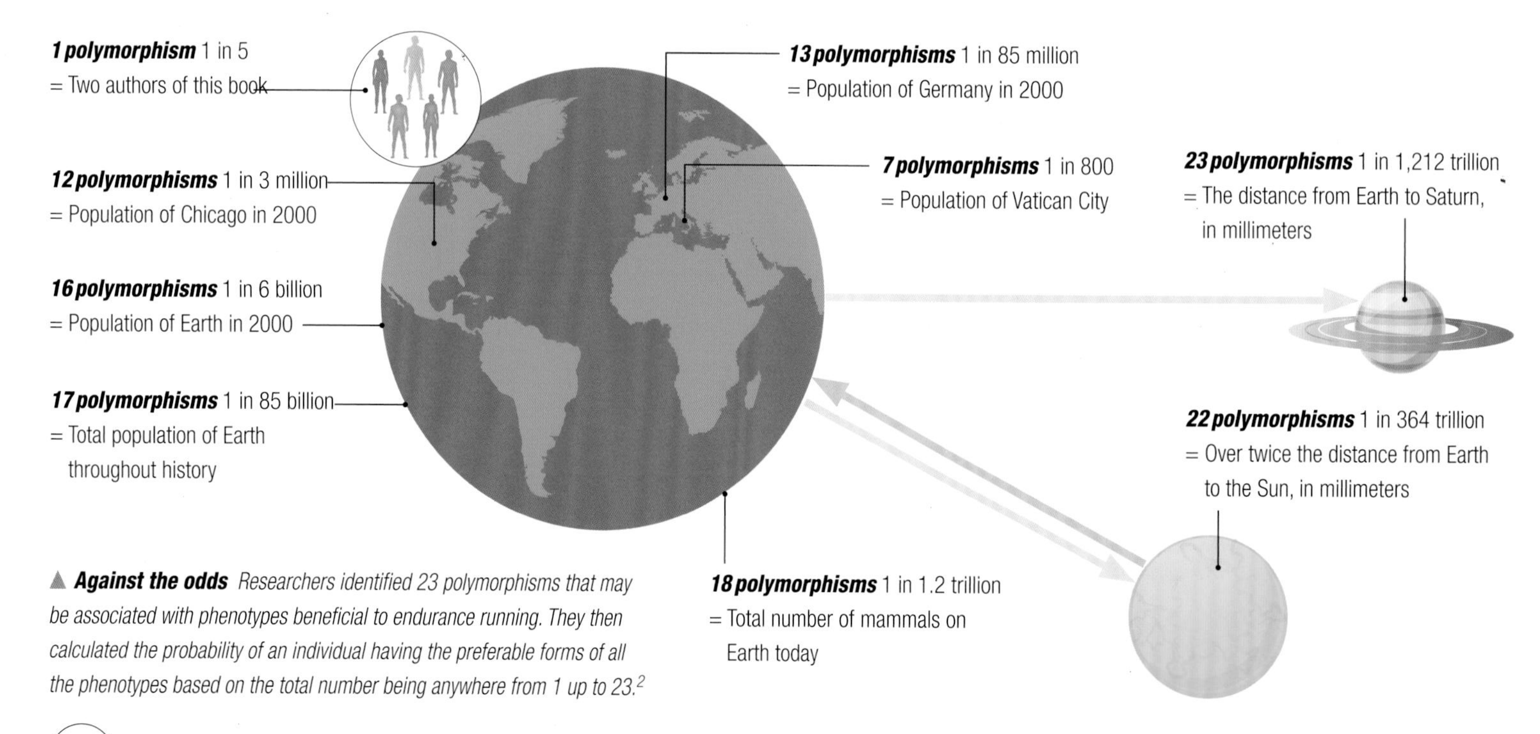

▲ **Against the odds** *Researchers identified 23 polymorphisms that may be associated with phenotypes beneficial to endurance running. They then calculated the probability of an individual having the preferable forms of all the phenotypes based on the total number being anywhere from 1 up to 23.*[2]

Optimizing the capacity that we have has more impact than the genetic starting point, and while recent scientific studies have suggested that some individuals are genetically more likely to "respond" to specific types of training than others,[1] for the vast majority of runners, genetics should not be seen as a barrier to adaptation and improved running performance.

▼ ***Getting to the top*** *While having genetic characteristics that are advantageous to running will separate the very elite runners from the rest, it is only just one piece of the jigsaw for the vast majority. Environmental factors, from having a supportive family and coaches to the money to pay for equipment or gym membership, and the dedication to train to maximize performance, are more important for the majority of runners, and factors such as not being spotted or not having the right environment or work ethic can prevent even those with a genetic advantage from fulfilling their potential.*

Maximizing your potential

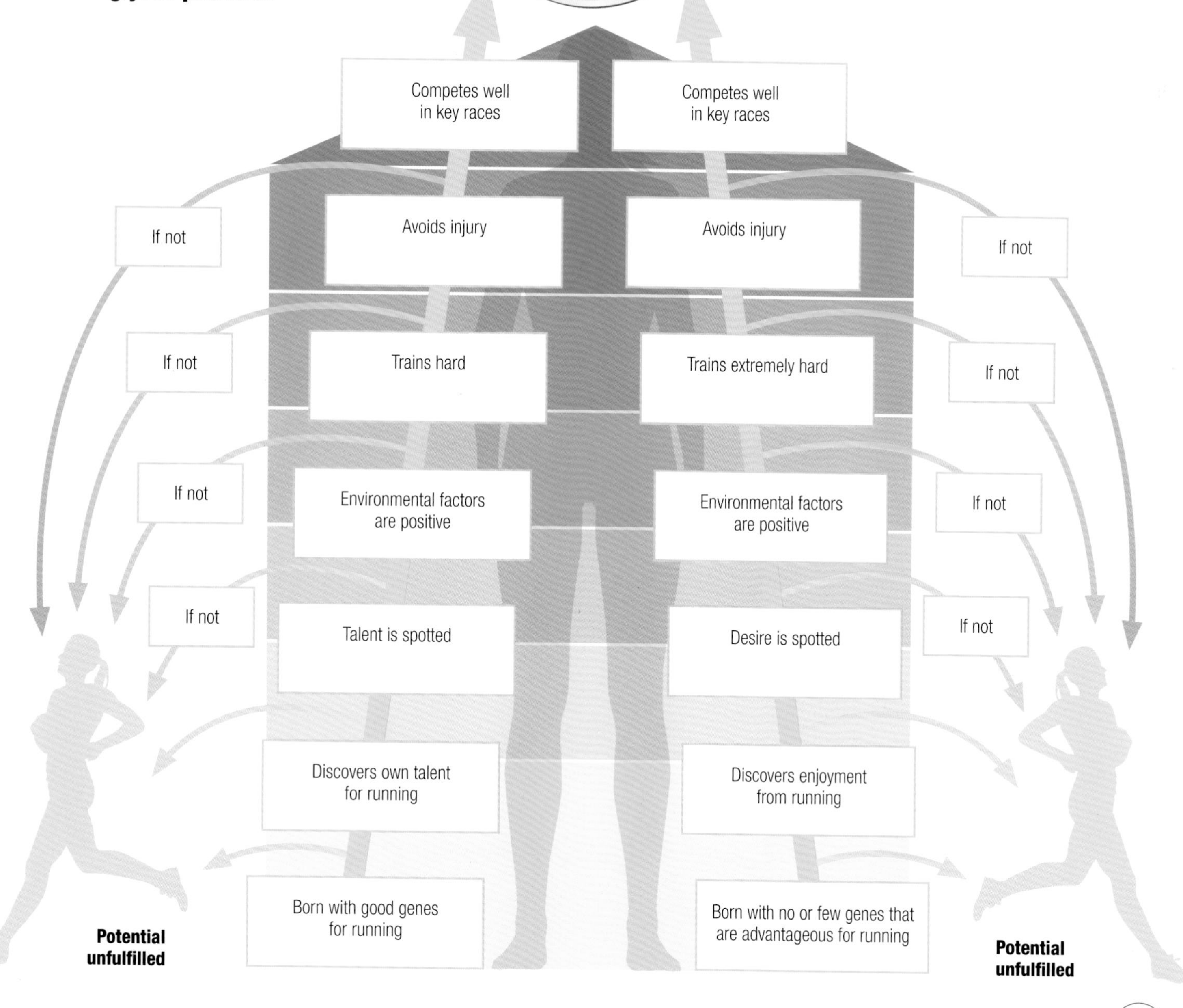

Are there physical limits to human marathon-running achievement?

Will there ever be a sub-two-hour marathon?

Ever since Roger Bannister ran a mile in under four minutes in 1954 runners have aspired to breaking running's next barrier—the sub-two-hour marathon. Some have questioned whether this will ever be possible, but the facts—and science—suggest that it is more a case of "when" than "if." While 5,000 m and 10,000 m times have plateaued since the mid-2000s, marathon speeds continue to increase.[1]

Completing 26.2 miles (42.2 km) in 1 hour 59 minutes 59 seconds equates to a pace of 4 minutes 34.8 seconds for each mile, or a speed of fractionally under 13.1 mph (5.86 m/s). We know from scientific studies that most elite runners can sustain a pace equivalent to around 80% of their maximum oxygen uptake (VO_2 max) during a marathon, and that running economy—the amount of oxygen used at any running speed—equates to approximately 5 ml/kg/min for each mile per hour of running speed. So on that basis, the oxygen uptake needed to run at 13.1 mph is approximately 65.5 ml/kg/min, and this in turns needs to represent 80% of VO_2 max, which elite runners can sustain for 26.2 miles. This therefore means that a VO_2 max of around 82 ml/kg/min would be required for an elite runner aiming to run a marathon in under two hours. While this is undoubtedly a high value, it is not unusual to see values of this kind when elite runners are assessed under laboratory conditions.

Men's marathon world record progression

▶ ***Marathon progression*** *Over the past century, the marathon record has been broken almost forty times, lowering it by over fifty minutes. However, forty of those minutes were knocked off in the first fifty years. Since the turn of the millennium the record has been lowered by 2 minutes 45 seconds.*

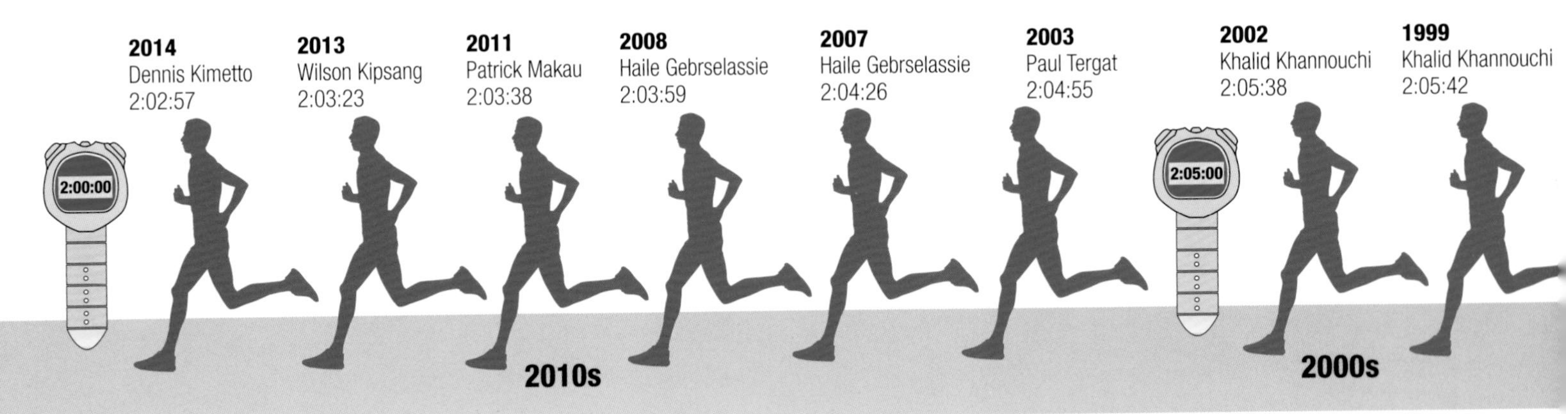

This suggests that the sub-two-hour marathon is feasible, but will certainly require an extraordinary run from a very talented individual, under conditions that are ideally suited to marathon running. The course will need to be flat with few sharp bends, at sea level, and with a climate that is cool enough not to over-stress the body's thermoregulation system, with no breeze to create resistance. Pacemakers will probably be needed to ensure the required speed is reached from the outset, and the successful runner will need to have combined "peaking" of their training with optimum nutrition, hydration, and, of course, a mental toughness that will help them to achieve the sub-two-hour goal.

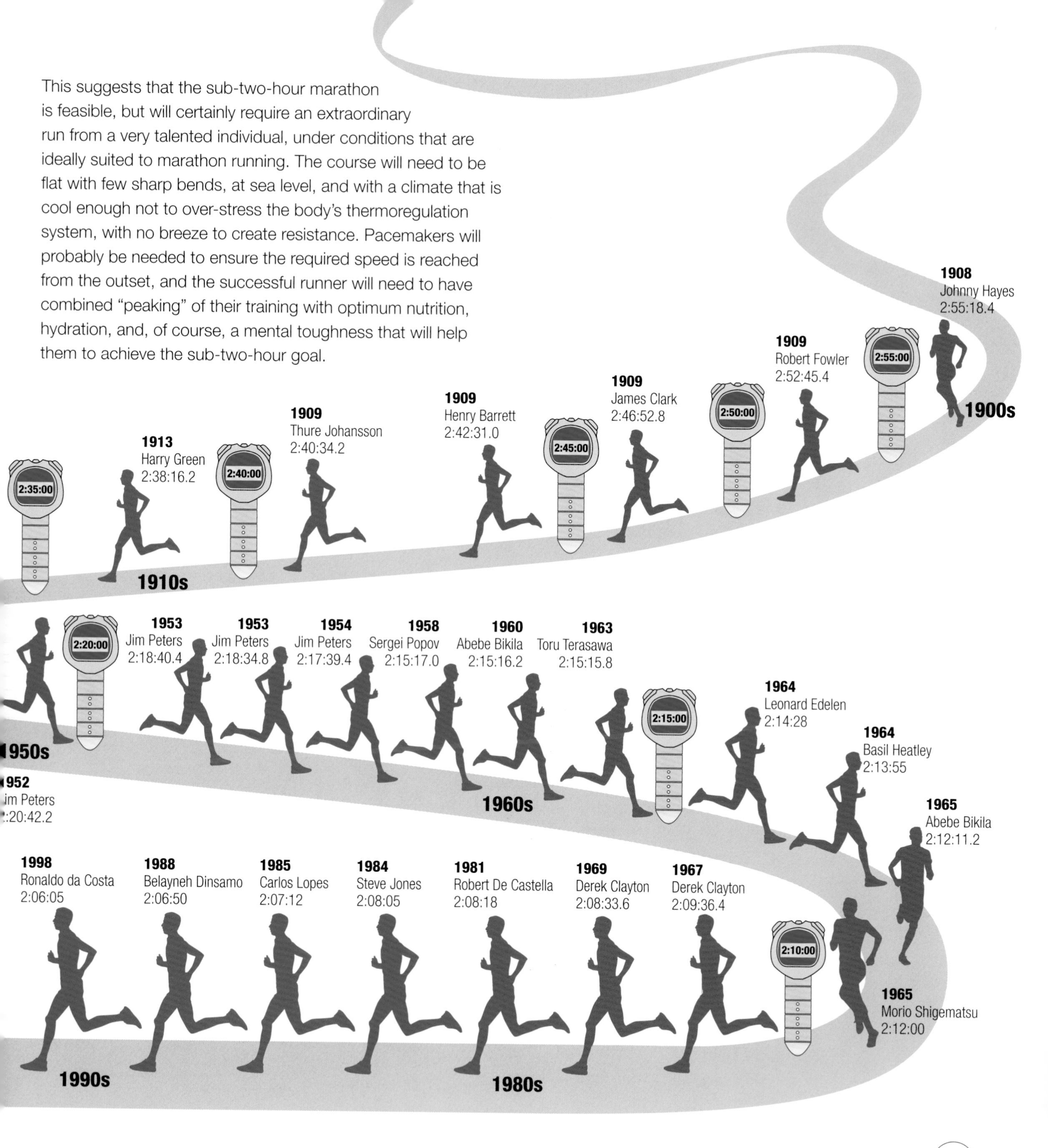

Is there a distance beyond which male and female runners are equal?

Will the best women ever beat the best men?

Scientific studies have shown that running long distances has to be at a lower intensity than when running shorter distances, otherwise fatigue quickly occurs. Scientists have also found that at lower intensities more fat is used as fuel (as opposed to carbohydrate) when compared with higher running intensities. We know that for genetic reasons females generally tend have a higher body fat percentage, and more body fat, than males. When these facts are considered alongside analysis of world-record times for the marathon, which shows that over the last 40 years the female record has steadily closed the gap on male best times, it is easy to see why it is possible to draw the conclusion that females may be better suited to long-distance running than males.

However, closer examination of the data and the science suggests that while good female runners will always be capable of beating many of their male counterparts, human physiology dictates that males will always have an advantage over females, regardless of distance. Even the leanest of males has enough calories from body fat reserves to fuel low-intensity, long-distance runs, and the greater lung volumes, hemoglobin concentrations, and muscle mass of males all result in a performance advantage. Some studies have suggested that females may have an advantage over males in warm conditions, since, at any given body weight, females have a slightly larger surface area than males. This can aid heat loss, but it is unlikely to result in significant performance gains.

One of the main reasons for the closing gap in world record times is the relatively late adoption of distance running among female athletes. The first female Olympic Games marathon was as recent as 1984, and this acceptance of females into major championships, and the growth of marathon running in general, has encouraged more female runners to take up, and excel at, endurance running. Times have consequently improved at a rapid rate, although in recent years this improvement has started to slow, and it is unlikely that the very best female times will ever beat those of males, regardless of distance.[1]

Women's and men's marathon record progression

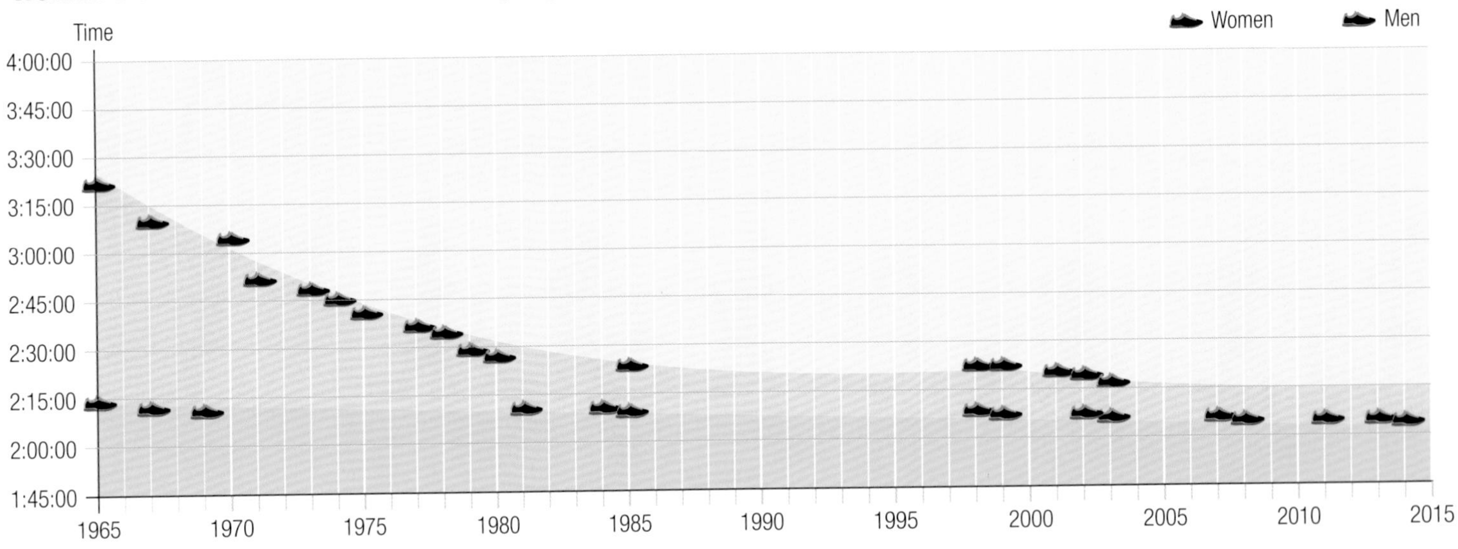

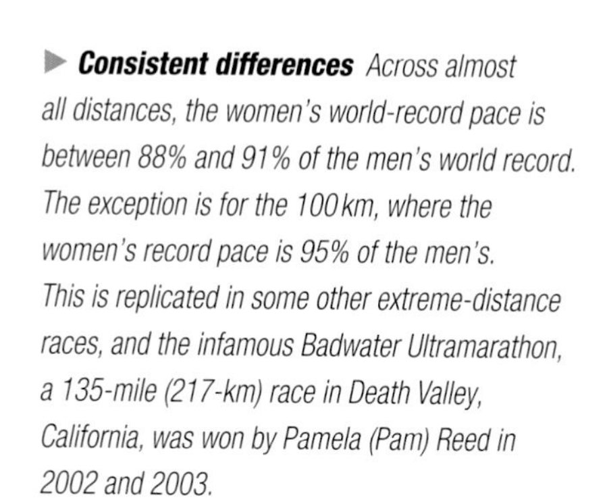

▶ ***Consistent differences*** *Across almost all distances, the women's world-record pace is between 88% and 91% of the men's world record. The exception is for the 100 km, where the women's record pace is 95% of the men's. This is replicated in some other extreme-distance races, and the infamous Badwater Ultramarathon, a 135-mile (217-km) race in Death Valley, California, was won by Pamela (Pam) Reed in 2002 and 2003.*

◀ ***Marathon records compared*** *The women's marathon record time improved dramatically following official recording of women's times in the 1960s, but while it was still closing in on the men's record at the start of the 1980s, it has not been improved upon since 2003.*

Women's record times related to men's

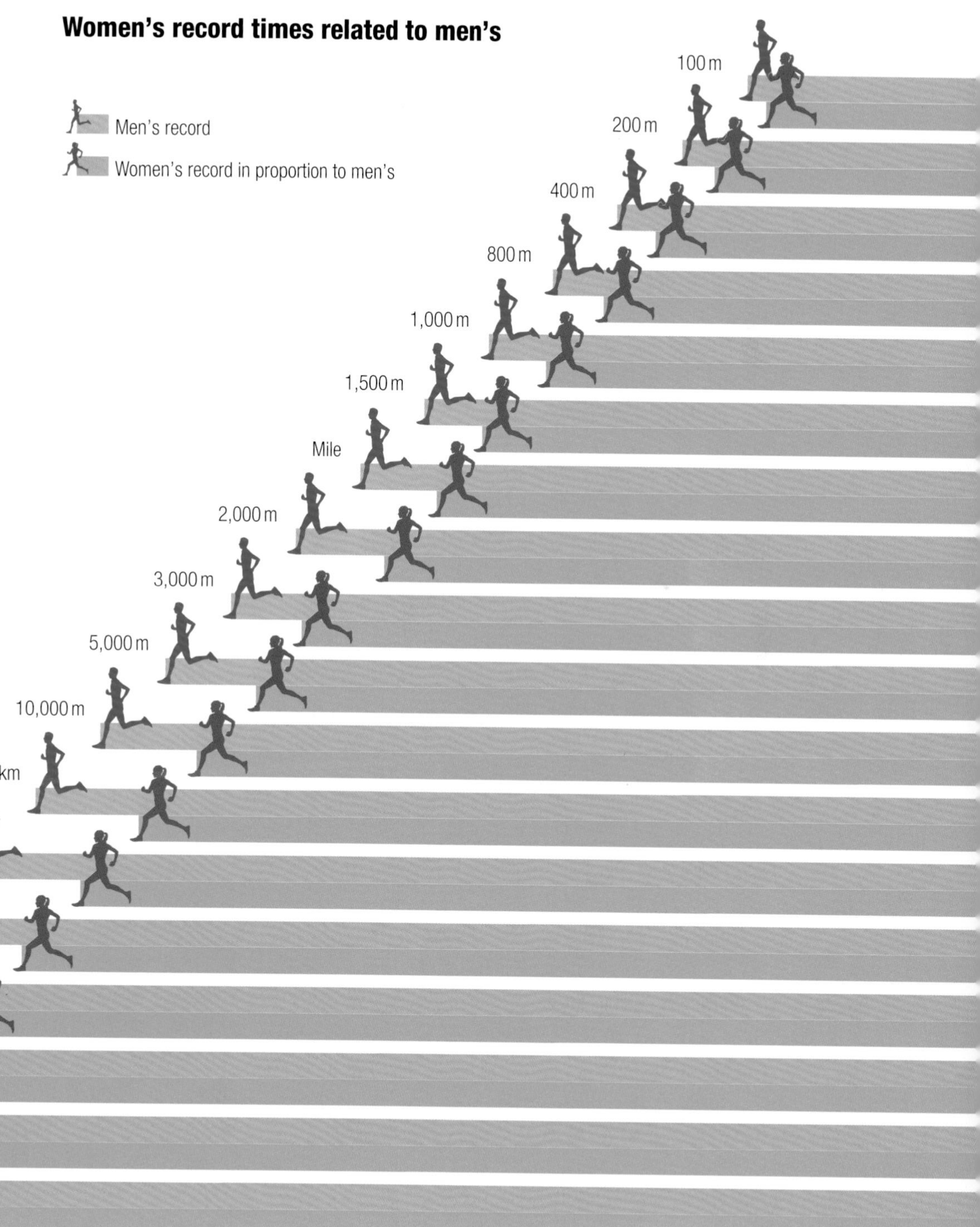

Is there evidence that running increases life expectancy?

Should I run for my life?

Numerous scientific studies over many years have shown that regular exercise such as running results in major benefits to health. When compared with sedentary individuals, runners tend to have lower body fat percentages, lower cholesterol, and lower blood pressure. We have also seen that lung capacity and the ability to transport oxygen to the muscles improves, and the heart—the body's most important muscle—becomes a more efficient pump. Muscle mass increases, making many daily tasks easier, and more dense capillary networks supply blood to the muscles. People who run also tend to have a more positive outlook on life—they are generally happy with their body image, and the regular release of endorphins, the body's natural opiate, supports a better sense of wellbeing.[1]

But does this mean that runners live longer than non-runners? This is a question that is very difficult to answer, not least because of the problems involved with conducting such a study. The anecdotal evidence suggests that this is the case—as a result of many of the benefits from exercise, we know that people who run regularly are almost certainly at a reduced risk of many life-threatening illnesses, including certain cancers, obesity, diabetes, and strokes. Unfortunately, when there are fatalities involving runners these hit the headlines, along with suggestions that running is harmful. Sad though these occurrences are, at some point they will be inevitable, especially in mass participation events where large numbers of people are running for a significant period of time. We should not overlook the fact that for every tragic fatality, there are many thousands—if not hundreds of thousands—of people who have gained huge health benefits from running that will have prevented, rather than created, life-threatening conditions. The Stanford Running Study suggested that longevity will be improved through running, after a study on 500 runners showed that runners had a reduced risk of cardio-vascular related deaths as well as reduced risk of death from some cancers, neurological diseases, and infections.[2]

While it is not possible to state definitively that runners will live longer than non-runners, there is every indication than running leads to a far better quality of life, and the capacity to retain our independence as we age. Running is an activity that can be continued for many years, and long into old age, enhancing physical and mental performance far better than any drug or medication.

Relative risk of developing cardiovascular disease

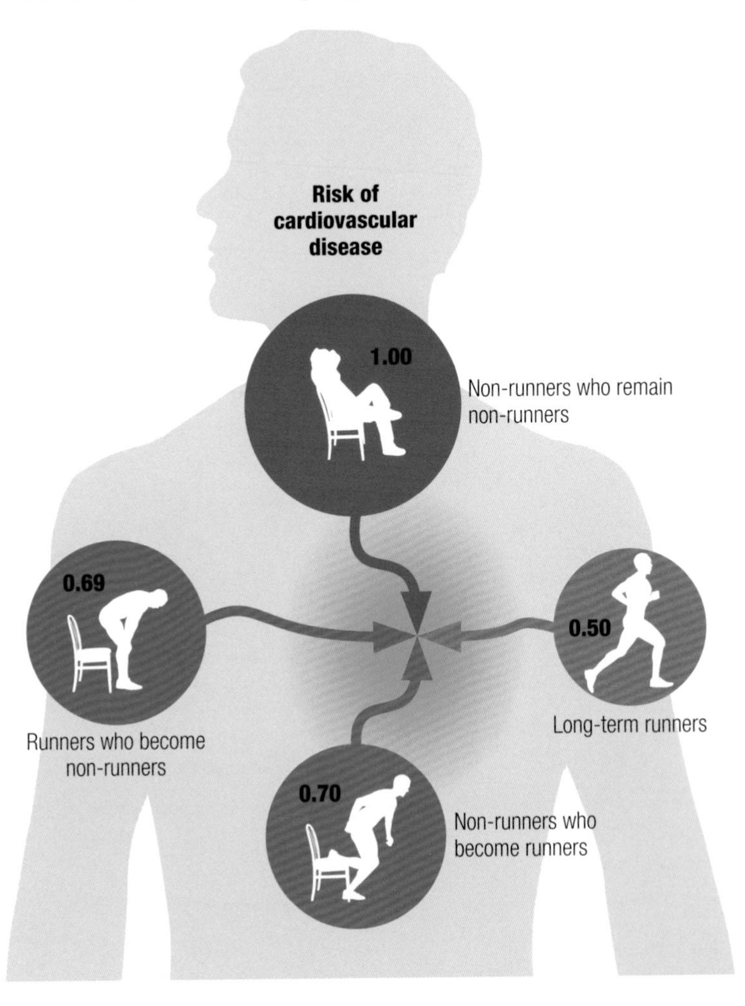

Is there a health risk from running too much?

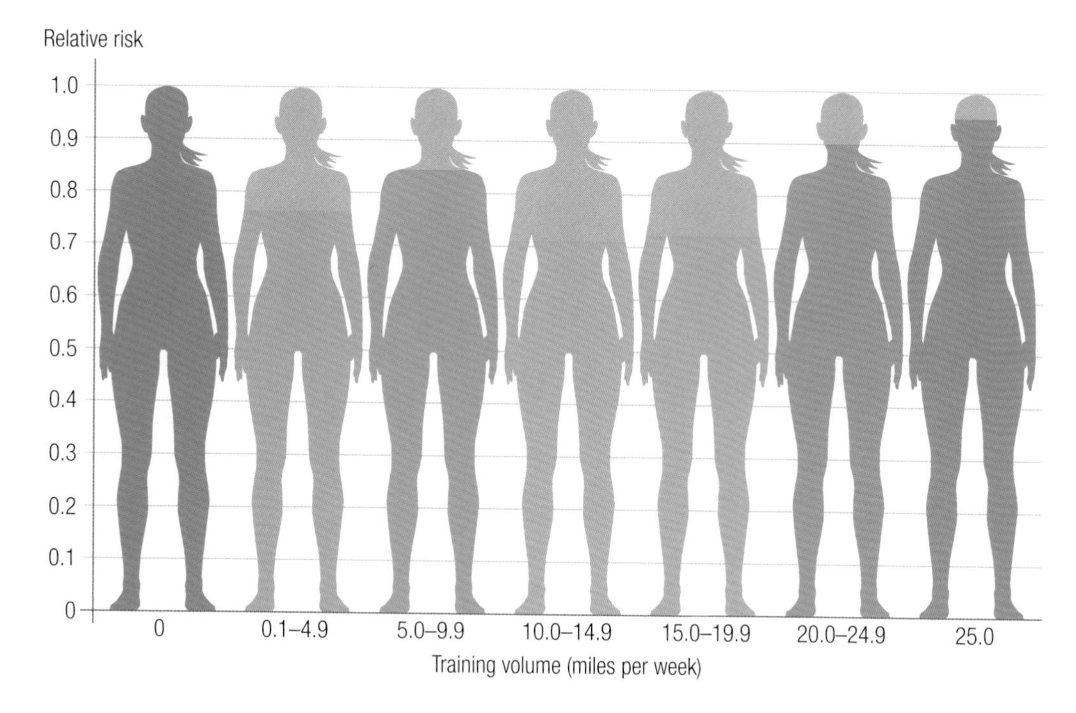

◀ ***Too much of a good thing?*** *That being fit is better than not being fit in terms of mortality risk is not in dispute, but what may seem more surprising is that there's much disagreement about whether it's counterproductive in terms of health to perform too much exercise. A survey of 52,600 runners and non-runners found that those who ran over 20 miles per week lost most of the reduced risk of death that runners had over non-runners.*[4]

▼ ***Keep on running?*** *In contrast, an analysis of data from over 660,000 men and women found that while there was a 20% benefit in terms of mortality risk from performing the recommended level of 7.5 metabolic-equivalent hours of exercise per week, more regular exercise of up to three to five times the recommended level lowered the risk by 39%. The risk increased at over ten times the recommended level of exercise, but it was still 31% lower than for the risk for the sedentary population.*[5]

◀ ***Run for your life*** *An analysis of over 55,000 subjects over a fifteen-year period revealed that long-term runners were half as likely to die from cardiovascular disease (CVD) than non-runners. Those who started as non-runners but took up running reduced their chance of dying from CVD to 70% of the previous risk, an almost identical risk to those who gave up running.*[3]

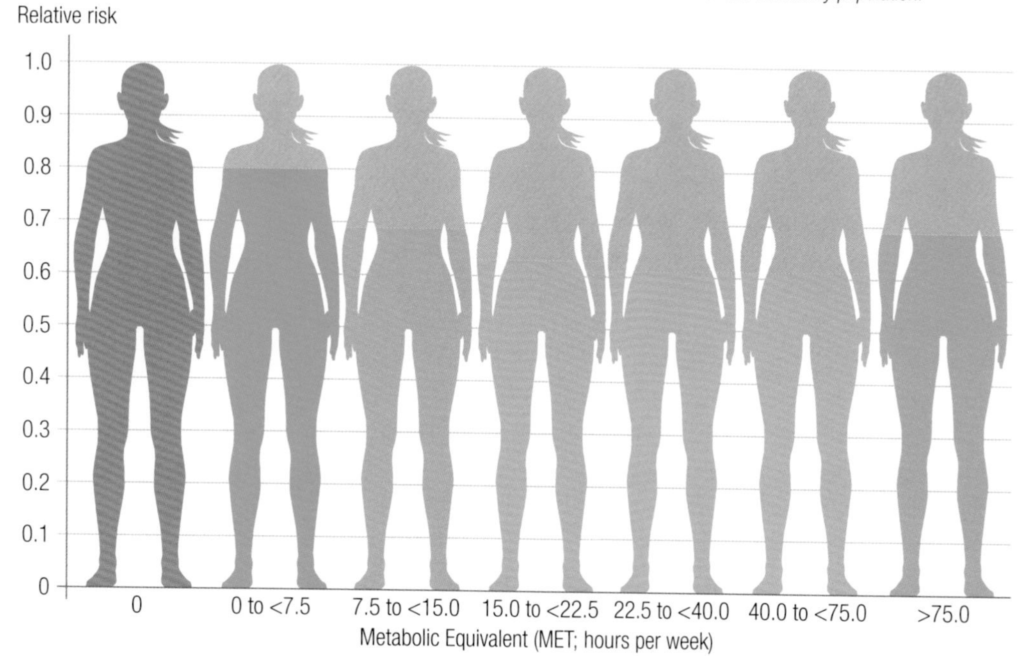

events:
relays

Beating personal goals and achieving individual success is important in running, but nothing is quite so rewarding and exciting as a relay. That could mean a 4x100 m or a 4x400 m race on the track, both of which feature in the major championships, but there are numerous other combinations on the track, road, and grass that are every bit as exciting.

On the road, relays often cover hundreds of miles and can be contested by big teams of a dozen or more runners. Some go from one landmark to another, others race along coastal paths or perhaps have the teams race the sun from sunrise to sunset such as the 93-mile (150-km) relay around Mont Blanc on the longest day.

Running clubs love the camaraderie of the 12-person road relay in Britain, which alternates 3- and 5-mile legs in a race that takes more than four hours, while every year in Portland, Oregon, hundreds of teams take part in 12-person, 195-mile (314-km) Hood to Coast relay hoping to beat the magical twenty-four hours.

Back on the track, it's about slick baton changes and great teamwork, which is perhaps why it's so popular with American college teams that compete in diverse events such as the sprint medley (200 m, 200 m, 400 m, 800 m), distance medley (1,200 m, 400 m, 800 m, 1,600 m), 4x1,500 m, and 4x800 m.

The method for timing any race varies with its nature, but we have moved a long way from a judge with a stopwatch and a tape across the finish line. Long-distance races may rely on the competitors wearing transponders that can be tracked around the course. On the track, where more accuracy and perhaps a photo-finish may be required, the favored system is Fully Automatic Time.

Fully automatic timing

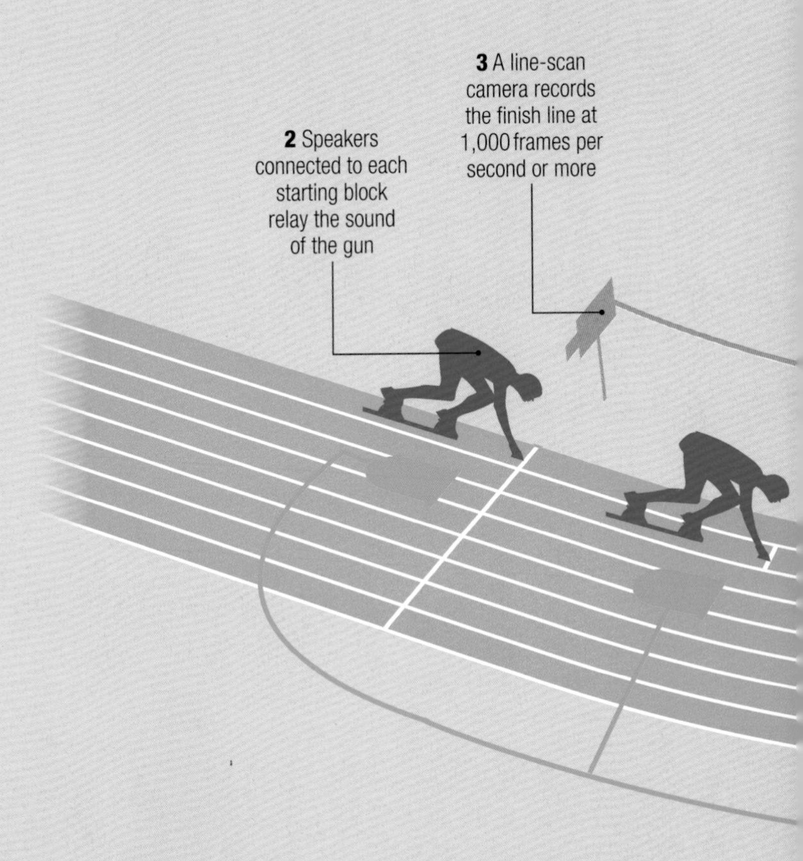

▲ ***All in the timing*** *Fully Automatic Time (FAT) is a form of race timing in which the clock is automatically activated by a starting device, and the finish time is automatically recorded by analysis of a photo finish. Electronic timing was introduced in 1977 as a method of ratifying world records after numerous studies indicated that hand timing was around 0.24 seconds slower. It's not as if it was a new phenomenon, however, given it had been in use at the 1912 Stockholm Olympics for the 100 m. A 1972 study at the Munich Olympics concluded that for events such as the 400 m, where the timekeepers are closer to the starter's gun, it was 0.18 seconds slower, the 100 m is 0.24 seconds, and further away events such as the 200 m meant a 0.26 second differential.*

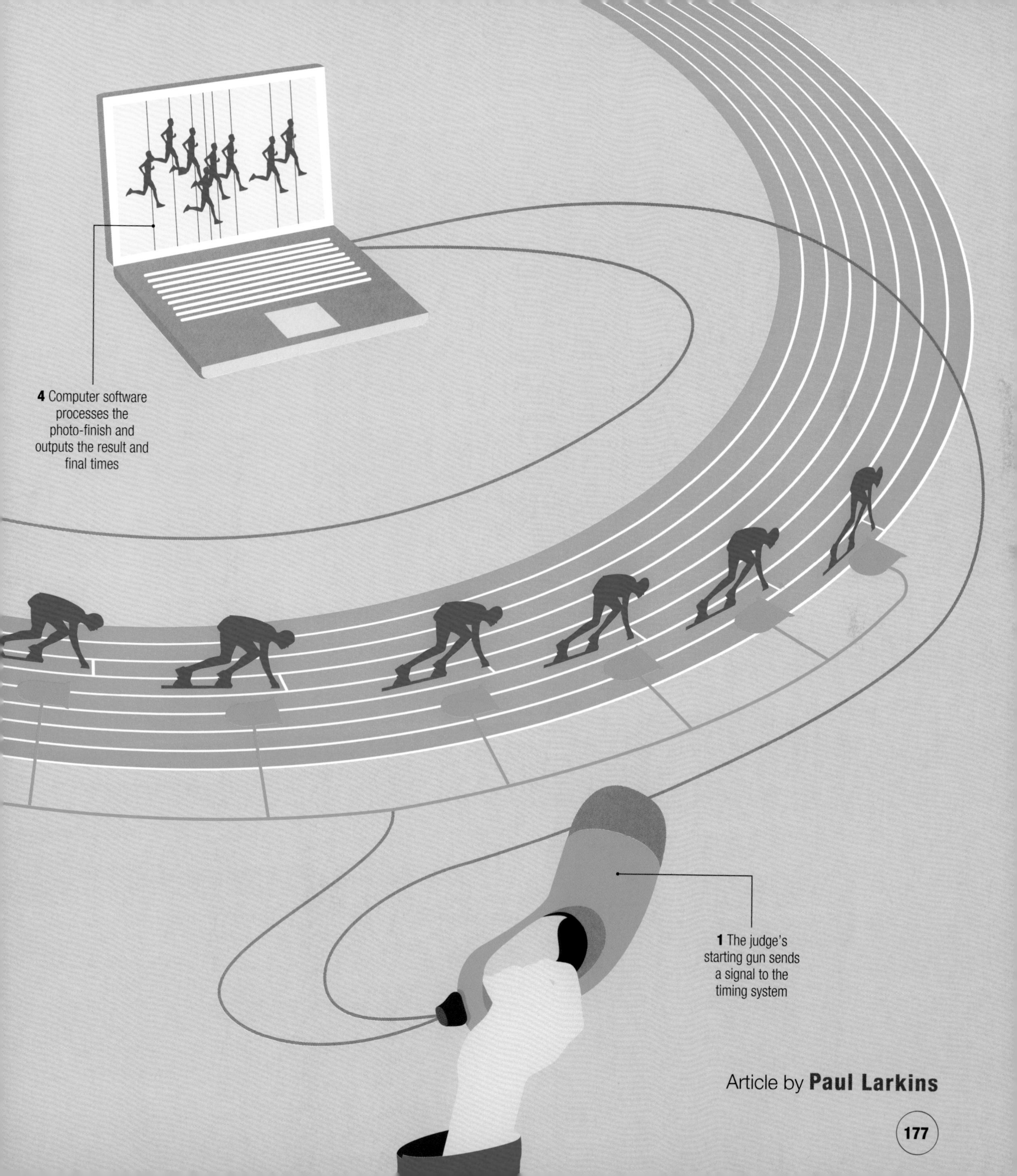

Article by **Paul Larkins**

Notes

CHAPTER 1
the runner's body

PAGES 16–17

1. Frederic N. Daussin, Joffrey Zoll, Stéphane. P. Dufour, Elodie Ponsot, Evelyne Lonsdorfer-Wolf, Stéphane Doutreleau, Bertrand Mettauer, François Piquard, Bernard Geny, and Ruddy Richard, "Effect of Interval Versus Continuous Training on Cardiorespiratory and Mitochondrial Functions: Relationship to Aerobic Performance Improvements in Sedentary Subjects," *American Journal of Physiology - Regulatory, Integrative and Comparative Physiology* 295 no. 1 (July 2008): 264–272. doi: 10.1152/ajpregu.00875.2007.

2. Jan Helgerud, Kjetil Høydal, Eivind Wang, Trine Karlsen, Pålr Berg, Marius Bjerkaas, Thomas Simonsen, Cecilies Helgesen, Ninal Hjorth, Ragnhild Bach, and Jan Hoff, "Aerobic High-Intensity Intervals Improve VO_2 max More Than Moderate Training," Medicine & Science in Sports & Exercise 39 no. 4 (April 2007): 665–671. doi: 10.1249/mss.0b013e3180304570.

PAGES 18–19

1. Rob Duffield and Brian Dawson, "Energy System Contribution in Track Running," *New Studies in Athletics* 18 no. 4 (2003): 47–56.

PAGES 30–31

1. Ralph F. Fregosi and Robert W. Lansing, "Neural Drive to Nasal Dilator Muscles: Influence of Exercise Intensity and Oralnasal Flow Partitioning," *Journal of Applied Physiology* 79 no. 4 (November 1995): 1330–1337.

2. Kellie M. Baker and David G. Behm, "The Ineffectiveness of Nasal Dilator Strips Under Aerobic Exercise and Recovery Conditions," *Journal of Strength and Conditioning Research* 13 no. 3 (August 1999): 206–209. doi: 10.1519/00124278-199908000-00004.

CHAPTER 2
perfect motion

PAGES 36–37

1. Milan Coh, Bojan Jošt, Branko Škof, Katja Tomažin, and Aleš Dolenec, "Kinematic and Kinetic Parameters of the Sprint Start and Start Acceleration Model of Top Sprinters," *Gymnica* 28 (April 1998): 33–42.

2. Neil E. Bezodis, Aki I. T. Salo, and Grant Trewartha, "Relationships Between Lower-Limb Kinematics and Block Phase Performance in a Cross Section of Sprinters," *European Journal of Sports Science* 15 no. 2 (June 2015): 118–124. doi: 10.1080/17461391.2014.928915.

3. Coh et al., "Kinematic and Kinetic Parameters of the Sprint Start and Start Acceleration Model of Top Sprinters."

4. Bezodis et al., "Relationships Between Lower-Limb Kinematics and Block Phase Performance in a Cross Section of Sprinters."

5. Antti Mero, Sami Kuitunen, Martin Harland, Heikki Kyröläinen, and Paavo V. Komi, "Effects of Muscle-Tendon Length on Joint Moment and Power During Sprint Starts," *Journal of Sports Sciences* 24 no 2. (February 2006): 165–173. doi: 10.1080/02640410500131753.

6. Tom E. Parry, Phillip Henson, and John Cooper, "Lateral Foot Placement Analysis of the Sprint Start," *New Studies in Athletics* 18 no. 1 (2003): 13–22.

7. Mitsuo Otsuka, Toshiyuki Kurihara, and Tadao Isaka, "Effect of a Wide Stance on Block Start Performance in Sprint Running," *PLoS ONE* 10 no. 11 (November 2015): e0142230. doi:10.1371/journal.pone.0142230.

8. Franklin M. Henry, "Force-Time Characteristics of the Sprint Start," *Research Quarterly* 23 (1952): 301–318. doi: 10.1080/10671188.1952.10624871.

9. Peter O. Sigerseth and Vernon F. Grinaker, "Effect of Foot Spacing on Velocity in Sprints," *Research Quarterly* 33 (1962): 599–606. doi: 10.1080/10671188.1962.10762113.

10. M. Arnold, "100 Metres Men," *Athletics Coach* 26 no. 4 (1992): 11–13.

11. Nathalie Guissard, Jacques Duchateau, and Karl Hainaut, "EMG and Mechanical Changes During Sprint Starts at Different Front Block Obliquities," *Medicine & Science in Sports & Exercise* 24 no. 11 (November 1992): 1257–1263.

12. Mero et al., "Effects of Muscle-Tendon Length on Joint Moment and Power During Sprint Starts."

13. Steven Cousins and Rosemary Dyson, "Forces at the Front and Rear Blocks During the Sprint Start," *Proceedings of XXII International Symposium on Biomechanics in Sports* (2004): 198–201.

14. Mero et al., "Effects of Muscle-Tendon Length on Joint Moment and Power During Sprint Starts."

15. Philip K. Schot and Kathleen M. Knutzen, "A Biomechanical Analysis of Four Sprint Start Positions," *Research Quarterly for Exercise and Sport* 63 no. 2 (June 1992): 137–147. doi: 10.1080/02701367.1992.10607573.

PAGES 38–39

1. Giovanni A. Cavagna, F. P. Saibene, and R. Margaria, "Mechanical Work in Running," *Journal of Applied Physiology* 19 (March 1964): 249–256.

2. Robert J. Butler, Harrison P. Crowell, and Irene M. Davis, "Lower Extremity Stiffness: Implications for Performance and Injury," *Clinical Biomechanics* 18 no. 6 (July 2003): 511–517. doi: 10.1016/S0268-0033(03)00071-8.

3. Mark L. Latash and Vladimir M. Zatsiorsky, "Joint Stiffness: Myth or Reality?," *Human Movement Science* 12 no. 6 (December 1993): 653–692. doi: 10.1016/0167-9457(93)90010-M.

4. Timothy E. Hewett, Amanda L. Stroupe, Thomas A. Nance, and Frank R. Noyes, "Plyometric Training in Female Athletes: Decreased Impact Forces and Increased Hamstring Torques," *American Journal of Sports Medicine* 24 no. 6 (December 1996): 765-773. doi: 10.1177/036354659602400611

5. Darren J. Stefanyshyn and Benno M. Nigg, "Dynamic Angular Stiffness of the Ankle Joint During Running and Sprinting," *Journal of Applied Biomechanics* 14 no. 3 (August 1998): 292–299. doi: 10.1123/jab.14.3.292.

6. Andre Seyfarth, Hartmut Geyer, Michael Gunther, and Reinhard Blickan, "A Movement Criterion for Running," *Journal of Biomechanics* 35 no. 5 (May 2002): 649–655. doi: 10.1016/S0021-9290(01)00245-7.

7. Daniel P. Ferris, Micky Louie, and Claire T. Farley, "Running in the Real World: Adjusting Leg Stiffness for Different Surfaces," *Proceedings of the Royal Society of London Series B Biological Sciences* 265 no. 1400 (June 1998): 989–994. doi: 10.1098/rspb.1998.0388.

8. Gerald Smith and P. Watanatada, "Adjustment to Vertical Displacement and Stiffness with Changes to Running Footwear Stiffness," *Medicine & Science in Sports & Exercise* 34 no. 5 (May 2002): 456–461. doi: 10.1097/00005768-200205001-00995

9. Ferris et al., "Running in the Real World: Adjusting Leg Stiffness for Different Surfaces."

10. Butler et al., "Lower Extremity Stiffness: Implications for Performance and Injury."

11. Michael B. Pohl, Joseph Hamill, and Irene S. Davis, "Biomechanical and Anatomic Factors Associated with a History of Plantar Fasciitis in Female Runners," *Clinical Journal of Sports Medicine* 19 no. 5 (September 2009): 37–376. doi: 10.1097/JSM.0b013e3181b8c270.

12. Clare E. Milner, Reed Ferber, Christine D. Pollard, Joseph Hamill, and Irene S. Davis, "Biomechanical Factors Associated with Tibial Stress Fracture in Female Runners," *Medicine & Science in Sports & Exercise* 38 no. 2 (February 2006): 323–328. doi: 10.1249/01.mss.0000183477.75808.92.

13. Anna V. Lorimer and Patria A. Hume, "Stiffness as a Risk Factor for Achilles Tendon Injury in Running," *Sports Medicine* (May 2016): 1–18. doi: 10.1007/s40279-016-0526-9.

PAGES 40–41

1. Joseph P. Hunter, Robert N. Marshall, and Peter J. McNair, "Interaction of Step Length and Step Rate During Sprint Running," *Medicine & Science in Sports & Exercise* 36 no. 2 (February 2004): 261–271. doi: 10.1249/01.MSS.0000113664.15777.53.

2. P.ekka Luhtanen and Paavo V. Komi, "Mechanical Factors Influencing Running Speed," *Biomechanics VI-B, International Series on Biomechanics* 2B (1978): 23–29.

3. Antti Mero and Paavo V. Komi, "Effects of Supramaximal Velocity on Biomechanical Variables in Sprinting," *International Journal of Sports Biomechanics* 1 no. 3 (August 1985):240–252.

4. Peter G. Weyand, Deborah B. Sternlight, Matthew J. Bellizzi, and Seth Wright, "Faster Top Running Speeds are Achieved With Greater Ground Forces Not More Rapid Leg Movements," *Journal of Applied Physiology* 89 no. 5 (November 2000): 1991–1999.

5. Ibid.

6. Sami Kuitunen, Paavo V. Komi, and Heikki Kyröläinen, "Knee and Ankle Joint Stiffness in Sprint Running," *Medicine & Science in Sports & Exercise* 34 no. 1 (January 2002): 166–173. doi: 10.1097/00005768-200201000-00025.

7. Hay, James G. *The Biomechanics of Sports Techniques.* 4th ed. University of Michigan: Prentice-Hall, 1993.

8. Hunter et al., "Interaction of Step Length and Step Rate During Sprint Running."

9. Aki Salo, Ian N. Bezodis, Alan M. Batterham, and David G. Kerwin, "Elite Sprinting: Are Athletes Individually Step Frequency or Step Length Reliant?," *Medicine & Science in Sports & Exercise* 43 no. 6 (June 2011): 1055–1062. doi: 10.1249/MSS.0b013e318201f6f8.

10. Laura Charalambous, David G. Kerwin, Gareth Irwin, Ian Bezodis, and Stephen Hailes, "Changes in Step Characteristics During Sprint Performance Development," *International Society of Biomechanics in Sport Conference* 11 no. S2 (2011): S467–S470.

11. Sofie Debaere, Ilse Jonkers, and Christophe H. Delecluse, "The Contribution of Step Characteristics to Sprint Running Performance in High-Level Male and Female Athletes," *Journal of Strength and Conditioning Research* 27 no. 1 (January 2013): 116–124. doi: 10.1519/JSC.0b013e31825183ef.

12. Deutscher Leichtathletik-Verband, "Biomechanical Analyses of Selected Events at the 12th IAAF World Championships in Athletics, Berlin 15–23 August 2009: Final Report, Sprint Men."

13. Yasushi Enomoto, Hirosuke Kadono, Yuta Suzuki, Tetsu Chiba, and Keiji Koyama, "Biomechanical Analysis of the Medalists in the 10,000 metres at the 2007 World Championships in Athletics," *New Studies in Athletics* 23 no. 3 (2008): 61–66.

PAGES 42–43

1. Carl Foster and Alejandro Lucia, "Running Economy: The Forgotten Factor in Elite Performance," *Sports Medicine* 37 no. 4–5 (February 2007): 316–319. doi: 10.2165/00007256-200737040-00011.

2. Ibid.

3. Masaki Ishikawa and Paavo V. Komi, "Muscle Fascicle and Tendon Behavior During Human Locomotion Revisited," *Exercise and Sport Sciences Reviews* 36 no. 4 (October 2008): 193–199. doi: 10.1097/JES.0b013e3181878417.

4. Dorsey S. Williams, Irene S. McClay, and Kurt T. Manal, "Lower Extremity Mechanics in Runners with Converted Forefoot Strike Pattern," *Journal of Applied Biomechanics* 16 no. 2 (May 2000): 210–218. doi: 10.1123/jab.16.2.210.

5. Adam I. Daoud, Gary J. Geissler, Frank Wang, Jason Saretsky, Yahya A. Daoud, and Daniel E. Lieberman, "Foot Strike and Injury Rates in Endurance Eunners: A Retrospective Study," *Medicine & Science in Sports & Exercise* 44 no. 7 (July 2012): 1325–1334. doi: 10.1249/MSS.0b013e3182465115.

PAGES 44–45

1. Masaki Ishikawa, Kanae Sano, Yoko Kunimasa, Toshiaki Oda, Caroline Nicol, Akira Ito, and Paavo V. Komi, "Economical Running Strategy for East African Distance Runners," *Journal of Physical Fitness and Sports Medicine* 2 no. 3 (2013): 361–363. doi: 10.7600/jpfsm.2.361.

2. Carl Foster and Alejandro Lucia, "Running Economy the Forgotten Factor in Elite Performance," *Sports Medicine* 37 no. 4–5 (February 2007): 316–319. doi: 10.2165/00007256-200737040-00011.

3. Bengt Saltin, Henrik Larsen, N. Terrados, Jens Bangsbo, T. Bak, Chang K. Kim, J. Svedenhag, and C. J. Rolf, "Aerobic Exercise Capacity at Sea Level and at Altitude in Kenyan Boys, Junior and Senior Runners Compared with Scandinavian Runners," *Scandinavian Journal of Medicine and Science in Sports* 5 no. 4 (August 1995): 209–221. doi: 10.1111/j.1600-0838.1995.tb00037.x.

4. Lida Mademli, Adamantios Arampatzis, and Mark Walsh, "Effect of Muscle Fatigue on the Compliance of the Gastrocnemius Medialis Tendon and Aponeurosis," *Journal of Biomechanics* 39 no 3. (2006): 426–434. doi: 10.1016/j.jbiomech.2004.12.016.

5. Masaki Ishikawa and Paavo V. Komi, "The Role of the Stretch Reflex in the Gastrocnemius Muscle During Human Locomotion at Various Speeds," *Journal of Applied Physiology* 103 no. 3 (September 2007): 1030–1036. doi: 10.1152/japplphysiol.00277.2007.

PAGES 46–47

1. Patrick O. Riley, Jay Dicharry, Jason Franz, Ugo Della Croce, Robert P. Wilder, and D. Casey Kerrigan, "A Kinematics and Kinetic Comparison of Overground and Treadmill Running," *Medicine & Science in Sports & Exercise* 40 no. 6 (June 2008): 1093–1100. doi: 10.1249/MSS.0b013e3181677530.

2. Jonathan Sinclair, Jim Richards, Paul J. Taylor, Christopher J. Edmundson, Darrell Brooks, and Sarah J. Hobbs, "Three-Dimensional Kinematic Comparison of Treadmill and Overground Running," *Sports Biomechanics* 12 no. 3 (September 2013): 272–282. doi: 10.1080/14763141.2012.759614.

3. Rebecca E. Fellin, Kurt Manal, and Irene S. Davis, "Comparison of Lower Extremity Kinematic Curves During Overground and Treadmill Running," *Journal of Applied Biomechanics* 26 no. 4 (November 2010): 407–414. doi: 10.1123/jab.26.4.407.

4. Riley et al., "A Kinematics and Kinetic Comparison of Overground and Treadmill Running."

5. B. C. Elliott and B. A. Blanksby, "A Cinematographic Analysis of Overground and Treadmill Running by Males and Females," *Medicine & Science in Sports & Exercise* 8 no. 2 (Summer 1976): 84–87.

6. Barry A. Frishberg, "An Analysis of Overground and Treadmill Sprinting," *Medicine & Science in Sports & Exercise* 15 no. 6 (1983): 478–485. doi: 10.1249/00005768-198315060-00007.

7. Benno M. Nigg, Rudd W. De Boer, and Veronica Fischer, "A Kinematic Comparison of Overground and Treadmill Running," *Medicine & Science in Sports & Exercise* 27 no. 1 (January 1995): 98–105. doi: 10.1249/00005768-199501000-00018.

8. Riley et al., "A Kinematics and Kinetic Comparison of Overground and Treadmill Running."

PAGES 48–49

1. Joseph P. Hunter, Robert N. Marshall, and Peter J. McNair, "Relationship Between Ground Reaction Force, Impulse and Kinematics of Sprint-Running Acceleration," *Journal of Applied Biomechanics* 21 no. 1 (February 2005): 31–43. doi: 10.1123/jab.21.1.31.

2. Ibid.

3. Matt Brughelli, John Cronin, and Anis Chaouachi, "Effects of Running Velocity on Running Kinetics and Kinematics," *Journal of Strength and Conditioning Research* 25 no. 4 (April 2011): 933–939. doi: 10.1519/JSC.0b013e3181c64308.

4. Hunter et al., "Relationship Between Ground Reaction Force, Impulse and Kinematics of Sprint-Running Acceleration."

PAGES 50–51

1. Richard N. Hinrichs, "Upper Extremity Function in Running II: Angular Momentum Considerations," *International Journal of Sport Biomechanics* 3 no. 3 (August 1987): 242–263. doi: 10.1123/ijsb.3.3.242.

2. Herman Pontzer, John H. Holloway, David A. Raichlen, and Daniel E. Lieberman, "Control and Function of Arm Swing in Human Walking and Running," *Journal of Experimental Biology* 212 (2008): 523–534. doi: 10.1242/jeb.024927.

3. Christopher J. Arellano and Rodger Kram, "The Metabolic Cost of Human Running: Is Swinging the Arms Worth It?," *Journal of Experimental Biology* 217 no. 14 (July 2014): 2456–2461. doi: 10.1242/jeb.100420.

4. Serge Gracovetsky, *The Spinal Engine,* rev. ed. (Sandy, UT: Aardvark Global Publishing, 2008).

PAGES 52–53

1. K. R. Williams and Peter R. Cavanagh, "Relationship Between Distance Running Mechanics, Running Economy, and Performance," *Journal of Applied Physiology* 63 no. 3 (September 1987): 1236–1245.

2. Hsiang-Ling Teng and Christopher M. Powers, "Sagittal Plane Trunk Posture Influences Patellofemoral Joint Stress During Running," *Journal or Orthopeadic and Sports Physical Therapy* 44 no. 10 (October 2014) 785–792. doi: 10.2519/jospt.2014.5249.

CHAPTER 3
fuel and fluid

PAGES 58–59

1. Louise M. Burke, John A. Hawley, Stephen H. Wong, and Asker E. Jeukendrup, "Carbohydrates for Training and Competition," *Journal of Sports Sciences* 29 no. S1 (May 2011): S17–27. doi: 10.1080/02640414.2011.585473.

2. Travis D. Thomas, K. A. Erdman, and Louise M. Burke, "Nutrition and Athletic Performance," *Medicine & Science in Sports & Exercise* 48 no. 3 (2016): 543–568.

3. Louise M. Burke, "Re-Examining High-Fat Diets for Sports Performance: Did We Call the "Nail in the Coffin" Too Soon?," *Sports Medicine* 45 no. S1 (November 2015): S33–S49. doi: 10.1007/s40279-015-0393-9.

4. James A. Betts and Clyde Williams, "Short-Term Recovery from Prolonged Exercise: Exploring the Potential for Protein Ingestion to Accentuate the Benefits of Carbohydrate Supplements," *Sports Medicine* 1 no. 40 (November 2010): 941–959. doi:10.2165/11536900-000000000-00000.

PAGES 60–61

1. John A. Hawley and Jill J. Leckey, "Carbohydrate Dependence During Prolonged, Intense Endurance Exercise," *Sports Medicine* 45 no. S1 (November 2015): S5–S12. doi: 10.1007/s40279-015-0400-1.

PAGES 62–63

1. Stuart M. Phillips, "A Brief Review of Critical Processes in Exercise-Induced Muscle Hypertrophy," *Sports Medicine* 44 S1 (May 2014): S71–S77. doi: 10.1007/s40279-014-0152-3.

2. Tyler A. Churchward-Venne, Nicholas A. Burd, and Stuart M. Phillips, "Nutritional Regulation of Muscle Protein Synthesis with Resistance Exercise: Strategies to Enhance Anabolism," *Nutritional Metabolism* 9 no. 1 (May 2012): 40–48. doi: 10.1186/1743-7075-9-40.

PAGES 64–65

1. Travis D. Thomas, Kelly Anne Erdman, and Louise M. Burke, "Nutrition and Athletic Performance," *Medicine & Science in Sports & Exercise* 48 no. 3 (March 2016): 543–568. doi: 10.1249/MSS.0000000000000852.

2. Ibid.

3. Ronald J. Maughan and Susan M. Shirreffs, "Nutrition for Sports Performance: Issues and Opportunities," *Proceedings of the Nutrition Society* 71 no. 1 (February 2012): 112–119. doi: 10.1017/S0029665111003211.

4. U.S. Department of Health and Human Services and U.S. Department of Agriculture. *2015–2020 Dietary Guidelines for Americans*. 8th ed. (2015).

PAGES 66–67

1. Lindsay B. Baker and Asker E. Jeukendrup, "Optimal Composition of Fluid-Replacement Beverages," *Comprehensive Physiology* 4 no. 2 (April 2014): 575–620. doi: 10.1002/cphy.c130014.

2. Ronald J. Maughan and Susan M. Shirreffs, "Nutrition for Sports Performance: Issues and Opportunities," *Proceedings of the Nutrition Society* 71 no. 1 (February 2012): 112–119. doi: 10.1017/S0029665111003211.

PAGES 68–69

1. Original graphic © Jonathan Savage 2016, http://fellrnr.com/wiki/Calories_burned_running_and_walking (accessed October 25, 2016).

PAGES 70–71

1. Lawrence L. Spriet, "Exercise and Sport Performance with Low Doses of Caffeine," *Sports Medicine* 44 no. S2 (November 2014): S175–S184. doi: 10.1007/s40279-014-0257-8.

2. U.S. Department of Health and Human Services and U.S. Department of Agriculture. *2015–2020 Dietary Guidelines for Americans*. 8th ed. (2015).

3. Mark Glaister, Stephen D. Patterson, Paul Foley, Charles R. Pedlar, John R. Pattison, and Gillian McInnes, "Caffeine and Sprinting Performance: Dose Responses and Efficacy," *Journal of Strength and Conditioning Research* 26 no. 4 (April 2012): 1001–1005. doi: 10.1519/JSC.0b013e31822ba300.

4. Jim D. Wiles, S. R. Bird, John Hopkins, and M. Riley, "Effect of Caffeinated Coffee on Running Speed, Respiratory Factors, Blood Lactate and Perceived Exertion During 1500-m Treadmill Running," *British Journal of Sports Medicine* 26 no. 2 (June 1992): 116–120. doi: 10.1136/bjsm.26.2.116.

5. Matthew P. O'Rourke, Brendan J. O'Brien, Wade L. Knez, and Carl D. Paton, "Caffeine Has A Small Effect On 5-Km Running Performance Of Well-Trained And Recreational Runners," *Journal of Science and Medicine in Sport* 11 no. 2 (April 2008): 231–233. doi: 10.1016/j.jsams.2006.12.118.

6. Craig A. Bridge and Melanie A. Jones, "The Effect of Caffeine Ingestion on 8 km Run Performance in a Field Setting," *Journal of Sports Sciences* 24 no. 4 (April 2006): 433–439. doi: 10.1080/02640410500231496.

PAGES 72–73

1. K. J. Cureton, P. B. Sparling, B. W. Evans, S. M. Johnson, U. D. Kong, and J. W. Purvis, "Effect of Experimental Alterations in Excess Weight on Aerobic Capacity and Distance Running Performance," *Medicine & Science in Sports & Exercise* 10 no. 3 (Fall 1978): 194–199.

2. Adrien Sedeaud, Andy Marc, Adrien Marck, Frédéric Dor, Julien Schipman, Maya Dorsey, Amal Haida, Geoffroy Berthelot, and Jean-François Toussaint, "BMI, A Performance Parameter for Speed Improvement," *PLoS ONE* 9 no. 2 (February 2014): e90183. doi:10.1371/journal.pone.0090183.

PAGES 74–75

1. Lindsay B. Baker and Asker E. Jeukendrup, "Optimal Composition of Fluid-Replacement Beverages," *Comprehensive Physiology* 4 no. 2 (April 2014): 575–620. doi: 10.1002/cphy.c130014.

2. P. R. Below, R. Mora-Rodriguez, J. Gonzalez-Alonso, and E. F. Coyle, "Fluid and Carbohydrate Ingestion Independently Improve Performance During 1 h of Intense Exercise," *Medicine & Science in Sports & Exercise* 27 no. 2 (February 1995): 200–210. doi: 10.1249/00005768-199502000-00009.

3. Johneric W. Smith, David D. Pascoe, Dennis H. Passe, Brent C. Ruby, Laura K. Stewart, Lindsay B. Baker, and Jeffrey J. Zachwieja, "Curvilinear Dose–Response Relationship of Carbohydrate ($0–120\,g{\cdot}h^{-1}$) and Performance," *Medicine & Science in Sports & Exercise* 45 no. 2 (February 2013): 336–341. doi: 10.1249/MSS.0b013e31827205d1.

4. Travis D. Thomas, Kelly Anne Erdman, and Louise M. Burke, "Nutrition and Athletic Performance," *Medicine & Science in Sports & Exercise* 48 no. 3 (March 2016): 543–568. doi: 10.1249/MSS.0000000000000852.

5. Kevin Currell and Asker E. Jeukendrup, "Superior Endurance Performance with Ingestion of Multiple Transportable Carbohydrates," *Medicine & Science in Sports & Exercise* 40 no. 2 (February 2008): 275–281. doi: 10.1249/mss.0b013e31815adf19.

PAGES 76–77

1. Freimut Schliess, Lee Richter, S. vom Dahl, and Dieter Haussinger, "Cell Hydration and mTOR-Dependent Signalling," *Acta Physiologica* 187 no. 1–2 (May–June 2006): 223–229. doi: 10.1111/j.1748-1716.2006.01547.x.

CHAPTER 4
running psychology

PAGES 82–83

1. Judy L. Van Raalte, Ruth Brennan Morrey, Britton W. Brewer, and Allen E. Cornelius, "Self-Talk of Marathon Runners," *Sport Psychologist* 29 no. 3 (September 2015): 258–260. doi: 10.1123/tsp.2014-0159.

2. Carl Foster, Jessica A. Florhaug, Jodi Franklin, Lori Gottschall, Lauri A. Hrovatin, Suzanne Parker, Pamela Doleshal, and Christopher Dodge, "A New Approach to Monitoring Exercise Training," *The Journal of Strength and Conditioning Research*, 15 no. 1 (February 2001): 109–115. doi: 10.1519/00124278-200102000-00019.

3. Francisco de Assis Manoel, Bruno P. Melo, Ramon Cruz, Cristóvão S. Villela, Danilo L. Alves, Sandro F. da Silva, and Fernando R. de Oliveira, "The Utilization of Perceived Exertion is Valid for the Determination of the Training Stress in Young Athletes," *Journal Of Exercise Physiology Online* 19 no. 1 (February 2016): 27–32. doi: 10.1007/s40279-014-0253-z.

PAGES 84–85

1. Noel E. Brick, Mark J. Campbell, Richard S. Metcalfe, Jacqueline L. Mair, and Tadhg E. Macintyre, "Altering Pace Control and Pace Regulation: Attentional Focus Effects During Running," *Medicine & Science in Sports & Exercise* 48 no. 5 (May 2016): 879–886. doi: 10.1249/MSS.0000000000000843.

PAGES 86–87

1. Costas I. Karageorghis, Peter C. Terry, Andrew M. Lane, Daniel T. Bishop, and David Priest, "The BASES Expert Statement on Use of Music in Exercise," *Journal of Sports Sciences* 30 no. 9 (May 2012): 953–956. doi: 10.1080/02640414.2012.676665.

2. Costas Karageorghis, David Priest, Peter Terry, Nikos Chatzisarantis, and Andrew Lane, "Redesign and Initial Validation of an Instrument to Assess the Motivational Qualities of Music in Exercise: The Brunel Music Rating Inventory-2," *Journal of Sports Sciences* 24 no. 8 (August 2006): 899–909. doi: 10.1080/02640410500298107.

3. Daniel T. Bishop, Costas I. Karageorghis, and Georgios Loizou, "A Grounded Theory of Young Tennis Players' Use of Music to Manipulate Emotional State," *Journal of Sport & Exercise Psychology* 29 no. 5 (October 2007): 584–607. doi: 10.1123/jsep.29.5.584.

PAGES 88–89

1. Daniel N. Ardoy, J. M. Fernández-Rodríguez, David Jiménez-Pavón, R. Castillo, J. R. Ruiz, and F. B. Ortega, "A Physical Education Trial Improves Adolescents' Cognitive Performance and Academic Achievement: The EDUFIT Study," *Scandinavian Journal of Medicine & Science in Sports* 24 no. 1 (February 2014): e52–e61. doi: 10.1111/sms.12093.

2. Noel Brick, Tadhg MacIntyre, and Michael Campbell, "Metacognitive Processes in the Self-regulation of Performance in Elite Endurance Runners," *Psychology of Sport & Exercise* 19 (July 2015): 1–9. doi: 10.1016/j.psychsport.2015.02.003.

3. Andrew M. Lane and Mathew Wilson, "Emotions and Trait Emotional Intelligence Among Ultra-Endurance Runners," *Journal of Science & Medicine in Sport* 14 no. 4 (July 2011): 358–362. doi: 10.1016/j.jsams.2011.03.001.

PAGES 90–91

1. Andrew M. Lane, Chris J. Beedie, Marc V. Jones, Mark Uphill, and Tracey J. Devonport, "The BASES Expert Statement on Emotion Regulation in Sport," *Journal of Sports Sciences* 30 no. 11 (June 2012): 1189–1195. doi:10.1080/02640414.2012.693621.

2. Andrew M. Lane, Tracey J. Devonport, Andrew P. Friesen, Christopher J. Beedie, Christopher L. Fullerton, and Damian M. Stanley, "How Should I Regulate My Emotions if I Want to Run Faster?," *European Journal of Sports Science* 16 no. 4 (September 2015): 465–472. doi: 10.1080/17461391.2015.1080305.

3. Damian M. Stanley, Andrew M. Lane, Christopher J. Beedie, and Tracey J. Devonport, "I Run to Feel Better; So Why Am I Thinking So Negatively?," *International Journal of Psychology and Behavioral Science* 2 no. 6 (December 2012): 208–213. doi: 10.5923/j.ijpbs.20120206.03.

4. Andrew M. Lane, Christopher J. Beedie, Tracey J. Devonport, and Damian M. Stanley, "Instrumental Emotion Regulation in Sport: Relationships Between Beliefs About Emotion and Emotion Regulation Strategies Used by Athletes," *Scandinavian Journal of Medicine & Science in Sports* 21 no. 6 (December 2011): e445–e451. doi: 10.1111/j.1600-0838.2011.01364.x.

5. Andrew M Lane. Peter Totterdell, Ian MacDonald, Tracey J. Devonport, Andrew P. Friesen, Christopher J. Beedie, Damian M. Stanley, and Alan Nevill, "Brief Online Training Enhances Competitive Performance: Findings of the BBC Lab UK Psychological Skills Intervention Study," *Frontiers in Psychology* 7 no. 413 (March 2016). doi: 10.3389/fpsyg.2016.00413.

PAGES 92–93

1. Andrew M. Lane, Tracey J. Devonport, Andrew P. Friesen, Christopher J. Beedie, Christopher L. Fullerton, and Damian M. Stanley, "How Should I Regulate My Emotions if I Want to Run Faster?," *European Journal of Sport Science* 16 no. 4 (September 2016): 465–472. doi: 10.1080/17461391.2015.1080305.

2. Andrew M. Lane, Paul A. Davis, and Tracey J. Devonport, "Effects of Music Interventions on Emotional States and Running Performance," *Journal of Sports Science and Medicine* 10 no. 2 (June 2011): 400–407.

3. Andrew M. Lane, Peter Totterdell, Ian MacDonald, Tracey J. Devonport, Andrew P. Friesen, Christopher J. Beedie, Damian M. Stanley, and Alan Nevill, "Brief Online Training Enhances Competitive Performance: Findings of the BBC Lab UK Psychological Skills Intervention Study," *Frontiers in Psychology* 7 no. 413 (March 2016). doi: 10.3389/fpsyg.2016.00413.

PAGES 94–95

1. Sheldon Hanton, Rich Neil, and Stephen D. Mellalieu, "Recent Developments in Competitive Anxiety Direction and Competition Stress Research," *International Review Of Sport & Exercise Psychology* 1 no. 1 (February 2008): 45–57. doi: 10.1080/17509840701827445.

2. J. Graham Jones, Austin Swain, and Andrew Cale, "Antecedents of Multidimensional Competitive State Anxiety and Self-Confidence in Elite Intercollegiate Middle-Distance Runners," *Sport Psychologist* 4 no. 2 (June 1990): 107–118. doi: 10.1123/tsp.4.2.107.

3. Andrew Lane, "Relationships Between Perceptions of Performance Expectations and Mood Among Distance Runners: The Moderating Effect of Depressed Mood," *Journal of Science & Medicine in Sport* 4 no. 1 (March 2001): 116–128. doi: 10.1016/S1440-2440(01)80013-X.

4. Sheldon Hanton and Graham Jones, "The Acquisition and Development of Cognitive Skills and Strategies: I. Making the Butterflies Fly in Formation," *Sport Psychologist* 13 no. 1 (March 1999): 1–21. doi: 10.1123/tsp.13.1.1.

5. Sheldon Hanton and Graham Jones, "The Effects of a Multimodal Intervention Program on Performers: II. Training the Butterflies to Fly in Formation," *Sport Psychologist* 13 no. 1 (March 1999): 22–41. doi: 10.1123/tsp.13.1.22.

CHAPTER 5
training and racing

PAGES 100–101

1. Michael K. Stickland, Brian H. Rowe, Carol H. Spooner, Ben Vandermeer, and Donna M. Dryden, "Effect of Warm-Up Exercise on Exercise-Induced Bronchoconstriction," *Medicine & Science in Sports & Exercise* 44 no. 3 (March 2012): 383–391. doi: 10.1249/MSS.0b013e31822fb73a.

2. Veronique L. Billat, Valery Bocquet, Jean Slawinski, L. Laffite, A. Demarle, P. Chassaing, and Jean Pierre Koralsztein, "Effect of a Prior Intermittent Run at vVO_2 Max on Oxygen Kinetics During an All-Out Severe Run in Humans," *Journal of Sports Medicine and Physical Fitness* 40 no. 3 (September 2000): 185–194.

3. Stephen A. Ingham, Barry W. Fudge, Jamie S. Pringle, and Andrew M. Jones, "Improvement of 800-m Running Performance with Prior High-Intensity Exercise," *International Journal of Sports Physiology and Performance* 8 no. 1 (January 2013): 77–83. doi: 10.1123/ijspp.8.1.77.

4. Karl B. Fields, Jeannie C. Sykes, Katherine M. Walker, and Jonathan C. Jackson, "Prevention of Running Injuries," *Current Sports Medicine Reports* 9 no. 3 (May–June 2010): 176–182. doi: 10.1249/JSR.0b013e3181de7ec5.

PAGES 102–103

1. Hans Seyle, "Stress and the General Adaptation Syndrome," *British Medical Journal* 1 no. 4667 (June 1950): 1383–1392. doi: 10.1136/bmj.1.4667.1383.

2. Ashleigh J. McNicol, Brendan J. O'Briena, Carl D. Paton, and Wade L. Knez, "The Effects of Increased Absolute Training Intensity on Adaptations to Endurance Exercise Training," *Journal of Science and Medicine in Sport* 12 no. 4 (July 2009): 485–489. doi: 10.1016/j.jsams.2008.03.001.

3. Jessica Hill, Glyn Howatson, Ken van Someren, Jonathan Leeder, and Charles Pedlar, "Compression Garments and Recovery from Exercise-Induced Muscle Damage: A Meta-Analysis," *British Journal of Sports Medicine* 47 no. 9 (June 2013). doi: 10.1136/bjsports-2013-092456.

4. Jonathan Leeder, Conor Gissane, Ken van Someren, Warren Gregson, and Glyn Howatson, "Cold Water Immersion and Recovery From Strenuous Exercise: A Meta-Analysis," *British Journal of Sports Medicine* 46 no. 4 (March 2012): 233–240. doi: 10.1136/bjsports-2011-090061.

5. Shona Halson, "Sleep in Elite Athletes and Nutritional Interventions to Enhance Sleep," *Sports Medicine* 44 no. S1 (May 2014): S13–S23. doi: 10.1007/s40279-014-0147-0.

6. Christina Haugaard Rasmussen, Rasmus Oestergaard Nielsen, Martin Serup Juul, and Sten Rasmussen, "Weekly Running Volume and Risk of Running-Related Injuries Among Marathon Runners," *The International Journal of Sports Physical Therapy* 8 no. 2 (April 2013): 111–120.

7. Tim J. Gabbett, "The Training—Injury Prevention Paradox: Should Athletes Be Training Smarter and Harder?" *British Journal of Sports Medicine* 50 no. 5 (2016): 273–280. doi:10.1136/bjsports-2015-095788.

PAGES 104–105

1. Thibaut Guiraud, Anil Nigam, Vincent Gremeaux, Philippe Meyer, Martin Juneau, and Laurent Bosquet, "High-Intensity Interval Training in Cardiac Rehabilitation," *Sports Medicine* 42 no. 7 (July 2012): 587–605. doi: 10.2165/11631910-000000000-00000.

2. Kyle Barnes and Andrew Kilding, "Strategies to Improve Running Economy," *Sports Medicine* 45 no. 1 (January 2015): 37–56. doi: 10.1007/s40279-014-0246-y.

3. Ibid.

4. L. Véronique. Billat, "Interval Training for Performance: A Scientific and Empirical Practice: Special Recommendations for Middle- and Long-Distance Running. Part I: Aerobic Interval Training," *Sports Medicine* 31 no. 1 (February 2001): 13–31. doi: 10.2165/00007256-200131020-00001.

5. Martin Buchheit and Paul Laursen, "High-Intensity Interval Training, Solutions to the Programming Puzzle: Part I: Cardiopulmonary Emphasis," *Sports Medicine* 43 no. 5 (May 2013): 313–338. doi: 10.1007/s40279-013-0029-x.

PAGES 106–107

1. Simon D. Angus, "Did Recent World Record Marathon Runners Employ Optimal Pacing Strategies?," *Journal of Sports Sciences* 32 no. 1 (July 2014): 31–45. doi: 10.1080/02640414.2013.803592.

2. A. St Clair Gibson and T. Noakes, "Evidence for Complex System Integration and Dynamic Neural Regulation of Skeletal Muscle Recruitment During Exercise in Humans," *British Journal of Sports Medicine* 38 (December 2004): 797–806. doi: 10.1136/bjsm.2003.009852.

3. Ross Tucker "The Anticipatory Regulation of Performance: The Physiological Basis for Pacing Strategies and the Development of a Perception-Based Model for Exercise Performance," *British Journal of Sports Medicine* 43 no. 6 (June 2009): 392–400. doi: 10.1136/bjsm.2008.050799.

4. Angus, "Did Recent World Record Marathon Runners Employ Optimal Pacing Strategies?."

5. Olaf Hoos, Tobias Boeselt, Martin Steiner, Kuno Hottenrott, and Ralph Beneke, "Long-Range Correlations and Complex Regulation of Pacing in Long-Distance Road Racing," *International Journal of Sports Physiology and Performance* 9 no. 3 (May 2014): 544–553. doi: 10.1123/IJSPP.2012-0334.

6. Andrew Renfree and Alan St Clair Gibson, "Influence of Difference Performance Levels on Pacing Strategy During the Women's World Championship Marathon Race," *International Journal of Sports Physiology and Performance* 8 no. 3 (May 2013): 279–285. doi: 10.1123/ijspp.8.3.279.

7. Robert Deaner, Rickey Carter, Michael Joyner, and Sandra Hunter, "Men are More Likely than Women to Slow in the Marathon," *Medicine & Science in Sport & Exercise* 47 no. 3 (March 2015): 607–616. doi: 10.1249/MSS.0000000000000432.

PAGES 108–109

1. C. T. M. Davies, "Effects of Wind Resistance and Air Resistance on the Forward Motion of a Runner," *Journal of Applied Physiology* 48 no. 4 (April 1980): 702–709.

2. Andrew M. Jones and Jonathan H. Doust, "A 1% Treadmill Grade Most Accurately Reflects the Energetic Cost of Outdoor Running," *Journal of Sports Sciences* 14 no. 4 (August 1996): 321–327. doi: 10.1080/02640419608727717.

3. François Péronnet, G. Thibault, and D. Cousineau, "A Theoretical Analysis of the Effect of Altitude on Running Performance," *Journal of Applied Physiology* 70 no. 1 (January 1991): 399–404.

4. Shinichiro Ito, "Aerodynamic Effects by Marathon Pacemakers on a Main Runner," *Transactions of the Japan Society of Mechanical Engineers* series B 73 no. 734 (2007): 1975–1980. doi: 10.1299/kikaib.73.1975.

PAGES 110–111

1. Alberto E. Minetti, Christian Moia, Giulio S. Roi, David Susta, and Guido Ferretti, "Energy Cost of Walking and Running at Extreme Uphill and Downhill Slopes," *Journal of Applied Physiology* 93 (September 2002) 1039–1046. doi: 10.1152/japplphysiol.01177.2001.

2. Johnny Padulo, Douglas Powell, Raffaele Milia, and Luca Ardihò, "A Paradigm of Uphill Running," *PLoS ONE* 8 no. 7 (July 2013): e69006. doi:10.1371/journal.pone.0069006.

3. Minetti et al., "Energy Cost of Walking and Running at Extreme Uphill and Downhill Slopes."

4. Christer Malm, Bertil Sjödin, Berit Sjöberg, Rodica Lenkei, Per Renström, Ingrid E. Lundberg, and Björn Ekblom, "Leukocytes, Cytokines, Growth Factors and Hormones in Human Skeletal Muscle and Blood After Uphill or Downhill Running," *Journal of Physiology* 556 no. 3 (May 2004): 983–1000. doi: 10.1113/jphysiol.2003.056598.

5. M. Giandolini, N. Horvais, J. Rossi, G. Millet, J-B. Morin, and P. Samozino, "Acute and Delayed Peripheral and Central Neuromuscular Alterations Induced by a Short and Intense Downhill Trail Run," *Scandinavian Journal of Medicine and Science in Sports* 19 no. 26 (November 2015): 1321–1333. doi: 10.1111/sms.12583.

6. Minetti et al., "Energy Cost of Walking and Running at Extreme Uphill and Downhill Slopes."

7. Ibid.

8. Giandolini et al., "Acute and Delayed Peripheral and Central Neuromuscular Alterations Induced by a Short and Intense Downhill Trail Run."

PAGES 112–113

1. Nathan Lewis, John Rogers, Dave Collins, and Charles Pedlar, "Can Clinicians and Scientists Explain and Prevent Unexplained Underperformance Syndrome in Elite Athletes: An Inter-Disciplinary Perspective and 2016 Update," *BMJ Open: Sport and Exercise Medicine* 1 (2015): e000063. doi: 10.1136/bmjsem-2015-000063.

2. Krista Howarth, Stewart Phillips, Maureen MacDonald, Douglas Richards, Natalie Moreau, and Martin Gibala, "Effect of Glycogen Availability on Human Skeletal Muscle Protein Turnover During Exercise and Recovery," *Journal of Applied Physiology* 109 no. 2 (August 2010): 431–438. doi: 10.1152/japplphysiol.00108.2009.

3. Lewis et al., "Can Clinicians and Scientists Explain and Prevent Unexplained Underperformance Syndrome in Elite Athletes: An Inter-Disciplinary Perspective and 2016 Update."

4. Andrew Philp, Matthew G. MacKenzie, Micah Y. Belew, Mhairi C. Towler, Alan Corstorphine, Angela Papalamprou, D. Grahame Hardie, and Keith Baar, "Glycogen Content Regulates Peroxisome Proliferator Activated Receptor-h (PPAR-h) Activity in Rat Skeletal Muscle," *PLoS ONE* 8 no. 10 (October 2013): e77200. doi:10.1371/journal.pone.0077200.

5. Giovanni Tanda and Beat Knechtle, "Effects of Training and Anthropometric Factors on Marathon and 100km Ultramarathon Race Performance," *Open Access Journal of Sports Medicine* 6 (April 2015): 129–136. doi: 10.2147/OAJSM.S80637.

6. Adrian Midgley, Lars McNaughton, and Andrew Jones, "Training to Enhance the Physiological Determinants of Long-Distance Running Performance," *Sports Medicine* 37 no. 10 (October 2007): 857–880. doi: 10.2165/00007256-200737100-00003.

7. Anaël Aubry, Christophe Hausswirth, Julien Louis, Aaron J. Coutts, and Yann Le Meur, "Functional Overreaching: The Key to Peak Performance During the Taper?," *Medicine & Science in Sports & Exercise* 46 no. 9 (2014): 1769–1777. doi: 10.1249/MSS.0000000000000301.

PAGES 114–115

1. Eileen Robertson, Philo Saunders, David Pyne, Robert Aughey, Judith Anson, and Christopher J. Gore, "Reproducibility of Performance Changes to Simulated Live High/Train Low Altitude," *Medicine & Science in Sports & Exercise* 42 no. 2 (February 2010): 394–401. doi: 10.1249/MSS.0b013e3181b34b57.

2. James Stray-Gundersen, Robert Chapman, and Benjamin Levine, ""Living High-Training Low" Altitude Training Improves Sea Level Performance in Male and Female Elite Runners," *Journal of Applied Physiology* 91 no. 3 (September 2001): 1113–1120.

3. Gregoire Millet, B. Roels, L. Schmitt, X. Woorons, and J. Richalet, "Combining Hypoxic Methods for Peak Performance," *Sports Medicine* 40 no. 1 (January 2010): 1–25. doi: 10.2165/11317920-000000000-00000.

4. Darrell Bonetti and Will Hopkins, "Sea-Level Exercise Performance Following Adaptation to Hypoxia: A Meta-Analysis," *Sports Medicine* 39 no. 2 (February 2009): 107–127. doi: 10.2165/00007256-200939020-00002.

5. Ben Holliss, Richard Burden, Andrew Jones, and Charles Pedlar, "Eight Weeks of Intermittent Hypoxic Training Improves Submaximal Physiological Variables in Highly Trained Runners," *Journal of Strength and Conditioning Research* 28 no. 8 (August 2014): 2195–2203. doi: 10.1519/JSC.0000000000000406.

6. James Stray-Gundersen, Robert F. Chapman, and Benjamin D. Levine, "'Living High-Training Low'" Altitude Training Improves Sea Level Performance in Male and Female Elite Runners," *Journal of Applied Physiology* 91 no. 3 (September 2001): 1113–1120. doi: 10.1034/j.1600-0838.2002.120111_1.x.

PAGES 116–117

1. Gregg Semenza, "Regulation of Oxygen Homeostasis by Hypoxia-Inducible Factor 1," *Physiology* 24 (April 2009): 97–106. doi: 10.1152/physiol.00045.2008.

2. Mujika, Iñigo. *Endurance Training – Science and Practice.* Vitoria-Gasteiz, 2012.

3. Iñigo Mujika and Sabino Padilla "Detraining: Loss of Training-Induced Physiological and Performance Adaptations. Part 1," *Sports Medicine* 30 no. 2 (August 2000): 79–87. doi: 10.2165/00007256-200030020-00002

4. Ibid.

5. Ibid.

CHAPTER 6
equipment

PAGES 126–127

1. E. C. Frederick, J. T. Daniels, and J. W. Hayes, "The Effect of Shoe Weight on the Aerobic Demands of Running," in: "Current Topics in Sports Medicine, Proceedings of the World Congress of Sports Medicine," ed. N. Bachl, L. Prokop and R. Suckert. Urban & Schwarzenberg: Vienna, (1984): 616–625.

2. Jason R. Franz, Corbyn M. Wierzbinski, and Rodger Kram, "Metabolic Cost of Running Barefoot Versus Shod: Is Lighter Better?," *Medicine & Science in Sports & Exercise* 44 no. 8 (August 2012): 1519–1525. doi: 10.1249/MSS.0b013e3182514a88.

3. Ibid.

4. Melissa A. Thompson, S. S. Lee, Jeff G. Seegmiller, and Craig P. McGowan, "Kinematic and Kinetic Comparison of Barefoot and Shod Running in Mid/Forefoot and Rearfoot Strike Runners," *Gait & Posture* 41 no. 4 (May 2015): 957–959. doi: 10.1016/j.gaitpost.2015.03.002.

5. Ibid.

6. Franz et al., "Metabolic Cost of Running Barefoot versus Shod: Is Lighter Better?."

PAGES 130–131

1. John R. Jakeman, Chris Byrne, and Roger G. Eston, "Lower Limb Compression Garment Improves Recovery from Exercise-Induced Muscle Damage in Young, Active Females," *European Journal of Applied Physiology* 109 no. 6 (August 2010): 1137–1144. doi: 10.1007/s00421-010-1464-0.

2. Rob Duffield, Jack Cannon, and Monique King, "The Effects of Compression Garments on Recovery of Muscle Performance Following High-Intensity Sprint and Plyometric Exercise," *Journal of Science and Medicine in Sport* 13 no. 1 (January 2010): 136–140. doi: 10.1016/j.jsams.2008.10.006.

3. Ajmol Ali, Mike P. Caine, and B. G. Snow, "Graduated Compression Stockings: Physiological and Perceptual Responses During and After Exercise," *Journal of Sports Sciences* 25 no. 4 (February 2007): 413–419. doi: 10.1080/02640410600718376.

4. Jakeman et al., "Lower Limb Compression Garment Improves Recovery from Exercise-Induced Muscle Damage in Young, Active Females."

5. Abigail S. L. Stickford, Robert F. Chapman, J. D. Johnston, and J. M. Stager, "Lower-Leg Compression, Running Mechanics, and Economy in Trained Distance Runners," *International Journal of Sports Physiology and Performance* 10 no. 1 (January 2015): 76–83. doi: 10.1123/ijspp.2014-0003.

6. Duffield et al., "The Effects of Compression Garments on Recovery of Muscle Performance Following High-Intensity Sprint and Plyometric Exercise."

7. Ajmol Ali, Mike P. Caine, and B. G. Snow, "Graduated Compression Stockings: Physiological and Perceptual Responses During and After Exercise," *Journal of Sports Sciences* 25 no. 4 (February 2007): 413–419. doi: 10.1080/02640410600718376.

8. John R. Jakeman, Chris Byrne, and Roger G. Eston, "Lower Limb Compression Garment Improves Recovery from Exercise-Induced Muscle Damage in Young, Active Females," *European Journal of Applied Physiology* 109 no. 6 (August 2010): 1137–1144. doi: 10.1007/s00421-010-1464-0.

PAGES 132–133

1. Ni Wang, Anxia Zha, and Jinxiu Wang, "Study on the Wicking Property of Polyester Filament Yarns," *Fibers and Polymers* 9 no. 1 (February 2008): 97–100. doi:10.1007/s12221-008-0016-2.

2. Bruce M. Latta, "Improved Tactile and Sorption Properties of Polyester Fabrics Through Caustic Treatment," *Textile Research Journal* 54 no. 11 (November 1984): 766–775. doi: 10.1177/004051758405401110.

PAGES 134–135

1. Samuel M. Fox III, John P. Naughton, and W. L. Haskell, "Physical Activity and the Prevention of Coronary Heart Disease," *Annals of Clinical Research* 3 (December 1971): 404–432.

2. W. C. Miller, J. P. Wallace, and K. E. Eggert, "Predicting Max HR and the HR-VO_2 Relationship for Exercise Prescription in Obesity," *Medicine & Science in Sports & Exercise* 25 no. 9 (September 1993): 1077–1081.

3. Hirofumi Tanaka, Kevin D. Monahan, and Douglas R. Seal, "Age-Predicted Maximal Heart Rate Revisited," *Journal of the American College of Cardiology* 37 no. 1 (January 2001):153–156. doi: 10.1016/S0735-1097(00)01054-8.

4. Greg P. Whyte, Keith George, Rob Shave, Natalie Middleton, and Alan M. Nevill, "Training Induced Changes in Maximum Heart Rate," *International Journal of Sports Medicine* 29 no. 2 (February 2008): 129–133. doi: 10.1055/s-2007-965783.

5. Ibid.

PAGES 138–139

1. C. Butler, "Sunglasses Should Be Worn All Year to Protect Eyes from UV Rays," *Washington Post* (July 4, 2011), accessed October 27, 2016, https://www.washingtonpost.com/national/health-science/sunglasses-should-be-worn-all-year-to-protect-eyes-from-uv-rays/2011/06/27/gHQAW005xH_story.html.

CHAPTER 7
running well

PAGES 144–145

1. Mark J. Comerford and Sarah L. Mottram, "Movement and Stability Dysfunction-Contemporary Developments," *Manual Therapy* 6 no. 1 (February 2001):15–26. doi: 10.1054/math.2000.0388.

2. Mark J. Comerford and Sarah L. Mottram, *Kinetic Control The Management of Uncontrolled Movement* (Australia: Elsevier, 2014).

3. G. Lorimer Mosely and Paul W. Hodges, "Reduced Variability of Postural Strategy Prevents Normalization of Motor Changes Induced by Back Pain: A Risk Factor for Chronic Trouble?," *Behavioral Neuroscience* 120 no. 2 (April 2006): 474–476. doi: 10.1037/0735-7044.120.2.474.

4. Peter O'Sullivan, "Diagnosis and Classification of Chronic Low Back Pain Disorders: Maladaptive Movement and Motor Control Impairments as Underlying Mechanism," *Manual Therapy* 10 no. 4 (November 2005): 242–255. doi: 10.1016/j.math.2005.07.001.

5. Wim Dankaerts and Peter O'Sullivan, "The Validity of O'Sullivan's Classification System (CS) for a Sub-group of NS-CLBP with Motor Control Impairment (MCI): Overview of a Series of Studies and Review of the Literature," *Manual Therapy* 16 no. 1 (February 2011): 9–14. doi: 10.1016/j.math.2010.10.006.

6. Comerford and Mottram, "Movement and Stability Dysfunction-Contemporary Developments."

7. Sahrmann, Shirley A. *Diagnosis and Treatment of Movement Impairment Syndromes.* United States of America: Mosby, 2005.

8. Comerford and Mottram, *Kinetic Control The Management of Uncontrolled Movement.*

9. Wim Dankaerts and Peter O'Sullivan, "The Validity of O'Sullivan's Classification System (CS) for a Sub-group of NS-CLBP with Motor Control Impairment (MCI): Overview of a Series of Studies and Review of the Literature."

10. Sahrmann, *Diagnosis and Treatment of Movement Impairment Syndromes.*

11. G. Lorimer Mosely and Paul W. Hodges, "Are the Changes in Postural Control Associated with Low Back Pain Caused by Pain Interference?," *Clinical Journal of Pain* 21 no. 4 (August 2005): 323–329. doi: 10.1097/01.ajp.0000131414.84596.99.

12. Marilyn R. Gossman, Shirley A. Sahrmann, and Steven J. Rose, "Review of Length-Associated Changes in Muscle," *Physical Therapy* 62 no. 12 (December 1982):1799–1808.

13. Peter B. O'Sullivan, Lance Twomey, and Garry T. Allison, "Altered Abdominal Muscle Recruitment in Patients with Chronic Back Pain Following a Specific Exercise Intervention," *Journal of Orthopaedic & Sports Physical Therapy* 27 no. 2 (February 1998): 114–124. doi: 10.2519/jospt.1998.27.2.114.

14. Comerford and Mottram, "Movement and Stability Dysfunction-Contemporary Developments."

15. Linda R. Van Dillen, K. S. Maluf, and Shirley A. Sahrmann, "Further Examination of Modifying Patient-Preferred Movement and Alignment Strategies in Patients with Low Back Pain During Symptomatic Tests," *Manual Therapy* 14 no. 1 (February 2009): 52–60. doi: 10.1016/j.math.2007.09.012.

16. Comerford and Mottram, *Kinetic Control The Management of Uncontrolled Movement.*

17. Ibid.

18. Nathalie A. Roussel, Jo Nijs, Sarah Mottram, Annouk Van Moorsel, Steven Truijen, and Gaetane Stassijns, "Altered Lumbopelvic Movement Control But Not Generalized Joint Hypermobility is Associated With Increased Injury in Dancers: A Prospective Study," *Manual Therapy* 14 no. 6 (December 2009): 630–635. doi: 10.1016/j.math.2008.12.004.

19. Sarah Mottram and Mark Comerford, "A New Perspective on Risk Assessment," *Physical Therapy in Sport* 9 no. 1 (February 2008): 40–51. doi: 10.1016/j.ptsp.2007.11.003.

20. Eyal Lederman, "The Myth of Core Stability," *CPDO Online Journal* 14 no. 1 (January 2007): 1–17. doi: 10.1016/j.jbmt.2009.08.001.

21. Mari Leppanen, Sari Aaltonen, Jari Parkkari, Ari Heinonen, and Urho M. Kujala, "Interventions to Prevent Sports Related Injuries: A Systematic Review and Meta-Analysis of Randomised Controlled Trials," *Journal of Sports Medicine* 44 no. 4 (April 2014): 473–486. doi: 10.1007/s40279-013-0136-8.

22. Jari Parkkari, Henri Taanila, Jaana Suni, Ville M. Mattila, Olli Ohrankammen, Petteri Vuorinen, Pekka Kannus, and Harri Pihlajamaki, "Neuromuscular Training with Injury Prevention Counselling to Decrease the Risk of Acute Musculoskeletal Injury in Young Men During Military Service: A Population-Based, Randomised Study," *BMC Medicine* 9 no. 35 (2011). doi: 10.1186/1741-7015-9-35.

23. Timothy E. Hewett, Thomas N. Lindenfeld, Jennifer V. Riccobene, and Frank R. Noyes, "The Effect of Neuromuscular Training on the Incidence of Knee Injury in Female Athletes: A Prospective Study," *The American Journal of Sports Medicine* 27 no. 6 (November–December 1999): 699–705.

24. Jeppe B. Lauersen, Ditte M. Bertelsen, and Lars B. Andersen, "The Effectiveness of Exercise Interventions to Prevent Sports Injuries: A Systematic Review and Meta-Analysis of Randomised Controlled Trials," *British Journal of Sports Medicine* 48 no. 11 (October 2013). doi: 10.1136/bjsports-2013-092538.

25. Leppanen et al., "Interventions to Prevent Sports Related Injuries: A Systematic Review and Meta-Analysis of Randomised Controlled Trials."

26. Kimitake Sato and Monique Mokha, "Does Core Strength Training Influence Running Kinetics, Lower-Extremity Stability, and 5000-M Performance in Runners?," *The Journal of Strength & Conditioning Research* 23 no. 1 (January 2009): 133–140. doi: 10.1519/JSC.0b013e31818eb0c5.

27. Comerford and Mottram, *Kinetic Control The Management of Uncontrolled Movement.*

28. W. Ben Kibler, "The Role of the Scapula in Athletic Shoulder Function," *The American Journal of Sports Medicine* 26 no. 2 (March–April 1998): 325–337. doi: 10.1177/03635465980260022801.

29. Carolyn Richardson, Paul W. Hodges, and Julie Hides. *Therapeutic Exercise for Lumbopelvic Stabilization—A Motor Control Approach for the Treatment and Prevention of Lower Back Pain.* Churchill Livingstone, 2014.

PAGES 146–147

1. Malachy P. Hugh and C. H. Cosgrave, "To Stretch or Not to Stretch: The Role of Stretching in Injury Prevention and Performance," *Scandinavian Journal of Medicine & Science in Sports* 20 no 2. (April 2010): 169–181. doi: 10.1111/j.1600-0838.2009.01058.x.

2. L. Simic, N. Sarabon and Goran Markovic, "Does Pre-Exercise Static Stretching Inhibit Maximal Muscular Performance? A Meta-Analytical Review," *Scandinavian Journal of Medicine & Science in Sports* 23 no. 2 (March 2013): 131–148. doi: 10.1111/j.1600-0838.2012.01444.x.

3. Ercole C. Rubini, André L. Costa, and Paulo Sergio Gomes, "The Effects of Stretching on Strength Performance," *Sports Medicine* 37 no. 3 (February 2007): 213–224. doi: 10.2165/00007256-200737030-00003.

4. Ian Shrier, "Stretching Before Exercise Does Not Reduce the Risk of Local Muscle Injury: A Critical Review of the Clinical and Basic Science Literature," *Clinical Journal of Sport Medicine* 9 no. 4 (November 1999): 191–251.

5. S. M. Weldon and R. H. Hill, "The Efficacy of Stretching for Prevention of Exercise Related Injury: A Systematic Review of the Literature," *Manual Therapy* 8 no. 3 (September 2003): 141–150. doi: 10.1016/S1356-689X(03)00010-9.

6. Duane V. Knudson, Peter Magnusson, and Malachy McHugh, "Current Issues in Flexibility Fitness," *President's Council on Physical Fitness and Sports* 3 no. 10 (June 2000): 1–6.

7. W. E. Garrett Jr., "Muscle Strain Injuries," *The American Journal of Sports Medicine* 24 no. S2 (1996): S2–S8.

8. Simon S. Yeung, Ella W. Yeung, and Lesley D. Gillespie, "Interventions for Preventing Lower Limb Soft-Tissue Running Injuries," *The Cochrane Database of Systematic Reviews* 6 no. 7 (July 2011): CD001256. doi: 10.1002/14651858.CD001256.pub2.

9. Rubini et al., "The Effects of Stretching on Strength Performance."

10. Hugh et al., "To Stretch or Not to Stretch: The Role of Stretching in Injury Prevention and Performance."

11. Simic et al., "Does Pre-Exercise Static Stretching Inhibit Maximal Muscular Performance? A Meta-Analytical Review."

12. Stephen B. Thacker, Julie Gilchrist, Donna F. Stroup, and C. Dexter Kimsey Jr., "The Impact of Stretching on Sports Injury Risk: A Systematic Review of the Literature," *Medicine & Science in Sports & Exercise* 36 no. 3 (March 2004): 371–378.

13. Ibid.

14. Timonthy E. Hewitt, Thomas N. Lindenfeild, Jennifer V. Riccobene, and Frank R. Noyes, "The Effects of Neuromuscular Training on the Incidence of Knee Injury in Female Athletes," *The American Journal of Sports Medicine* 27 no. 6 (November 1999): 699–704.

15. Hugh et al., "To Stretch or Not to Stretch: The Role of Stretching in Injury Prevention and Performance."

16. Nicholas Caplan, Rebecca Rogers, Michael K. Parr, and Philip R. Hayes, "The Effect of Proprioceptive Neuromuscular Facilitation and Static Stretch Training on Running Mechanics," *The Journal of Strength & Conditioning Research* 23 no. 4 (July 2009): 1175–1180. doi: 10.1519/JSC.0b013e318199d6f6.

17. Gro Jamtvedt, Robert D. Herbert, Signe Flottorp, Jan Odgaard-Jensen, Karl Havelsrud, Alex Barratt, Erin Mathieu, Amanda Burls, and Andrew Oxman, "A Pragmatic Randomised Trial of Stretching Before and After Physical Activity to Prevent Injury and Soreness," *British Journal of Sports Medicine* 44 no. 14 (2010): 1002–1009. doi: 10.1136/bjsm.2009.062232.

18. Katie Small, Lars Mc Naughton, and Martyn M Matthews, "A Systematic Review into the Efficacy of Static Stretching as Part of a Warm-Up for the Prevention of Exercise-Related Injury," *Research in Sports Medicine* 16 no. 3 (2008): 213–231. doi: 10.1080/15438620802310784.

19. Phil Page, "Current Concepts in Muscle Stretching for Exercise and Rehabilitation," *International Journal of Sports Physical Therapy* 7 no. 1 (February 2012): 109–119.

20. C. A. Smith, "The Warm-Up Procedure: To Stretch or Not to Stretch. A Brief Review," *The Journal of Orthopaedic and Sports Physical Therapy* 19 no. 1 (January 1994): 12–17. doi: 10.2519/jospt.1994.19.1.12.

21. Daniel Cipriani, Bobbie Abel, and Dayna Pirrwitz, "A Comparison of Two Stretching Protocols on Hip Range of Motion: Implications for Total Daily Stretch Duration," *Journal of Strength & Conditioning Research* 17 no. 2 (May 2003): 274–278.

22. Malachy P. McHugh, S. Peter Magnusson, G. W. Gleim, and J. A. Nicholas, "Viscoelastic Stress Relaxation in Human Skeletal Muscle," Medicine & Science in Sports & Exercise 24 no. 12 (December 1992): 1375–1382. doi: 10.1111/j.1600-0838.1996.tb00101.x.

23. J. C. Tabary, C. Tabary, C. Tardieu, G. Tardieu, and G. Goldspink, "Physiological and Structural Changes in the Cat's Soleus Muscle Due to Immobilization at Different Lengths by Plaster Casts," *The Journal of Physiology* 224 no. 1 (July 1972): 231–244. doi: 10.1113/jphysiol.1972.sp009891.

24. P. E. Williams and G. Goldspink, "Changes in Sarcomere Length and Physiological Properties in Immobilized Muscle," *Journal of Anatomy* 127 no. 3 (December 1978): 459–468.

25. Richard L. Gajdosik, "Passive Extensibility of Skeletal Muscle: Review of the Literature with Clinical Implications," *Clinical Biomechanics* 16 no. 2 (February 2001): 87–101. doi: 10.1016/S0268-0033(00)00061-9.

26. Cynthia Holzman Wepplera and S. Peter Magnusson, "Increasing Muscle Extensibility: A Matter of Increasing Length or Modifying Sensation?," *Physical Therapy* 90 (March 2010): 438–449. doi: 10.2522/ptj.20090012.

27. Ulrike H. Mitchell, J. William Myrer, J. Ty Hopkins, Ian Hunter, J. Brent Feland, and Sterling C. Hilton, "Acute Stretch Perception Alteration Contributes to the Success of the PNF "Contract-Relax" Stretch," *Journal of Sport Rehabilitation* 16 no. 2 (May 2007): 85–92.

28. Caplan et al., "The Effect of Proprioceptive Neuromuscular Facilitation and Static Stretch Training on Running Mechanics."

29. Mitchell et al., "Acute Stretch Perception Alteration Contributes to the Success of the PNF "Contract-Relax" Stretch."

PAGES 148–149

1. Karoline Cheung, Patria A. Hume, and Linda Maxwell, "Delayed Onset Muscle Soreness Treatment Strategies and Performance Factors," *Sports Medicine* 33 no. 2 (February 2003): 145–164. doi: 10.2165/00007256-200333020-00005.

2. Robert D. Herbert, Marcos de Noronha, and Steven J. Kamper, "Stretching to Prevent or Reduce Muscle Soreness After Exercise, Editorial Group: Cochrane Bone, Joint and Muscle Trauma Group," *Cochrane Database of Systematic Reviews* (July 2011). doi: 10.1002/14651858.CD004577.pub3.

3. J. Friden, M. Sjostrom, and B. Ekblom, "Myofibrillar Damage Following Intense Eccentric Exercise in Man," *International Journal of Sports Medicine* 4 no. 3 (1983): 170–176. doi: 10.1055/s-2008-1026030.

4. D. J. Newham, G. Mc Phail, K. R. Mills, and R. H. Edwards, "Ultrastructural Changes after Concentric and Eccentric Contractions of Human Muscle," *Journal of Neurological Sciences* 61 no. 1 (September 1983): 109–122. doi: 10.1016/0022-510X(83)90058-8.

5. D. L. Morgan and D. G. Allen, "Early Events in Stretch-Induced Muscle Damage," *Journal of Applied Physiology* 87 no. 6 (December 1999): 2007–2015.

6. U. Proske and D. L. Morgan, "Muscle Damage from Eccentric Exercise: Mechanism, Mechanical Signs, Adaptation and Clinical Applications," *The Journal of Physiology* 538 no. 2 (December 2001): 333–345. doi: 10.1111/j.1469-7793.2001.00333.x.

7. S. J. Brown, R. B. Child, S. H. Day, and A. E. Donelly, "Exercise-Induced Skeletal Muscle Damage and Adaptation Following Repeated Bouts of Eccentric Muscle Contractions," *Journal of Sports Sciences* 15 no. 2 (April 1997): 215–222. doi: 10.1080/026404197367498.

8. Malachy P. McHugh, "Recent Advances in the Understanding of the Repeated Bout Effect: The Protective Effect Against Damage from a Single Bout of Exercise," *Scandinavian Journal of Medicine & Science in Sports* 13 no. 2 (April 2003): 88–97. doi: 10.1034/j.1600-0838.2003.02477.x.

9. Priscilla M. Clarkson and Monica J. Hubal, "Exercise Induced Muscle Damage in Humans," *American Journal of Physical Medication & Rehabilitation* 81 no. 11 (November 2002): 52–69. doi: 10.1097/01.PHM.0000029772.45258.43.

10. Kazunori Nosaka, Mike Newton, and Paul Sacco, "Delayed-Onset Muscle Soreness Does Not Reflect the Magnitude of Eccentric Exercise Induced Muscle Damage," *Scandinavian Journal of Medicine & Science in Sports* 12 no. 6 (December 2002): 337–346. doi: 10.1034/j.1600-0838.2002.10178.x.

11. Brad J. Schoenfeld and Bret Contreras, "Is Post Exercise Muscle Soreness a Valid Indicator of Muscular Adaptations?," *Strength and Conditioning Journal* 35 no. 5 (October 2013): 16–21. doi: 0.1519/SSC.0b013e3182a61820.

12. Camilla L. Brockett, David L. Morgan, and Uwe Proske, "Human Hamstring Muscles Adapt to Eccentric Exercise by Changing Optimum Length," *Medicine & Science in Sports & Exercise* 33 no. 5 (May 2001): 783–790.

13. John A. Faulkner, Susan V. Brooks, and Julie A. Opiteck, "Injury to Skeletal Muscle Fibers During Contractions: Conditions of Occurrence and Prevention," *Physical Therapy* 73 no. 12 (December 1993): 911–921.

14. Uwe Proske and David L. Morgan, "Muscle Damage from Eccentric Exercise: Mechanism, Mechanical Signs, Adaptation and Clinical Applications," *Journal of Physiology* 537 no. 2 (2001): 333–345. doi: 10.1111/j.1469-7793.2001.00333.x.

15. Ibid.

16. Ji-Guo Yu, Dieter O. Furst, and Lars-Eric Thornell, "The Mode of Myofibril Remodelling in Human Skeletal Muscles Affected by DOMs Induced by Eccentric Contractions," *Histochemistry and Cell Biology* 119 no. 5 (May 2003): 383–393. doi: 10.1007/s00418-003-0522-7.

17. Ji-Guo Yu, L. Carlsson, and Lars-Eric Thornell, "Evidence for Myofibril Remodelling as Opposed to Myofibril Damage in Human Muscles with DOMs: An Ultrastructural and Immunoelectron Microscopic Study," *Histochemistry and Cell Biology* 121 (March 2004): 219–227. doi: 10.1007/s00418-004-0625-9.

18. McHugh, "Recent Advances in the Understanding of the Repeated Bout Effect: The Protective Effect Against Damage from a Single Bout of Exercise."

19. Camilla L. Brockett, David L. Morgan, and Uwe Proske, "Human Hamstring Muscles Adapt to Eccentric Exercise by Changing Optimum Length," *Medicine & Science in Sports & Exercise* 33 no. 5 (May 2001): 783–790. doi: 10.1097/00005768-200105000-00017.

20. Newham et al., "Ultrastructural Changes after Concentric and Eccentric Contractions of Human Muscle."

21. Proske and Morgan, "Muscle Damage from Eccentric Exercise: Mechanism, Mechanical Signs, Adaptation and Clinical Applications."

22. Fredrik Lauritzen, Gøran Paulsen, Truls Raastad, Linde H. Bergersen, and Simen G. Owe, "Gross Ultrastructural Changes and Necrotic Fiber Segments in Elbow Flexor Muscles After Maximal Voluntary Eccentric Action in Humans," *Journal of Applied Physiology 1985* 107 no. 6 (December 2009): 1923–1934. doi: 10.1152/japplphysiol.00148.2009.

23. Andrew J. Mc Kune, Stuart J. Semple, and E. M. Peters-Futre, "Acute Exercise Induced Muscle Injury," *Biology of Sport* 29 no. 1 (January 2012): 3–10. doi: 10.5604/20831862.978976.

24. Clarkson and Hubal, "Exercise Induced Muscle Damage in Humans."

25. Mc Kune et al., "Acute Exercise Induced Muscle Injury."

PAGES 150–151

1. Brian Hemmings, Marcus Smith, Jan Graydon, and Rosemary Dyson, "Effects of Massage on Physiological Restoration, Perceived Recovery, and repeated sports performance," *British Journal of Sports Medicine* 34 no. 2 (2000): 109–114. doi: 10.1136/bjsm.34.2.109.

2. Brian Hemmings, "Physiological, Psychological and Performance Effects of Massage Therapy in Sport: A Review of the Literature," *Physical Therapy in Sport* 2 no. 4 (2001): 165–170. doi: 10.1054/ptsp.2001.0070.

3. I. M. Fletcher, "The Effects of Pre-Competition Massage on the Kinematic Parameters of 20-m Sprint Performance," *Journal of Strength and Conditioning Research* 24 no. 5 (May 2010): 1179–1183. doi: 10.1519/JSC.0b013e3181ceec0f.

4. Jason Brummit, "The Role of Massage in Sports Performance and Rehabilitation: Current Evidence and Future Direction," *North American Journal of Sports Physical Therapy* 3 no. 1 (February 2008): 7–21.

5. Pornratshanee Weerapong, Patria A. Hume, and Gregory S. Kolt, "The Mechanisms of Massage and Effects on Performance, Muscle Recovery and Injury Prevention," *Sports Medicine* 35 no. 3 (2005): 235–256.

6. J. Hilbert, G. A. Sforzo, and Thomas Swenson, "The Effects of Massage on Delayed Onset Muscle Soreness," *British Journal of Sports Medicine* 37 no. 1 (2003): 72–75. doi: 10.1136/bjsm.37.1.72.

7. Zainal Zainuddin, Mike Newton, Paul Sacco, and Kazunori Nosaka, "Effects of Massage on Delayed Onset Muscle Soreness, Swelling and Recovery of Muscle Function," *Journal of Athletic Training* 40 no. 3 (July–September 2005): 174–180.

8. C. Haas, T. Butterfield, S. Abshire, Y. Zhao, X. Zhang, D. Jarjoura, and T. M. Best, "Massage Timing Affects Postexercise Muscle Recovery and Inflammation in a Rabbit Model," *Medicine & Science in Sports & Exercise* 45 no. 6 (June 2003): 1105–1112. doi: 10.1249/MSS.0b013e31827fdf18.

9. Ibid.

10. T. M. Best, R. Hunter, A. Wilcox, and F. Haq, "Effectiveness of Sports Massage for Recovery of Skeletal Muscle From Strenuous Exercise," *Clinical Journal of Sports Medicine* 18 no. 5 (September 2008): 446–460. doi: 10.1097/JSM.0b013e31818837a1.

11. Attila Szabo, Mária Rendi, Tamás Szabó, Attila Velenczei, and Árpád Kovács, "Psychological Effects of Massage on Running," *Journal of Social, Behavioral, and Health Sciences* 2 no. 1 (2008): 1–7. doi: 10.5590/JSBHS.2008.02.1.01.

12. Hemmings et al., "Effects of Massage on Physiological Restoration, Perceived Recovery, and repeated sports performance."

13. Hemmings, "Physiological, Psychological and Performance Effects of Massage Therapy in Sport: A Review of the Literature."

14. Weerapong et al., "The Mechanisms of Massage and Effects on Performance, Muscle Recovery and Injury Prevention."

15. Manuel Arroyo-Morales, Nicolas Olea, Concepción Ruíz, Juan de Dios Luna del Castilo, Manuel Martínez, Carmen Lorenzo, and Lourdes Díaz-Rodríguez, "Massage After Exercise—Responses of Immunologic and Endocrine Markers: A Randomized Single-Blind Placebo-Controlled Study," *Journal of Strength and Conditioning Research* 23 no. 2 (March 2009): 638–644. doi: 0.1519/JSC.0b013e318196b6a6.

16. Mark H. Rapaport, Pamela Schettler, and Catherine Bresee, "A Preliminary Study of the Effects of Repeated Massage on Hypothalamic–Pituitary–Adrenal and Immune Function in Healthy Individuals: A Study of Mechanisms of Action and Dosage," *The Journal of Alternative and Complementary Medicine* 18 no. 8 (August 2012): 789–797. doi: 10.1089/acm.2011.0071.

17. J. Y. Ang, J. L. Lua, A. Mathur, R. Thomas, B. I. Asmar, S. Savasan, S. Buck, M. Long, and S. Shankaran, "A Randomized Placebo-Controlled Trial of Massage Therapy on the Immune System of Preterm Infants," *Pediatrics* 130 no. 6 (December 2012): e1549–e1558. doi: 10.1542/peds.2012-0196.

18. Alberta Moraska, Robin A. Pollini, Karen Boulanger, Marissa Z. Brooks, and Lesley Teitlebaum, "Physiological Adjustments to Stress Measures Following Massage Therapy: A Review of the Literature," *Evidence-Based Complementary and Alternative Medicine* 7 no. 4 (December 2008): 409–418. doi: 10.1093/ecam/nen029.

19. Robert Schleip, Thomas W. Findley, Leon Chaitow, and Peter Huijung, *Fascia: The Tensional Network of the Human Body*, (Australia: Elsevier, 2012).

20. Leon Chaitow, *Fascial Dysfunction: Maunual Therapy Approaches*, (United Kingdom: Handspring Publishing, 2014).

21. Ibid.

22. B. Bramah, "Swedish Massage in a Sporting Context: Part One," *Sportex Dynamics Journal* 44 (April 2015).

23. Giampietro L. Vairo, Sayers J. Miller, Nicole M. McBrier, and William E. Buckley, "Systematic Review of Efficacy for Manual Lymphatic Drainage Techniques in Sports Medicine and Rehabilitation: An Evidence-Based Practice Approach," *Journal of Manual and Manipulative Therapy* 17 no. 3 (2009): e80–e89. doi: 10.1179/jmt.2009.17.3.80E.

24. Zainuddin et al., "Effects of Massage on Delayed Onset Muscle Soreness, Swelling and Recovery of Muscle Function."

PAGES 152–153

1. C. Bleakley, S. McDonough, E. Gardner, G. D. Baxter, J. T. Hopkins, and G. W. Davison, "Cold Water Immersion (Cryotherapy) for Preventing and Treating Muscle Soreness After Exercise," *The Cochrane Database of Systematic Reviews* 15 no. 2 (February 2012): 1–22. doi: 10.1002/14651858.CD008262.pub2.

2. Llion A. Roberts, Truls Raastad, James F. Markworth, Vandre C. Figueiredo, Ingrid M. Egner, David Cameron-Smith, Jeff S. Coombes, and Jonathan M. Peake, "Post-Exercise Cold Water Immersion Attenuates Acute Anabolic Signalling and Long-Term Adaptations in Muscle to Strength Training," *Journal of Physiology* 593 no. 18 (September 2015): 4285–4301. doi: 10.1113/JP270570.

3. S. L. Halson, J. Bartram, N. West, J. Stephens, C. K. Argus, M. W. Driller, C. Saerent, M. Lastella, W. G. Hopkins, and D. T. Martin, "Does Hydrotherapy Help or Hinder Adaptation to Training in Competitive Cyclists?," *Medicine & Science in Sports & Exercise* 46 no. 8 (August 2014): 1631–1639. doi: 10.1249/MSS.0000000000000268.

4. Alan Dunne, David Crampton, and Michael Egaña, "Effect of Post-Exercise Hydrotherapy Water Temperature on Subsequent Exhaustive Running Performance in Normothermic Conditions," *Journal of Science and Medicine in Sport* 16 no. 5 (September 2013): 466–471. doi: 10.1016/j.jsams.2012.11.884.

5. Marc J. Quod, David Martin, and Paul B. Laursen, "Cooling Athletes Before Competition in the Heat: Comparison of Techniques and Practical Considerations," *Sports Medicine* 36 no. 8 (August 2006): 671–682. doi: 10.2165/00007256-200636080-00004.

6. J. Vaile, S. Halson, and N. Gill, "Effect of Cold Water Immersion on Repeat Cycling Performance and Thermoregulation in the Heat," *Journal of Sports Sciences* 26 no. 5 (March 2008): 431–440. doi: 10.1080/02640410701567425.

7. Nathan G. Versey, Shona L. Halson, and Brian T. Dawson, "Water Immersion Recovery for Athletes: Effect on Exercise Performance and Practical Recommendations," *Sports Medicine* 43 no. 11 (November 2013): 1101–1130. doi: 10.1007/s40279-013-0063-8.

8. S. W. Yeargin, D. J. Casa, and J. M. McClung, "Body Cooling Between Two Bouts of Exercise in the Heat Enhances Subsequent Performance," *Journal of Strength and Conditioning Research* 20 no. 2 (May 2006): 383–389. doi: 10.1519/R-18075.1.

9. D. C. Mac Auley, "Ice Therapy: How Good is the Evidence?," *International Journal of Sports Medicine* 22 no. 5 (July 2001): 379–384.

10. Versey et al., "Water Immersion Recovery for Athletes: Effect on Exercise Performance and Practical Recommendations."

11. J. M. Peake, L. A. Roberts, V. C. Figueiredo, I. Egner, S. Krog, S. N. Aas, K. Suzuki, J. F. Markworth, J. S. Coombes, D. Cameron-Smith, and T. Raastad, "The Effects of Cold Water Immersion and Active Recovery on Inflammation and Cell Stress Responses in Human Skeletal Muscle After Resistance Exercise," *Journal of Physiology* (2016). doi: 10.1113/JP272881.

12. François Bieuzen, Chris M. Bleakley, and Joseph T. Costello, "Contrast Water Therapy and Exercise Induced Muscle Damage: A Systematic Review and Meta-Analysis," *PLoS ONE* 4 (April 2013). doi: 10.1371/journal.pone.0062356.

13. Daryl J. Cochrane, "Alternating Hot and Cold Water Immersion for Athlete Recovery: A Review," *Physical Therapy in Sport* 5 no. 1 (February 2004): 26–32. doi: 10.1016/j.ptsp.2003.10.002.

14. Bleakley et al., "Cold Water Immersion (Cryotherapy) for Preventing and Treating Muscle Soreness After Exercise."

15. Lindy L. Washington, Stephen J. Gibson, and Robert D. Helme, "Age-Related Differences in the Endogenous Analgesic Response to Repeated Cold Water Immersion in Human Volunteers," *Pain* 89 no. 1 (December 2000): 89–96. doi: 10.1016/S0304-3959(00)00352-3

16. W. Freund, F. Weber, C. Billich, F. Birklien, M. Breimhorst, and U. H. Schuetz, "Ultra-Marathon Runners Are Different: Investigations Into Pain Tolerance and Personality Traits of Participants of the TransEurope FootRace 2009," *Pain* 13 no. 7 (September 2013): 524–532. doi: 10.1111/papr.12039.

17. Versey et al., "Water Immersion Recovery for Athletes: Effect on Exercise Performance and Practical Recommendations."

18. Ian M. Wilcock. John B. Cronin, and Wayne A. Hing, "Physiological Response to Water Immersion. A Method for Sport Recovery?," *Sports Medicine* 36 no. 9 (September 2006): 747–765. doi: 10.2165/00007256-200636090-00003.

19. Cochrane, "Alternating Hot and Cold Water Immersion for Athlete Recovery: A Review."

20. Wilcock et al., "Physiological Response to Water Immersion. A Method for Sport Recovery?."

21. Joel A. DeLisa, Bruce M. Gans, and Nicholas E. Walsh, *Physical Medicine and Rehabilitation. Principles and Practice*, (United States of America: Lippincott, Williams and Wilkins, 2005).

22. Kwon Sik Park, Jang Kyu Choi, and Yang Saeng Park, "Cardiovascular Regulation During Water Immersion," *Journal of Physiological Anthropology and Applied Human Science* 18 no. 6 (November 1999): 233–241. doi: 10.2114/jpa.18.233.

23. K. J. Hayter, K. Doma, M. Schumann, and G. B. Deakin, "The Comparison of Cold-Water Immersion and Cold Air Therapy on Maximal Cycling Performance and Recovery Markers Following Strength Exercises," *PeerJ* 4 (March 2016): e1841. doi: 10.7717/peerj.1841.

PAGES 154–155

1. D. M. Urquhart, J. F. Tobing, F. S. Hanna, P. Berry, A. E. Wluka, C. Ding, and F. M. Cicuttini, "What is the Effect of Physical Activity on the Knee Joint? A Systematic Review," *Medicine & Science in Sports & Exercise* 43 no. 3 (March 2011): 432–442. doi: 10.1249/MSS.0b013e3181ef5bf8.

2. Ibid.

3. A Van Ginckel, N. Baelde, K. F. Almgvist, P. Roosen, P. McNair, and E. Witvrouw, "Functional Adaptation of Knee Cartilage in Asymptomatic Female Novice Runners Compared to Sedentary Controls. A Longitudinal Analysis Using Delayed Gadolinium Enhanced Magnetic Resonance Imaging of Cartilage," *Osteoarthritis and Cartilage* 18 no. 12 (December 2010): 1564–1569. doi: 10.1016/j.joca.2010.10.007.

4. Tim D. Spector and Alex J. MacGregor, "Risk Factors for Osteoarthritis: Genetics 2001," *Osteoarthritis and Cartilage* 12 (2004): 39–44. doi: 10.1016/j.joca.2003.09.005.

5. S. P. Messier, D. J. Gutekunst, C. Davis, and P. DeVita, "Weight Loss Reduces Knee-Joint Loads in Overweight and Obese Older Adults With Knee Osteoarthritis," *Arthritis & Rheumatism* 52 no. 7 (July 2005): 2026–2032. doi: 10.1002/art.21139.

6. P. T. Williams, "Effects of Running and Walking on Osteoarthritis and Hip Replacement Risk," *Medicine & Science in Sports & Exercise* 45 no. 7 (July 2013): 1292–1297. doi: 10.1249/MSS.0b013e3182885f26.

7. Heidari Behzad, "Knee Osteoarthritis Prevalence, Risk Factors, Pathogenesis and Features: Part I," *Caspian Journal of International Medicine* 2 no. 2 (Spring 2011): 205–212.

8. Robert A. Magnussen, Alfred A. Mansour, James L. Carey, and Kurt P. Spindler, "Meniscus Status at Anterior Cruciate Ligament Reconstruction Associated with Radiographic Signs of Osteoarthritis at 5- to 10-Year Follow-Up: A Systematic Review," *The Journal of Knee Surgery* 22 no. 4 (October 2009): 347–357.

9. A. Van Ginckel, N. Baelde, K. F. Almqvist, P. Roosen, P. McNair, and E. Witvrouw, "Functional Adaptation of Knee Cartilage in Asymptomatic Female Novice Runners Compared to Sedentary Controls," *Osteoarthritis Cartilage* 18 no. 12 (December 2010): 1564–1569. doi: 10.1016/j.joca.2010.10.007.

10. R. H. Miller, W. B. Edwards, S. C. Brandon, A. M. Morton, and K. J. Deluzio, "Why Don't Most Runners Get Knee Osteoarthritis? A Case for Per-Unit-Distance Loads," *Medicine & Science in Sports & Exercise* 46 no. 3 (March 2014): 572–579. doi: 10.1249/MSS.0000000000000135.

11. Willem van Mechelen, "Running Injuries. A Review of the Epidemiological Literature," *Sports Medicine* 14 no. 5 (November 1992): 320–335. doi: 10.2165/00007256-199214050-00004.

12. Scott F. Dye, "The Pathophysiology of Patellofemoral Pain: A Tissue Homeostasis Perspective," *Clinical Orthopaedics and Related Research* 436 (July 2005): 100–110. doi: 10.1097/01.blo.0000172303.74414.7d.

13. B. T. Zazulak, T. E. Hewett, N. P. Reeves, B. Goldberg, and J. Cholewicki, "Deficits in Neuromuscular Control of the Trunk Predict Knee Injury Risk: A Prospective Biomechanical-Epidemiologic Study," *American Journal of Sports Medicine* 35 no. 7 (July 2007): 1123–1130. doi: 10.1177/0363546507301585.

14. Richard B. Souza, Christie E. Draper, Michael Fredericson, and Christopher M. Powers, "Femur Rotation and Patellofemoral Joint Kinematics; A Weight-Bearing Magnetic Resonance Imaging Analysis," *Journal of Orthopaedic & Sports Physical Therapy* 40 no. 5 (2010): 277–285. doi: 10.2519/jospt.2010.3215.

15. Simon Lack, "Proximal Intervention for the Management of Patella Femoral Pain," *SportEX medicine* 63 (January 2015): 22–26.

PAGES 156–157

1. W. Freund, F. Weber, C. Billich, F. Birklien, M. Breimhorst, and U. H. Schuetz, "Ultra-Marathon Runners Are Different: Investigations Into Pain Tolerance and Personality Traits of Participants of the TransEurope FootRace 2009," *Pain* 13 no. 7 (September 2013): 524–532. doi: 10.1111/papr.12039.

2. Ibid.

3. K. G. Silbernagel, R. Thomeé, B. I. Eriksson, and J. Karlsson, "Continued Sports Activity, Using a Pain-Monitoring Model, During Rehabilitation in Patients With Achilles Tendinopathy: A Randomized Controlled Study," *American Journal of Sports Medicine* 35 no. 6 (June 2007): 897. doi: 10.1177/0363546506298279.

4. David Butler and G. Lorimer Moseley, *Explain Pain* (Adelaide: NOI Group Publishing, 2003).

5. G. Lorimer Moseley, "Reconceptualising Pain According to Modern Pain Science," *Physical Therapy Reviews* 12 no. 3 (2007): 169–178. doi: 10.1179/108331907X223010.

6. C. Daly, U. Persson, R. Twycross-Lewis, R. C. Woledge, and D. Morrissey, "The Biomechanics of Running in Athletes with Previous Hamstring Injury: A Case-Control Study," *Scandinavian Journal of Medicine & Science in Sports* 26 no. 4 (April 2016): 413–420. doi: 10.1111/sms.12464.

7. Michael J. Mueller, and Katrina S. Maluf, "Tissue Adaptation to Physical Stress: A Proposed "Physical Stress Theory" to Guide Physical Therapist Practice, Education, and Research," *Physical Therapy* 82 no. 4 (April 2002): 383–403.

8. Marc R. Safran, Douglas McKeag, and Steven P. Van Camp, *Manual of Sports Medicine*, (Lippincott Williams & Wilkins, 1998).

9. Peter Brucker and Karim Khan, *Clinical Sports Medicine*, (Australia: McGraw-Hill, 2012).

PAGES 158–159

1. C. M. Bleakley, P. Glasgow, and M. J. Webb, "Cooling an Acute Muscle Injury: Can Basic Scientific Theory Translate into the Clinical Setting?," *British Journal of Sports Medicine* 46 (2012): 296–298. doi: 10.1136/bjsm.2011.086116.

2. C. M. Bleakley and J. T. Hopkins, "Is it Possible to Achieve Optimal Levels of Tissue Cooling in Cryotherapy?," *Physical Therapy Reviews* 15 no. 4 (2010): 344–350. doi: 10.1179/174328810X12786297204873.

3. C. M. Bleakley, P. Glasgow, and D. C. MacAuley, "PRICE Needs Updating, Should We Call the POLICE?," *British Journal of Sports Medicine* 46 (2012): 220–221. doi: 10.1136/bjsports-2011-090297.

4. Ibid.

5. Kenneth Knight, *Cryotherapy in Sport Injury Management*, (United States of America: Human Kinetics, 1995).

6. C. M. Bleakley, P. D. Glasgow, N. Phillips, L. Hanna, M. J. Callaghan, G. W. Davison, T. L. Hopkins, and E. Delahunt, "Management of Acute Soft Tissue Injury Using Protection Rest Ice Compression and Elevation: Recommendations from the Association of Chartered Physiotherapists in Sports and Exercise Medicine (ACPSM)," *SKIPP* (2010).

7. J. A. Paoloni, C. Milne, J. Orchard, and B. Hamilton, "Non-Steroidal Anti-Inflammatory Drugs in Sports Medicine: Guidelines for Practical but Sensible Use," *British Journal of Sports Medicine* 43 no. 11 (October 2009): 863–865. doi: 10.1136/bjsm.2009.059980.

8. Jay Hertel, "The Role of Nonsteroidal Anti-Inflammatory Drugs in the Treatment of Acute Soft Tissue Injuries," *Journal of Athletic Training* 32 no. 4 (October–December 1997): 350–358.

9. R. A. Hauser, E. E. Dolan, H. J. Phillips, A. C. Newlin, R. E. Moore, and B. A. Bentham, "Ligament Injury and Healing: A Review of Current Clinical Diagnostics and Therapeutics," *The Open Rehabilitation Journal* 6 (2013): 1–20. doi: 10.2174/1874943701306010001.

10. Robert Schleip, Thomas W. Findley, Leon Chaitow, and Peter Huijung, *Fascia: The Tensional Network of the Human Body*, (Australia: Elsevier, 2012).

11. Leon Chaitow, *Fascial Dysfunction: Maunual Therapy Approaches*, (Handspring Publishing, 2014).

12. K. M. Khan and A. Scott, "Mechanotherapy: How Physical Therapists' Prescription of Exercise Promotes Tissue Repair," *British Journal of Sports Medicine* 43 no. 4 (2009): 247–252. doi: 10.1136/bjsm.2008.054239.

13. Chenyu Huang and Rei Ogwana, "Mechanotransduction in Bone Repair and Regeneration," *The FASEB Journal* 24 no. 10 (October 2010): 3625–3632. doi: 10.1096/fj.10-157370.

14. Donald E. Ingber, "Cellular Mechanotransduction: Putting All The Pieces Together Again," *The FASEB Journal* 20 no. 7 (May 2006): 811–827. doi: 10.1096/fj.05-5424rev.

15. Ashton Acton, *Issues in Biological and Life Sciences Research: 2013 Edition*, (United States of America: ScholarlyEditions, 2013).

16. Khan and Scott, "Mechanotherapy: How Physical Therapists' Prescription of Exercise Promotes Tissue Repair."

17. Bleakley et al., "Management of Acute Soft Tissue Injury Using Protection Rest Ice Compression and Elevation: Recommendations from the Association of Chartered Physiotherapists in Sports and Exercise Medicine (ACPSM)."

PAGES 160–161

1. Jonathan Yeung, Andrew Cleves, Hywell Griffiths, and Len Nokes, "Mobility, Proprioception, Strength and FMS as Predictors of Injury in Professional Footballers," *BMJ Open Sport & Exercise Medicine* 2 (2016): e000134. doi: 10.1136/bmjsem-2016-000134.

2. Kyle Kiesel, Phillip J. Plisky, and Michael L. Voight, "Can Serious Injury in Professional Football Be Predicted by a Preseason Functional Movement Screen," *North American Journal of Sports Physical Therapy* 2 no. 3 (August 2007): 147–158.

3. Kyle Kiesel, P. Plisky, and R. Butler, "Functional Movement Test Scores Improve Following a Standardized Off-Season Intervention Program in Professional Football Players," *Scandinavian Journal of Medicine & Science in Sports* 21 no. 2 (April 2009): 287–292. doi: 10.1111/j.1600-0838.2009.01038.x.

4. Takayuki Hotta, Shu Nishiguchi, Naoto Fukutani, Yuto Tashiro, Daiki Adachi, Saori Morino, Hidehiko Shirooka, Yuma Nozaki, Hinako Hirata, Moe Yamaguchi, and Tomoki Aoyama, "Functional Movement Screen for Predicting Running Injuries in 18- to 24-Year-Old Competitive Male Runners," *Journal of Strength & Conditioning Research* 29 no. 10 (October 2015): 2808–2815. doi: 10.1519/JSC.0000000000000962.

5. Gregory R. Waryasz and Ann Y. McDermott, "Patellofemoral Pain Syndrome (PFPS): A Systematic Review of Anatomy and Potential Risk Factors," *Dynamic Medicine* 7 no. 9 (June 2008). doi: 10.1186/1476-5918-7-9.

6. Bradley S. Neal, Ian B. Griffiths, Geoffrey J. Dowling, George S. Murley, Shannon E. Munteanu, Melinda E. Franettovich Smith, Natalie J. Collins, and Christian J. Barton, "Foot Posture as a Risk Factor for Lower Limb Overuse Injury: A Systematic Review and Meta-analysis," *Journal of Foot and Ankle Research* 7 no. 1 (December 2014): 55. doi: 10.1186/s13047-014-0055-4.

7. Jeppe B. Lauersen, Ditte M. Bertelsen, and Lars B. Andersen, "The Effectiveness of Exercise Interventions to Prevent Sports Injuries: A Systematic Review and Meta-Analysis of Randomised Controlled Trials," *British Journal of Sports Medicine* 48 no. 11 (October 2014): 871–877 doi: 10.1136/bjsports-2013-092538.

8. Alan Hreljac, Robert N. Marsall, and Patria A. Hume, "Evaluation of Lower Extremity Overuse Injury Potential in Runners," *Medicine & Science in Sports & Exercise* 32 no. 9 (September 2000):1635–1641. doi: 10.1097/00005768-200009000-00018.

9. Brian Noerhen, J. Scholz, and I. Davis, "The Effect of Real-time Gait Retraining on Hip Kinematics, Pain and Function in Subjects with Patellofemoral Pain Syndrome," British Journal of Sports Medicine 45 no. 9 (July 2011): 691–696. doi: 10.1136/bjsm.2009.069112.

PAGES 162–163

1. A. N. Schroder and T. M. Best, "Is Self Myofascial Release an Effective Preexercise and Recovery Strategy? A Literature Review," *Current Sports Medicine Reports* 14 no. 3 (May–June 2015): 200–208. doi: 10.1249/JSR.0000000000000148.

2. K. Jay, E. Sundstrup, D. S. Sondergaard, D. Behm, M. Brandt, C. A. Saervoll, M. D. Jakobsen, and L. L. Andersen, "Specific and Cross Over Effects of Massage for Muscle Soreness: Randomized Controlled Trial," *International Journal of Sports Physical Therapy* 9 no. 1 (February 2014): 82–91.

3. G. Z. Macdonald, D. C. Button, E. J. Drinkwater, and D. G. Behm, "Foam Rolling as a Recovery Tool After an Intense Bout of Physical Activity," *Medicine & Science in Sports & Exercise* 46 no. 1 (January 2014): 131–142. doi: 10.1249/MSS.0b013e3182a123db.

4. Patrick F. Curran, Russell D. Fiore, and Joseph J. Crisco, "A Comparison of the Pressure Exerted on Soft Tissue by 2 Myofascial Rollers," Journal of Sport Rehabilitaion 17 no. 4 (January 2008): 432–442. doi: 10.1123/jsr.17.4.432.

5. Leon Chaitow, *Fascial Dysfunction: Maunual Therapy Approaches*, (Handspring Publishing, 2014).

6. Jeppe B. Lauersen, Ditta M. Bertelsen, and Lars B. Andersen, "The Effectiveness of Exercise Interventions to Prevent Sports Injuries: A Systematic Review and Meta-Analysis of Randomised Controlled Trials," *British Medical Journal* 48 (2013): 871–877. doi: 10.1136/bjsports-2013-092538.

7. Robert Schleip, Thomas W. Findley, Leon Chaitow, and Peter Huijung, *Fascia: The Tensional Network of the Human Body*, (Australia: Elsevier, 2012).

8. Myroslava Kumka and Jason Bonar, "Fascia: A Morphological Description and Classification Based on a Literature Review," *The Journal of the Canadian Chiropractic Association* 56 no. 3 (September 2012): 1–13.

9. L. Zheng, Y. Huang, W. Song, X. Gong, M. Liu, X. Jia, G. Zhou, L. Chen, A. Li, and Y. Fan, "Fluid Shear Stress Regulates Metalloproteinase-1 and 2 in Human Periodontal Ligament Cells: Involvement of Extracellular Signal-Related Kinase (ERK) and P38 Signalling Pathways," *Journal of Biomechanics* 45 no. 14 (September 2012): 2368–2375. doi: 10.1016/j.jbiomech.2012.07.013. Epub 2012 Aug 3.

10. Chaitow, *Fascial Dysfunction: Maunual Therapy Approaches.*

11. R. A. Brand, "Autonomous Informational Stability in Connective Tissues," *Medical Hypotheses* 37 no. 2 (February 1992): 107–114. doi: 10.1016/0306-9877(92)90050-M.

12. Schleip et al., *Fascia: The Tensional Network of the Human Body.*

13. Kumka and Bonar, "Fascia: A Morphological Description and Classification Based on a Literature Review."

14. Schleip et al., *Fascia: The Tensional Network of the Human Body.*

15. T. A. Mc Guine and J. S. Keene, "The Effect of a Balance Training Program on the Risk of Ankle Sprains in High School Athletes," *American Journal of Sports Medicine* 34 no. 1 (July 2006): 1003–1111. doi: 10.1177/0363546505284191.

16. C. A. Emery, D. J. Cassidy, T. P. Klassen, R. J. Rosychuk, and B. H. Rowe, "Effectiveness of a Home-Based Balance-Training Program in Reducing Sports-Related Injuries Among Healthy Adolescents: A Cluster Randomized Controlled Trial," *Canadian Medical Association Journal* 172 no. 6 (March 2015): 749–754. doi: 10.1503/cmaj.1040805.

17. H. C. Heitkamn, T. Horstmann, F. Mayer, J. Weller, and H. H. Dickhuth, "Gain in Strength and Muscular Balance After Balance Training," *International Journal of Sports Medicine* 22 no. 4 (May 2001): 285–290. doi: 10.1055/s-2001-13819.

18. Lauersen et al., "The Effectiveness of Exercise Interventions to Prevent Sports Injuries: A Systematic Review and Meta-Analysis of Randomised Controlled Trials."

19. Maarten D. W. Hupperets, Evert A. L. M. Verhagen, and Willem van Mechelen, "Effect of Unsupervised Home Based Proprioceptive Training on Recurrences of Ankle Sprain: Randomised Controlled Trial," *British Medical Journal* 339 (July 2009). doi: 10.1136/bmj.b2684.

20. E. Holme, S. P. Magnusson, K. Becher, T. Bieler, P. Aagaard, and M. Kjaer, "The Effect of Supervised Rehabilitation on Strength, Postural Sway, Position Sense and Re-Injury Risk After Acute Ankle Ligament Sprain," Scandinavian Journal of Medicine and Science in Sports 9 no. 2 (April 1999): 104–109. doi: 10.1111/j.1600-0838.1999.tb00217.x.

21. Schleip et al., *Fascia: The Tensional Network of the Human Body.*

22. H. Langevin, "Connective Tissue: A Body-Wide Signaling Network?," *Medical Hypotheses* 66 no. 6 (2006): 1074–1077. doi: 10.1016/j.mehy.2005.12.032.

23. H. Su, N. J. Chang, W. L. Wu, L. Y. Guo, and I. H. Chu, "Acute Effects of Foam Rolling, Static Stretching, and Dynamic Stretching During Warm-Ups on Muscular Flexibility and Strength in Young Adults," *Journal of Sport Rehabilitation* 13 (October 2016): 1–24. doi: 10.1123/jsr.2016-0102.

24. Graham Z. MacDonald, Michael D. Penney, Michelle E. Mullaley, Amanda L. Cuconato, Corey D. Drake, David G. Behm, and Duane C. Button, "An Acute Bout of Self-Myofascial Release Increases Range of Motion without a Subsequent Decrease in Muscle Activation or Force," *Journal of Strength and Conditioning Research* 27 no. 3 (March 2013): 812–821. doi: 10.1519/JSC.0b013e31825c2bc1.

25. D. H. Junker and T. L. Stoggl, "The Foam Roll as a Tool to Improve Hamstring Flexibility," *Journal of Strength and Conditioning Research* 29 no. 12 (December 2015): 3480–3485. doi: 10.1519/JSC.0000000000001007.

26. Schleip et al., *Fascia: The Tensional Network of the Human Body.*

27. Schroder and Best, "Is Self Myofascial Release an Effective Preexercise and Recovery Strategy? A Literature Review."

28. Macdonald et al., "Foam Rolling as a Recovery Tool After an Intense Bout of Physical Activity."

CHAPTER 8
the big questions

PAGES 168–169

1. Joshua Denham, Francine Z. Marques, Brendan J. O'Brien, and Fadi J. Charchar, "Exercise: Putting Action into Our Epigenome," *Sports Medicine* 44 no. 2 (February 2014): 189–209. doi: 10.1007/s40279-013-0114-1.

2. Alun G. Williams and Jonathan P. Folland, "Similarity of Polygenic Profiles Limits the Potential for Elite Human Physical Performance," *The Journal of Physiology* 586 no. 1 (2008): 113–121. doi: 10.1113/jphysiol.2007.141887.

PAGES 170–171

1. Timothy N. Kruse, Rickey E. Carter, Jordan K. Rosedahl, and Michael J. Joyner, "Speed Trends in Male Distance Running," *PLoS ONE* 9 no. 11 (November 2014): e112978. doi: 10.1371/journal.pone.0112978.

PAGES 172–173

1. Valérie Thibault, Marion Guillaume, Geoffroy Berthelot, Nour E. Helou, Karine Schaal, Laurent Quinquis, Hala Nassif, Muriel Tafflet, Sylvie Escolano, Olivier Hermine, and Jean-François Toussaint, "Women and Men in Sport Performance: The Gender Gap has not Evolved since 1983," *Journal of Sports Science and Medicine* 9 no. 2 (June 2010): 214–231. eCollection 2010. doi: 10.1123/ijspp.8.1.99.

PAGES 174–175

1. C. Knoepfli-Lenzin, C. Sennhauser, M. Toigo, U. Boutellier, Jens Bangsbo, Peter Krustrup, A. Junge, and Jiri Dvorak, "Effects of a 12-Week Intervention Period with Football and Running for Habitually Active Men with Mild Hypertension," *Scandinavian Journal of Medicine & Science in Sports* 20 S1 (April 2010): S72–S79. doi: 10.1111/j.1600-0838.2009.01089.x. Epub 2010 Feb 2.

2. Eliza F. Chakravarty, Helen B. Hubert, Vijaya B. Lingala, and James F. Fries, "Reduced Disability and Mortality Among Aging Runners: A 21-Year Longitudinal Study," *Archives of Internal Medicine* 168 no. 15 (August 2008): 1638-1646. doi:10.1001/archinte.168.15.1638

3. Duck-chul Lee, Russell R. Pate, Carl J. Lavie, Xuemei Sui, Timothy S. Church, and Steven N. Blair, "Leisure-Time Running Reduces All-Cause and Cardiovascular Mortality Risk," *Journal of The American College of Cardiology* 64 no. 5 (August 2014): 472–481. doi: 10.1016/j.jacc.2014.04.058.

4. D. C. Lee, R. R. Pate, and C. J. Lavie, "Running and All-Cause Mortality Risk—Is More Better?," *Medicine & Science in Sports & Exercise* 44 no. 5 (2012): S699. Quoted in James H. O'Keefe and Carl J. Lavie, "Run For Your Life . . . at a Comfortable Speed and Not Too Far," *Heart* 95 (2013): 516–519. doi:10.1136/heartjnl-2012-302886.

5. Hannah Arem, Steven C. Moore, Alpa Patel, Patricia Hartge, Amy Berrington de Gonzalez, Kala Visvanathan, Peter T. Campbell, Michael Freedman, Elisabete Weiderpass, Hans O. Adami, Martha S. Linet, I-Min Lee and Charles E. Matthews, "Leisure Time Physical Activity and Mortality: A Detailed Pooled Analysis of the Dose-Response Relationship," *JAMA Internal Medicine* 175 no. 6 (June 2015): 959–967. doi: 10.1001/jamainternmed.2015.0533.

Notes on contributors

John Brewer is currently Professor of Applied Sports Science and Head of School at St. Mary's University, London. Previously he has been Director of Sports Science at GlaxoSmithKline and Director of the Lilleshall Sports Injury and Human Performance Centre. John has acted as a consultant for the London Marathon and been on the board of UK Anti-Doping. His areas of expertise include endurance running, sports nutrition, drugs in sport, fitness, and training. An eighteen-time London Marathon runner, he wrote the official *London 2012 Training Guide to Athletics (Track)* and is a regular writer for *Running Magazine*. John has written numerous articles for a range of publications, and has made regular appearances on UK TV and radio.

Iain Fletcher is a principle lecturer in biomechanics at the University of Bedfordshire, where he runs the MSc in Sports Performance. He is an accredited strength and conditioning (S&C) coach with the UK Strength and Conditioning Association and the British Olympic Association. He has 25 years of experience as an S&C coach working with a wide range of sports. Formally an English Institute of Sport S&C coach, Iain now works as a consultant with a number of professional rugby clubs and track and field athletes, assessing biomechanical function and training movement efficiency.

Laura Charalambous is a biomechanics lecturer at the University of Bedfordshire. Her PhD research developed methods of providing integrated kinematic and kinetic feedback for sprinters and evaluated the effectiveness of different feedback interventions. Laura spent a year with the International Rugby Board and Cardiff Metropolitan University, investigating the use of artificial pitches and their effects on surface and player mechanics. She is a Fellow of the Higher Education Academy and the current Chair of the Ethics Committee for the ISPAR Research Institute.

Bob Murray is Managing Principal of Sports Science Insights, a consulting group that assists companies and organizations in need of targeted expertise in exercise science and sports nutrition. Bob was the cofounder of the Gatorade Sports Science Institute and its director from 1985 to 2008, and prior to that he served on the faculties of Boise State University (1980–1985; Associate Professor), Ohio State University (1979–1980; Lecturer), and Oswego State University (1974–1977; Assistant Professor and Men's Swimming and Diving Coach). Bob is an honorary member of the Academy of Nutrition and Dietetics.

Daniel Craighead is a PhD candidate in kinesiology at the Pennsylvania State University at University Park. Working at Noll Laboratory, Dan performs research on microvascular function in men and women with essential hypertension, and on the vasoactive properties of topical analgesic gels. Dan is also interested in how unique exercise interventions can improve cardiovascular health in aging men and women. For his work at Penn State, he has been awarded the Joseph and Jean Britton Distinguished Graduate Fellowship. Dan is an active member of the American College of Sports Medicine and the American Physiological Society. He is also a former collegiate runner and an avid marathoner, having qualified for the Boston Marathon multiple times.

Andy Lane is Professor of Sport and Learning at the University of Wolverhampton in the UK. Andy's areas of expertise include psychological factors such as mood and emotion in sport. A sub-three-hour marathon London marathon runner himself, a goal that took 15 marathons and 24 years to attain, he wrote the webpages for music and motivation that featured on the London Marathon website from 2011 to 2014. He is a regular contributor for *Running World* magazine and has written numerous articles for leading sports science journals.

Charles Pedlar is an applied sport scientist with extensive experience of working with elite-distance runners, and a Fellow of the British Association of Sport and Exercise Sciences (BASES). Currently he is a visiting research associate in the Cardiovascular Performance Program at Massachusetts General Hospital/Harvard Medical School in Boston and a Reader of Applied Sport and Exercise Science at St. Mary's University. Charlie has 15 years' experience working with Olympic-standard and pro-league athletes in Europe and the USA and was awarded BASES Practitioner of the Year in 2014.

James Earle is an applied sports scientist at St. Mary's University and has recently taken the role as Head of Strength and Conditioning at Culford School and Tennis Academy. He has a Master's degree in exercise physiology and is an accredited strength and conditioning coach with the UK Strength and Conditioning Association. His areas of interest are applied exercise physiology and strength training for sports performance.

Paul Larkins is a content manager on the *Cambridge News* (UK). Paul is also a former international athlete, representing Great Britain at the World Indoor Championship and the World Road Relays, and has won three national titles in the mile and 3000 m. At Oklahoma State University he was NCAA Mile Champion and has run the mile under four minutes more than ten times. He is a regular event reporter for *Athletics Weekly*, covering two Olympic Games.

Jess Hill is a senior lecturer and a research scientist and sports physiologist at St. Mary's University. She teaches physiology within the Sport Science Program, specializing in research methods and recovery from exercise, and is also a BASES accredited physiologist working with a range of elite athletes.

Anna Barnsley works from her own practice based in London, where she specializes in sports injuries and back and neck issues. She has over 21 years' of experience from working at Crystal Palace Sports Injuries center (a National center of excellence) and six years with the England Under-18 rugby team, to various media roles including physio for Sky's charity football programme "The Match" and Talk Sport radio. Anna wrote a series for *Running Fitness Magazine* and teaches physios/osteopaths and GPs. She runs herself for fitness and pleasure and has guided many a runner through injury to marathon fitness.

Index

N

O

P

R

S

Acknowledgments

The Ivy Press would like to thank the following for permission to reproduce copyright material:

Getty Images: Lars Baron/Staff: 35; Fabrice Coffrini/Staff: 61; Elsa/Staff: 137; Victor Fraile/Contributor: 9; Sean Garnsworthy/Staff: 13; Rene Johnston/Contributor: 87; Zak Kaczmarek/Stringer: 4–5; Keystone/Stringer: 119; Philippe Lopez/Staff: 85; Andy Lyons/Staff: 81; David McNew/ Stringer: 29; Gonzalo Arroyo Moreno/Stringer: 167; New York Times Co./Contributor: 89; Doug Pensinger/Staff: 99; Ryan Pierse/Staff: 57; Oli Scarff/Staff: 95; Wally Skalij/Contributor: 143; Michael Steele/Staff: 123; Boris Streubel/Stringer: 45.

REX Shutterstock: ddp USA/REX/Shutterstock: 82–83, 92–93; Robert Perry/REX/Shutterstock: 91.

Science Photo Library: 161.

Every effort has been made to trace copyright holders and to obtain their permission for the use of copyright material. The publisher apologizes for any errors or omissions in the lists above and will gratefully incorporate any corrections in future reprints if notified.